Phenomenological Creep Models of Composites and Nanomaterials

Deterministic and Probabilistic Approach

Leo Razdolsky

L.R. Structural Engineering Inc.
Lincolnshire, Illinois, USA

CRC Press
Taylor & Francis Group
Boca Raton London New York

CRC Press is an imprint of the
Taylor & Francis Group, an **informa** business
A SCIENCE PUBLISHERS BOOK

CRC Press
Taylor & Francis Group
6000 Broken Sound Parkway NW, Suite 300
Boca Raton, FL 33487-2742

Printed on acid-free paper
Version Date: 20181106

International Standard Book Number-13: 978-1-138-50601-5 (Hardback)

Visit the Taylor & Francis Web site at
http://www.taylorandfrancis.com

and the CRC Press Web site at
http://www.crcpress.com

Preface

The mechanics of composites as an independent branch of the mechanics of deformable media is at the early stage of growth, development and formation. Its science and technology is a broad and interdisciplinary area of research and developmental activity that has been growing very fast worldwide in the past few years. Composites have emerged as a natural response to the needs of modern technology. They are based on the unique idea of simplicity of reinforcement, when they combine "polar" properties of materials—a compliant matrix and a rigid and strong reinforcement. It is important to emphasize that the idea of reinforcement is deeper than just strength and manufacturability. It also increases the reliability of the material. Apparently, composites are the only materials in which the increase in strength is accompanied by an increase in the fracture toughness. Composites, strictly speaking, are not materials in the classical sense; final product, for example, metallurgy, with given properties and practically unchanged during processing. They constitute a vast family of materials created from semi-finished products together with the construction. It is the mechanics of composites that is the scientific basis for understanding, describing, predicting and controlling the structural properties of the entire variety of materials and the technology. When creating structural analyses and designs from composites, design issues (understood in the traditional sense), optimal reinforcement and the development of the technological process are three sides of a single problem and cannot be considered in isolation. It is this approach to the construction of the phenomenological models of creep of composites and nanocomposites that is proposed in this book and as a consequence, the classical Volterra integral equation of the second kind, describing the creep process of a material under a uniaxial stressed state, undergoes appropriate changes and additions.

The book contains a large amount of original research material (as well as substantially modified and increased) from previously published articles by the author. At the same time the book contains many other results obtained by other researches, primarily reflecting the most important data in the area of statistical structural analysis of engineering creep.

The majority of the book is devoted to the deterministic and applied probabilistic methods and simplifications that are specifically tailored to the creep deformation of composites and the problems related to nanocomposites. The text is divided into seven chapters; each of which begins with a short theoretical introduction, followed by relevant formulas and examples. The authors have paid particular attention to the fact that the statistical data bases are presented in dimensionless form (the original deterministic creep constitutive integral equations as well as equations of energy and mass conservation are dimensionless!), therefore the applied probability based results (such as, for instance, mean and standard deviation values and other numerical characteristics of stochastic stress-strain creep deformation process) that makes it possible to solve a number of problems with exceptional simplicity.

This book is intended for graduate students, professors, scientists, and engineers. Thus, the book should be considered not only as a graduate textbook, but also as a reference handbook to those working or interested in areas of Deterministic and Applied Probability Methods in Creep Mechanics, Stress Analysis, and Mechanical Properties of Composites and Nanocomposites. The scope of the work is broad enough to be useful to practicing and research engineers as well as serve as a teaching tool in colleges and universities. In addition, the book provides extensive coverage of a great many practical problems and numerous references to the literature.

Contents

Notations

k—The thermal conductivity [W/m*K] or [J/m*s*K]

Time: $t = \frac{h^2}{a}\tau$ [sec]

Temperature: $T = \frac{RT_*^2}{E}\theta + T_*$ [K], where $T_* = 600^o$ K is the base line temperature

Coordinates: $\overline{x} = x/h$ and $\overline{z} = z/h$—"x" and "z"—dimensionless coordinates

ν—kinematic viscosity [m^2/sec]; "u" and "w"—dimensionless velocities

$\beta = \frac{RT_*}{E}$—Dimensionless parameter

$\overline{\gamma} = \frac{c_p RT_*^2}{QE}$—Dimensionless parameter

$P = \frac{e\sigma K_v (\beta T_*)^3 h}{\lambda}$—Thermal radiation dimensionless coefficient

$\sigma = 5.67(10^{-8})$ [watt/m^2K^4]—Stefan-Boltzmann constant

$\delta = (\frac{E}{RT_*^2})Qz(\exp(-\frac{E}{RT_*}))$—Frank-Kamenetskii's parameter

$C = [1 - P(t)/P_o]$—Concentration of the material

L and h—Length (width) and thickness accordingly

W; U—dimensionless velocities

A is set of outcomes (events) to which a probability is assigned

$P(E_2|E_1)$—conditional probability

$\Phi^*(.)$ Denotes the cumulative distribution function of standard normal distribution $\Phi^*(z) = \frac{1}{\sqrt{2\pi}} \int_{-\infty}^{z} e^{-\frac{z^2}{2}} dz$

μ_A, μ_B, σ_A, σ_B are mean and standard deviation of A and B, respectively

J(t, t') is the creep compliance function

T_m is the melting point of the metal matrix material

ε (t)—strain

σ (t)—stress

$\bar{\sigma}$ (t) = E(t)ε(t)—instantaneous stress

ε_e—instantaneous (elastic) strain

ε_c—creep strain

ε_T—thermal expansion due to temperature effect

K (t, t') =—∂J(t, t')/∂t'—retardation function (memory function)

R (t, t')—relaxation function (also called the relaxation modulus)

P_f—probability of failure

P_{rel}—reliability

θ—dimensionless temperature

θ_g—transitional dimensionless temperature (Glass Transition Temperature)

θ_{gm}—transitional dimensionless temperature of matrix material

θ_{gf}—transitional dimensionless temperature of filler material

τ—dimensionless time

ω—frequency

S(ω)—spectral density

D—Diffusion rate (Flick's law)

η—*viscosity* parameter of the material

E—Modulus of elasticity

n = η/E—relaxation time

n—Power law exponent

α_i—material property parameter

k—The thermal conductivity [W/m*K] or [J/m*s*K]

T—Temperature

E_a—activation energy

R is the ideal gas constant

e—Emissivity factor

σ—Stefan-Boltzmann constant (σ = 5.6703(10^{-8}) watt/m^2K^4)

T_o—ambient temperature

t—Time

a—thermal diffusivity parameter [m^2/sec]

G_{tot}—Total free energy

G—Gibbs energy

1
Introduction and Overview

1.1 Definition of structured composites and nanomaterials

The most recent definition of structural composites is given in [1]:

Composites are considered to be combinations of materials differing in composition or form on a macro scale. The constituents retain their identities in the composite; that is, they do not dissolve or otherwise merge completely into each other although they act in concert. Normally, the components can be physically identified and exhibit an interface between one another.

1.1.1 *The use of nanocomposites in space technology*

The success of astronautics and aeronautics cannot be imagined without the development of materials that are able to withstand high temperatures during operation of engines, as well as those that arise when a spacecraft reenters the atmosphere during descent, carries high mechanical loads, faces extreme conditions of outer space to ensure the containment inhabited compartments of the ship and the fulfillment of many other requirements that are vital for the guaranteed safety of crew members and functioning of space vehicles. It is impossible to forget the losses that humanity suffered on the way to mastering space, due to problems associated with the defects and shortcomings in the properties of the materials used. Having learnt lessons from past debacles and the continuing need to pursue ambitious tasks of projects dictate the need to develop new materials. One direction could be the creation of composite materials with a polymer matrix and carbon nanotubes (CNTs) as fillers.

In order to ensure long-term trouble-free operation of space vehicles (satellites), the durability of equipment on-board—materials and elements—and its resistance to the debilitating effects of the surrounding space environment is vital. According to estimates [2] more than half of the failures and malfunctions in the operation of on-board spacecraft equipment are due to the adverse impact of space factors (ISF). The spacecraft is influenced by an extensive ISF complex: high energy electron and ion fluxes, cold and hot cosmic plasma, solar electromagnetic radiation, meteoric matter, artificial particles (so-called "Space debris"), etc. Analysis of the statistics and the conditions for the occurrence of anomalies in the operation of spacecraft equipment shows that in most cases, failures are due to the impact of cosmic radiation on materials and equipment elements (radiation effects on spacecraft are caused by electrons and Ions with energy above ~ $10^5 - 10^6$ eve ($1.6(10^{-14})$ J – $1.6(10^{-13})$ J), capable of penetrating into the materials to a depth of more than a few tens of micrometers causing, amongst others, ionization of atoms in the composites, the formation of local electrical charges, the formation of defects, nuclear transformations, etc. In space conditions, the spacecraft surfaces are in a state of repeated heating cycles of up to 420° K and cooling cycles of up to 120° K. The absorbed energy of ionizing radiation is expended on irreversible chemical changes in the material and radiative processes, but most of it passes into heat, which leads to increase in the temperature of the material (radiation heating). When irradiation under conditions close to the adiabatic conditions occurs the temperature of the materials can be increased by several tens and even hundreds of degrees.

Nanocomposites, currently being implemented in many different areas of human activity, will, of course find wide application in the creation of promising samples of aerospace engineering. Created with the help of nanotechnology, polymer-matrix composites due to their properties are able to protect the spacecraft from many of the space effects, which makes them suitable for use as structural and functional materials. Till date, a rather limited volume of data has been obtained on the radiation resistance of nanocomposites materials. Most nanocomposites have a higher radiation resistance to various constituents of cosmic ionizing radiation in comparison with traditional composite materials. The high radiation resistance of CNTs (carbon nanotubes transistors) can be explained using the physical model, which is based on the notion of being less dense in comparison with bulk solid bodies, anisotropic packing of atoms in nanotubes and adsorption of the surface of a tube with carbon atoms knocked out of it. Migration processes occurring on the surface of the nanotubes lead to "healing" of radiation defects. Thus, for sufficiently high strength properties, CNTs (with tunable diodes) have the property of shielding electromagnetic radiation, and also account for high thermal and electrical conductivity, being able to shoot thermal load and static electricity charges.

The most important direction of CNT-PN application is providing thermal and mechanical protection to the spacecraft. By the developing technologies for creating composite materials with the required properties, we will be able to solve the problem in the long term with a single screen made of nanocomposites offering triple protection—thermal, radiation and shockproof. In the latter case, we mean impact on the particle's meteor matter and space debris.

1.1.2 Classification of polymers

A large number of polymers can be divided into three main classes, underlying the classification that is now adopted. *The first class* includes a large group of carbohydrate polymers whose macromolecules have a skeleton constructed from carbon atoms. Typical representatives of polymers of this class can be called polyethylene, polypropylene, polyisobutylene, poly (methyl methacrylate) (PMMA), polyvinyl alcohol and many others. The fragment of the macromolecule of the first of them has the following structure: [-CH_2 -CH_2-]n. *The second class* includes an equally large group, heterochain polymers whose macromolecules in the main chain contain heteroatom's (for example, oxygen, nitrogen, sulfur, etc.) in addition to carbon atoms. To polymers of this class belong numerous simple and complex polyesters, polyamides, polyurethanes, natural proteins, etc., as well as large group element organic polymers: polyethylene oxide (simple polyester); Polyethylene terephthalate (polyester) polyamide; polydimethylsiloxane. *The third class* of polymers is composed of high-molecular compounds with conjugate system of constraints. These include various polyacetylenes, polyphenylenes, polyoxadiazoles and many others connection. Examples of such polymers are: polyacetylene; polyphenylene; polyoxadiazole.

An interesting group of chelating polymers also belong to this class, the composition of which includes various elements capable of forming coordination links (they are usually denoted by arrows). The elementary link of such polymers often has complex structure. Among numerous polymeric materials the most practical application so far is based on materials founded on representatives of the first class of polymers—carbohydrate high-molecular compounds. Of carbon-chain polymers, one can get the most valuable materials—synthetic rubbers, plastics, fibers, films, etc., and historically these polymers are the first to have found practical application (obtaining phenol-formaldehyde resins, synthetic rubber, organic glass, etc.). Many of the carbon-chain polymers are subsequently classical objects for research and for creation of the theory of mechanical behavior of polymeric bodies (for example, polyisobutylene, polymethylmethacrylate, polypropylene, phenol-formaldehyde resin, etc.). Polymers are divided into thermoplastics and thermoset by their ability to recycle. Consider the first, thermoplastic materials or thermoplastics (thermoplastic) include polymers, which, when heated during processing, pass from solid state of aggregation in liquid: highly elastic or viscous (molded thermoplastics pass into a viscous flow state). When the material is cooled, the reverse transition to a solid state occurs. The behavior of thermoplastics differs from thermoset materials, which are cured during processing and are not capable of then going into a liquid aggregate state.

1.2 What is the main difference between the composites and ordinary solid bodies?

If in a crystalline solid the energy of interaction between its constituent atoms or molecules bound by covalent bonds that are much higher than thermal energy, $U >> kT$, then for composites $U \approx kT$. For polymers, this is the energy of interaction of macromolecules, interchange interactions. The consequence of this is that the forces binding the molecules are small and the configuration of the molecular and macroscopic structures is easily changed under the action of external forces and temperature. The physical properties of such substances are also easily changed [3]. Thus, one of the main features of composites is their sensory properties, and therefore the phenomenological creep model must be built on the energy basis of the Gibbs theory. As a result of the analysis of these thermodynamic models and phase chemical compositions, it is possible to explain not only the rheological properties of materials, but also to predict the optimization of the technological conditions for their synthesis (temperature regime, initial composition of components).

One can see from Table 1.1 [4] that in modern composites the strength and stiffness of fibers is far above those of traditional bulk materials.

Primary characteristics for reinforcement fibers in matrix composites are high stiffness and strength. The fibers must maintain these characteristics in hostile environments such as elevated temperatures, exposure to fluids, and environmental loads. These characteristics and requirements have substantial implications for the physical, chemical and mechanical properties of the fiber, which in turn implies processing and acceptance parameters.

Table 1.1: Properties of composite reinforcing fibers material [4].

Material	E [GPa]	σ_b [GPa]	ρ [kg/m^3]	E/ρ [MJ/kg]	σ_b/ρ [MJ/kg]	C (cost) $/kg
E-glass	72.4	2.4	2,540	28.5	0.95	1.1
S-glass	85.5	4.5	2,490	34.3	1.8	22–33
Aramid	124	3.6	1,440	86	2.5	22–33
boron	400	3.5	2,450	163	1.43	330–440
HS graphite	253	4.5	1,800	140	2.5	66–110
HM graphite	520	2.4	1,850	281	1.3	220–660

Composites, by definition, are heterogeneous material; however the structural analysis proceeds on the assumption that the composite material is homogeneous, because on the macroscopic scale, they appear homogeneous and respond homogeneously when tested. Therefore the analysis of composite materials usually uses effective properties (for instance modulus of elasticity) which are based on the average stress and average strain.

Out of various definitions of composite materials the following is used in this book. Composites are materials consisting of chemically or physically dissimilar components which when combined produce a material with characteristics different from the individual components. Components are divided into two groups (phases): filler (or reinforcement) in the form of discrete particles or fibers—forms a discrete phase; binder (matrix)—in the form of any material that fills the space between the reinforcement and forms a continuous medium. A variety of analytical procedures may be used to determine the different properties of laminate composites from volume fractions and fiber and matrix properties. The derivations of these procedures may be found in [5, 6].

1.3 Composite structural systems

Composites have emerged as a natural response to the needs of modern techniques. They are based on a unique idea of simplicity of reinforcement, when connecting "polar" properties of materials—a compliant matrix and rigid and strong reinforcement. It is important to emphasize that the idea of reinforcement is deeper than just strength and manufacturability. This also enhances the reliability of the composite structures. Apparently, composites are the only materials in which the increase in strength is accompanied by an increase in viscosity of a failure process. Composites, strictly speaking, are not materials in the classical sense where the final product has, for example, like metallurgy, with preset and practically unchanged in the process of processing properties. They constitute a vast family of materials created from semi-finished products that together form the design. It is the mechanics of composites that is the scientific basis for understanding, describing, predicting and controlling structural engineering properties of the entire variety of materials and the technology of casting products from them. When creating structures from composites, design issues (understood in the traditional sense), optimal reinforcement and the development of the technological process are the three sides of a unified problems and cannot be considered in isolation, as can occur in the creation of metal structures. Modern fibrous composites are heterogeneous anisotropic materials. Elasticity and inelasticity of fibrous composites are determined by the type of reinforcement (glass, boron, carbon and organic fibers) and matrices (polymer, carbon, metal, and ceramic), the degree of their interaction in the composite, and also the angle of loading application with respect to directions of reinforcement.

Composites have two levels of heterogeneity, micro—no homogeneity with a monolayer composed of fibers and a binder and macro—no homogeneity within a multi-layered structural system composed of layers that are arbitrary packed (with possibly different thicknesses) into the whole package. Hence there are two directions in the mechanics of composites: micro- and macro-mechanics. Structurally heterogeneous medium in its physical and mechanical behavior is much richer than homogeneous material. A variety of possible situations in the process of deformation and fracture of composites makes studying these materials attractive for specialists from different areas of solid state mechanics. For example, in fibrous composite elements there are always micro-defects–cracks, caused not only by the imperfection of technology, but also by the deviation from idealized material model. The central point in the mechanics of fibrous composites is the significant accounting of the material structure at the level of reinforcing elements, a circumstance uncharacteristic of the classical mechanics of solids. Depending on the level of reinforcing elements, the mechanical properties of the engineered composite material are created; by managing the ply or fiber stacking sequence, it is possible within certain limits to control the resistance fields of the material, "adjusting" them to existing efforts. In this way opportunities open up the development of principles for the optimal design of the material itself. It is the composites that materialized this branch of the mechanics of the solid, which have recently been intensively developing. It should be emphasized that the formation of physical properties and the origins of the theoretical foundations of technological processes of composites molding—current and future—is also based on the level of reinforcement of composites. Depending on the percentage of reinforcement, numerous features of fracture are also explained (for example, delamination, stratification of layers, destruction of fibers, etc.) and features of properties of composites, such as viscoelasticity for a polymer and plasticity for metal matrices. On the physical and mechanical phenomena occurring at the structural design level, prediction of the life span and reliability of composite structures is based on assumptions from [7].

When describing a monolayer and materials composed of homogeneous thickness of the layers, heterogeneity could be excluded from consideration by the transformation of fibrous composites to the reduced anisotropic medium; the definition of effective stress-strain diagrams was the subject of the reinforced media theory [8, 9].

The method of replacing an inhomogeneous composite with an effective homogeneous anisotropic body is successfully used for solving the problems of rigidity, stability, and beams, plates and shells oscillation. In case of refusal of traditional kinematic hypotheses of Kirchhoff-Love type and searches for refined solutions that can "feel" bad resistance composites shear and transverse separation, attention was mainly paid to study of the resistance of fibrous composites to shear, especially in planes, where it is essentially determined by the matrix (the so-called interlayer shear deformation).

The classic works of Kelly [10–12] that gave rise to modern compositional studies were devoted to the study of the mechanical behavior of metal matrix composite (MMC). Modern space systems are inconceivable without the use of composites. The use of such materials in structural design of the airframe of modern aircraft reaches about 50% by weight, and this is probably close to the limit for the existing design of the aircraft. Unlike the use of fiber reinforced plastics (FRP) in large-sized structures, their use in small structural elements does not lead to a significant weight loss due to rather massive connecting elements. Moreover, special design requirements, such as high electro-thermal conductivity, strength at high temperatures, low anisotropy of elastic and strength characteristics, limit FRP applications. In these cases advanced composites are the best solution. Along with composites with a ceramic matrix, they are the material basis for a new generation of gas turbines of any kind, high-temperature nuclear reactors, hypersonic aircrafts, and so on. Previous experience of some research groups has formed a solid scientific and technological basis for the development of new effective composites, lightweight composite structures, as well as high-temperature resistance composites with operating temperatures up to 1300°C.

In the phenomenological approach, an inhomogeneous composite is considered as a continuous medium (homogeneous anisotropic material), mathematical model of which takes into account the experimental data on strength and strain. Many varieties of phenomenological criteria (limit state design) have been proposed for anisotropic materials. The most popular among them are the criterion of maximum stresses (or strains) and quadratic polynomial criteria, such as the Hill-Mises criteria [13], Tsai-Wu [14], and Chamis-Hoffman [15, 16]. For multi-layered composites, a structural-phenomenological model is often used. The phenomenological approach is used to describe the behavior of a unidirectional monolayer, and the structural one for consideration of a multilayer material composed of differently oriented plies. This requires a theory of layered media, allowing to carry out the transition from the stresses and deformations of the composite structural member to stresses and strains in any of its layers. Micromechanical models of polymer composites allow us to identify analytical dependences showing the influence of the properties of fibers, matrices, their adhesion interaction, material structure and fracture mechanisms on the macroscopic elastic strength characteristics of a unidirectional layer. Most successfully they describe the ultimate effective modulus of elasticity and the strength of the composite under tension. In the event that the deformations of the fibers and the matrix are the same, the following additive relationships hold, which show the contribution of each component in proportion to its volume content

$$E_{eff} = E_f \varphi_f + E_m (1-\varphi_f); \sigma = \sigma_f \varphi_f + \sigma_m (1-\varphi_f) \quad (1.1a)$$

These equations are called the "rule of the mixture." Since the contribution of polymer matrix to the overall stiffness usually does not exceed 2–5% (since $E_m << E_f$), it can be ignored

$$E_{eff} (\Leftrightarrow) = E_f \varphi_f \text{ and } \sigma_{eff} (\Leftrightarrow) = \sigma_f \varphi_f \quad (1.2a)$$

Composite elongation in the transverse direction is composed from deformation of fibers and binder. The elastic modulus E^{tr}_{eff} can be calculated according to the formula

$$1/E^{tr}_{eff} = \varphi_f/E_f + \varphi_M/E_M \quad (1.3a)$$

The strength of composites in the case of transverse tension-compression and shear depends on many factors, primarily on the properties of the matrix, adhesion interaction, the structure of the material—the presence of pores and other defects. Analytical dependencies in this case can only have an approximate character. It is generally accepted that reinforcement reduces the strength of the composite in the transverse direction approximately 2 times compared with the strength of a homogeneous matrix. The redistribution of stresses between matrix and reinforcement is depended on the fiber size. For instance, because of the large diameters of boron fibers, their number per unit volume of plastic is 15–20 times less than in glass-carbon fibers. Consequently, the role of each fiber increases sharply. The failure of several fibers in the boron composite dramatically increases the stress on the remaining fibers. Because the strength of the fiber is determined by the ratio of length to diameter, which is necessary for the effective hardening of plastic, the length of boron fibers is hundreds of times the length of fine glass or carbon fibers, scale dependence of the strength of boron fibers on length determines less strength of the material on their basis. Stress σ and modulus of elasticity E in case of unidirectional arrangement of fibers increases with the growth of the volume fiber content. The optimum filling degree is 65–75%, slightly higher than that of glass and carbon fiber. In this case, in contrast to glass- and carbon fiber–fibers σ under compression is greater than under tension. The influence of the arrangement of the fibers on the modulus of elasticity

[17] under compression and the degree of anisotropy of epoxy-fibrous fiber ($\varphi = 0.42$) is presented in Table 1.2.

The reinforcement absorbs the main stresses arising in the composite under the influence of external loads, and determines the basic mechanical characteristics. The matrix ensures the joint operation of discrete elements of the reinforcement, combining them into a monolith, due to its own rigidity and adhesion at the matrix-reinforcement interface. This phase to a lesser extent determines the mechanical properties, but it has a decisive influence on the technological characteristics, for example, the possibility of formation and technological regimes. It should be noted that the concept of "filler" is general and does not only combines discrete elements with high strength and rigidity. Under this concept, the inclusion of gases in the form of small-sized bubbles in the volume of the matrix material is also suitable. Such a conventional reinforcement provides not strength, but other functional characteristics, for example, high sound and heat insulation capabilities. Composites should be distinguished from mixtures and solutions. In the first case, the material is not monolithic, in the second case there is no interface between the components. In the technical literature, composites are divided into classes. Each class was named by the type of matrix: composite material (CM) with a polymer matrix is called polymer (PMC), with metal matrix it is known as metal (MMC), and with ceramic–ceramic composite materials (CMC). The first class of composite materials, PMC is most commonly used in technology. The matrix is based on various polymer resins. The combination of a polymer matrix with a metal reinforcement is indicated as PM. As filler, boron fibers and, more rarely, other metals, such as aluminum, are used. The name of CM with a polymer matrix is formed from the name of the type of reinforcement with the addition of the word "plastic". For example: fiberglass, boroplastics, carbon fiber, organoplastic, etc. The second class of composites has a metal matrix: light metals—aluminum, magnesium and heavy metals—steel and nickel alloys. The nomenclature and volume of use of metal composites is much smaller than that of polymeric ones. The MM group (matrix and reinforcement of metals) is represented in the industry, mainly by boroalumination, that

Table 1.2: The influence of fibers on the modulus of elasticity.

Mutual arrangement of fibers	The angle between the direction of the fibers in adjacent layers	The modulus of elasticity under compression, kg/mm^2				Anisotropy indices		
		E_x	E_y	E_{45}	E_z	E_x/E_y	E_x/E_z	E_x/E_{45}
Unidirectional	0	16200	1860	1540	1860	8.7	8.7	10.8
Cross-plane	90	8600	8500	1620	1870	1.02	4.5	5.3
Cross-spatial	90	5400	5300	–	5800	1.02	0.96	–

is, by the combination of an aluminum matrix with boron fiber. In the third class, various types of ceramics are used as the matrix. The information on the materials included in this group is very limited. Basically, they are in the development and research stage now. In the SM group, various metallic powders are used as the reinforcing filler. Such materials are called cermets. The fourth class contains composites with a matrix that does not belong to polymers, metals and ceramics. The most known composition here is carbon–carbon (UCMM). The reinforcement in it is represented by carbon rods or threads. The matrix is also carbon, but with much less strength.

1.3.1 Classification according to a structured feature

The first group includes composites with reinforcement in the form of discrete particles. Here we can distinguish two subgroups: CM, reinforced with solid particles and gas-filled. As solid particles, powders, flakes, chopped fibers, and microsphere is used. Composites with a polymer matrix and discrete filler are called "filled plastics." An effective disperse filler is a microsphere, which is a small diameter ball (d = 0.05–0.2 mm) hollow inside and made, most often, of glass. The second subgroup includes materials filled with small gas bubbles. Depending on the type of matrix, they have received the corresponding name: foams (polymer matrix), foam metals, foam ceramics, foam rubber and the like. If the gas bubbles are connected with each other, then in the name of materials the phrase "foam" is replaced by "porous", for example, "poroplast". The second group of CM is characterized by the fact that the reinforcement has a fibrous structure. This type of CM can be divided into four subgroups: unidirectional composites, woven layered fibrous volumetric weaving, and nonwoven. Unidirectional reinforcement can be: in the form of a primary thread (obtained immediately after manufacture); Filament (twisted) thread; Roving, consisting of strands of complex filaments; Tape strings connected together in a transverse direction through a certain distance and forming an elongated flat surface; A bundle of interlocked yarns with a cross-section close to the circumference. The second subgroup includes CM consisting of layers of tissue joined by a matrix. Reinforcing filler is characterized by the type of weaving and the thickness of the threads.

The third subgroup is characterized by an armature, which is a three-dimensional weave of threads or rods. The spatial interlacing can be organized by a different number of threads—from 3 to 11. The fourth subgroup contains materials with non woven filler (for instance, "pile fabric"). For using composite materials in the creation of modern structural elements and structural systems it is necessary to take into account their specific features, such as: anisotropy of stiffness and strength, viscoelastic properties, heterogeneity of elastic and strength parameters, and risk of failure, which determine the bearing capacity of the structure. The solution of this problem is impossible without complex theoretical and experimental research aimed at elucidating the physical pictures of the processes taking place in the structure and in the material, subjected to expected operational loads. An important link in such studies is the development of mathematical models of the phenomena under consideration that satisfy the requirements of accuracy on the one hand, economic and practical applicability to engineering calculations and design, on the other. Composites, strictly speaking, are not materials in the classical sense; for example, metallurgy, in which the final product is preset and practically unchanged in the process of processing properties. They constitute a vast family of materials created from semi-finished products together with the design. It is the mechanics of composites that is the scientific basis for understanding, describing, predicting and controlling structural properties of the entire variety of materials and technology of producing products from them. When creating structures from composites, design issues (understood in the traditional sense), optimal reinforcement and the development of the technological process are the three sides of unified problems and cannot be considered in isolation. Modern fibrous composites are heterogeneous anisotropic materials. Elasticity and inelasticity of fibrous composites are determined by the type of reinforcement (glass, boron, carbon and organic fibers) and matrices (polymer, carbon, metal, and ceramic), the degree of their interaction in the composite, and also the angle of load application with respect to directions of reinforcement. Composites have two levels of heterogeneity: micro—no homogeneity (a monolayer composed of fibers and a binder) and macro—no homogeneity (a multilayered structure composed of a given number of different layers with arbitrary packing by the thicknesses). Hence there are two directions in the mechanics of composites: micro- and macro-mechanics. Structurally heterogeneous medium in its physical and mechanical behavior is much richer than homogeneous material. A variety of possible situations in the process of deformation and fracture of composites makes studying these materials attractive for specialists from different areas of solid state mechanics. For example, in fibrous composites of level of reinforcing elements there are always micro-defects—cracks, caused not only by the imperfection of technology, but also by the deviation from the idealized material model. The central point in the mechanics of fibrous composites is the significant accounting of the material structure at the level of reinforcing elements—a circumstance uncharacteristic of the classical mechanics of solids. At the level of reinforcing elements, the mechanical properties of the composites are created; by managing and creating the layout of fibers (within certain limits control) the required resistance of the composite structure can be obtained. Moreover, in this way opportunities for the development of principles of the optimal design of the material itself open up. It is the composites that materialized in this branch of the mechanics of the solid that have been intensively developing recently. It should be emphasized that the formation of physical properties and the origins theoretical foundations of technological processes of producing composites—modern and future, also lie at the structural level of reinforcement elements. Similarly, numerous features, such as prediction of the life span and reliability of composite structures, are based on a probabilistic approach [18, 19].

1.3.2 *Reinforced media theory*

The definition of effective deformation characteristics was the subject of the reinforced media theory. The method of replacing an inhomogeneous composite with an effective homogeneous anisotropic body is successfully used for solving the rigidity problems, stability, and oscillations of beams, plates and shells. A characteristic moment—refusal of traditional kinematic hypotheses of Kirchhoff-Love type and searches for refined solutions that can "feel" bad resistance of composites, shear and transverse separation are presented in [20–22]. The main attention was paid to study of the resistance of fibrous composites to shear, especially in planes, where it is mainly determined by the matrix (the so-called interlayer shear deformation). The specific properties of modern composites are heterogeneity, anisotropy, and nonlinear elasticity which require the search for ways to construct theories more specialized in nature than the theory of elasticity of an anisotropic body, taking into account the real composition of the material. Difficulties facing the structural engineer are obvious. The depth of models should correspond to the accuracy of information about the properties of components, their manufacturing process and interaction in composite material. For problems occurring when the direction of reinforcing fibers and principal stresses do not coincide, the effect of viscoelasticity of the material becomes essential. The fact is that the polymer matrix makes composites extremely sensitive to the stress and temperature hierarchy of the structural element, manufacturing technology and the regime of subsequent operation. Technological limitations play a much more important role in the design of composite structures than metals. More research is needed in areas, such as, the creation of the theory of a "growing" body of nanomaterials, the macro-mechanics of nanocomposites and the subsequent phenomenological models of creep deformations.

Models of deformation and failure of composites rely heavily on initial information obtained from mechanical tests of a monolayer or multilayered structure as a whole. Hence the increased requirements for accuracy and reliability of this information are demanding. The mechanical properties of composites in contrast to metals are characterized by a large number of experimental constants. Their definition is associated with significant methodological difficulties. In practice the structural engineer is using many different shapes, sizes and technologies in preparation of samples for the experimental procedures. This leads to incomparability of results and creates an environment of contradictory judgments on the structural capabilities of composites. A realistic approach to the problem of mechanical testing requires a clear regulation of the number determined characteristics and methods for determining strength and rigidity composites, and reinforces the need for a critical analysis of existing methods, their evaluation and validation.

Developing models of elastic deformation of unidirectional and cross-reinforced composite materials is based on the structural-phenomenological approach to the analysis of defining relations.

The averaged stiffness characteristics of the multi-layered packages are determined by integrating over the thickness of the element. The analysis takes into account the orientation of elementary layers and shell hypotheses, superimposed on the entire stack of layers. In the case of solving dynamic problems in the framework of the concept of effective modules, the real heterogeneous material is allowed to be replaced by a hypothetical. There are additional difficulties arising from the need to analyze the wave processes occurring in a given structural system. Obviously, such an approach is justified only when the wavelength is an order of magnitude greater than the characteristic size of the structure.

1.4 Model of elastic deformation of a unidirectional multilayered composite material

Orthotropic is the condition expressed by variation of mechanical properties as a function of orientation. Lamina exhibit orthotropic characteristic because large difference in properties between the 0° and 90° directions is present. If a material is orthotropic, it contains planes of symmetry and can be characterized by four independent elastic constants. The stress-strain curves for composite materials are frequently assumed to be linear to simplify the analysis. For purposes of laminate analysis, it is important to consider the plane stress version (x1 and x2 coordinates) of the effective stress-strain relations. Let x3 be the normal to the plane [x1; x2] of a unidirectional reinforced lamina (see Figs. 1.1 and 1.2 below).

Suppose that each element of the multilayered structure is in a plane stressed state. Since the coordinate axes 1, 2 are the orthotropic axes, Hooke's law for a layer can be written in the form:

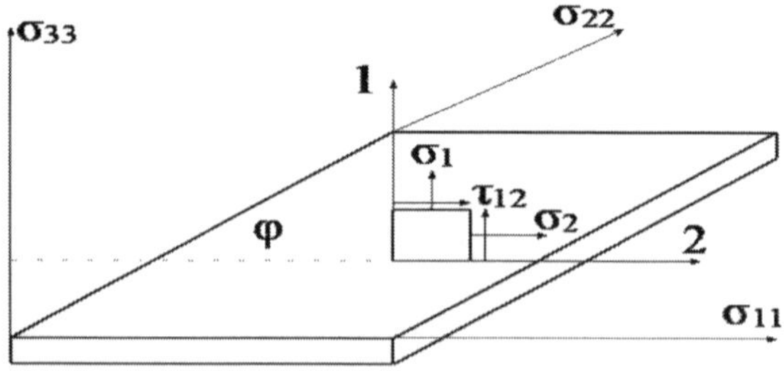

Figure 1.1: Unidirectional reinforced lamina coordinates.

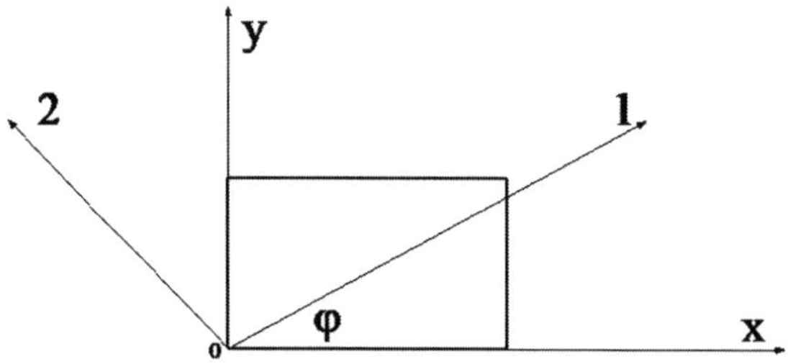

Figure 1.2: Axes rotation.

$$\begin{aligned}
&\sigma_1 = \overline{E}_1(\varepsilon_1 + v_{12}\varepsilon_2)\\
&\sigma_2 = \overline{E}_2(\varepsilon_2 + v_{21}\varepsilon_1)\\
&\tau_{12} = G_{12}\gamma_{12}\\
&\overline{E}_1 = \frac{E_1}{1 - v_{12}v_{21}};\quad \overline{E}_2 = \frac{E_2}{1 - v_{12}v_{21}}
\end{aligned} \tag{1.1}$$

where ε_1 ε_2 γ_{12} deformations in the directions 1, 2 and the plane 12, respectively, E_1, E_2 and G_{12} are the elastic module in directions 1, 2 and the shear modulus in the plane of the layer, and v_{12} and v_{21} are the Poisson coefficients. Here the symmetry condition of the elastic constants: $E_1\, v_{12} = E_2\, v_{21}$.

Elastic constants in (1.1) are determined either experimentally, or are calculated within the framework of the theory of reinforcement. Since the actual distribution of stresses and strains in the composite body, are where the main material has more rigid inclusions, it is very difficult, to obtain any practically useful dependencies for determination of the elastic constants E_1, E_2, v_{12}, G_{12}. In order to do so it is necessary to make some assumptions.

1. One direction of the reinforced material is a continuous macroscopically homogeneous (transversely isotropic) body.

2. The main material (hereinafter referred to as the binder) and the reinforcement material are linearly elastic, isotropic and homogeneous: the connection between deformations and the stresses in the binder and reinforcement follow Hooke's law.
3. There is an ideal grip between the binder and the reinforcement.
4. Additional stresses perpendicular to the direction of reinforcement, which in this case have a different Poisson's ratio (between the reinforcement and the binder), can be considered negligibly small.
5. When the sample is loaded across the direction of stress reinforcement in the binder and reinforcement are the same, and the proportion of deformations of the components materials are calculated with the assumption that in a general deformation, composite material is proportional to the volume content of each component.
6. The composite material is reinforced with straight-line fibers.

With sufficiently thick uniform saturation of the main material (binder) with fibers, the first premise is completely acceptable. All researchers who study the mechanical properties dispersed-reinforced media have come to such conclusion. In order to be able to use this premise with sufficient accuracy for practical purposes, it is necessary to know what should be the smallest number of fibers in z unit cross-sectional area of unidirectional reinforced material. The second premise is introduced because of the need to simplify the solution of the problem and keep it limited only to the elastic stage of operation, which is true for many composite materials. From the third premise it follows that the component materials in the direction of the reinforcement are deformed together. The use of this assumption almost excludes errors, if the stresses acting along the reinforcing fibers are not varying along its length. The fourth premise is very close to the real situation and stresses (even if the Poisson's ratio for the reinforcement and binder is the same) are rapidly damped. When discussing the fifth premise, it should be remembered that actual stresses in the main material near the reinforcing bar increase, because for reinforcement more rigid material is used, but this increase has a local character and it gradually disappears as a result of elastic redistribution. Actual stress distribution is a very complex issue requiring special consideration. When determining the averaged elasticity of the deformation characteristics of the composite material, we will operate with the average stresses for the entire volume of the considered element.

Assuming that the distribution of fibers in an elementary volume is uniform, and the fiber diameter is small in comparison with the distances over which the averaged stress and strain fields that vary markedly. With sufficient accuracy of effective stiffness characteristics unidirectional composite material can be calculated from the following formulas [23, 24].

$$E_1 = \mu E_f + (1-\mu)E_m$$
$$E_2 \approx \frac{E_f E_m}{\mu E_f + (1-\mu)E_m} \tag{1.2}$$
$$G_{12} = \frac{(1+\mu)G_f + (1-\mu)G_m}{(1+\mu)G_m + (1-\mu)G_f} G_m$$

where the shear module of the isotropic fiber and the matrix are respectively equal

$$G_f = \frac{E_f}{2(1+\nu_f)}; \quad G_m = \frac{E_m}{2(1+\nu_m)}$$
$$\nu_{12} = \mu\nu_f + (1-\mu)\nu_m; \quad \nu_{21} = \frac{\nu_{12}E_2}{E_1} \tag{1.3}$$

Formulas for the elastic modulus E_1 and Poisson's ratio ν_{12} are sufficiently accurate for use in engineering calculations. The relations (1.2) for modules E_2 and G_{12} can be recommended for approximate calculations. It is expedient to refine them by the results of experiments using given material.

1.4.1 *Elastic deformation model of cross-reinforced composite material*

Let's introduce the orthogonal coordinates (1, 2) and assume that the axis 1 (the reinforced layer) makes an angle φ with the axis σ_{11}. Static relationships, the connecting stresses in the coordinate systems σ_{11}. σ_{22} and 1, 2 are as follows:

$$\sigma_{11} = \sigma_1 \cos^2\varphi + \sigma_2 \sin^2\varphi - \tau_{12}\sin 2\varphi$$
$$\sigma_{22} = \sigma_1 \sin^2\varphi + \sigma_2 \cos^2\varphi - \tau_{12}\sin 2\varphi \tag{1.4}$$
$$\sigma_{12} = (\sigma_1 - \sigma_2)\sin\varphi\cos\varphi + \tau_{12}\cos 2\varphi$$

Geometric relationships that allow expressing deformations in the system coordinate 1, 2 through deformations in the system coordinate σ_{11}. σ_{22}, can be written as follows:

$$\varepsilon_1 = \varepsilon_{11}\cos^2\varphi + \varepsilon_{22}\sin^2\varphi + \varepsilon_{12}\sin\varphi\cos\varphi$$
$$\varepsilon_2 = \varepsilon_{11}\sin^2\varphi + \varepsilon_{22}\cos^2\varphi - \varepsilon_{12}\sin\varphi\cos\varphi \tag{1.5}$$
$$\gamma_{12} = (\varepsilon_{22} - \varepsilon_{11})\sin 2\varphi + \varepsilon_{12}\cos 2\varphi$$

Now we obtain the relations connecting the stresses σ_{11} σ_{22} σ_{12} with deformations ε_{11} ε_{22} ε_{12}. For this purpose we substitute the deformations

ε_1; ε_2 and γ_{12} into Hooke's law, and the stresses obtained as a result of this substitution in the relation (1.5). After some simplification and by using symmetry conditions, we write the physical relationships for the layer reinforced at an angle φ to the axis α_1.

$$\begin{aligned}\sigma_{11} &= A_{11}\varepsilon_{11} + A_{12}\varepsilon_{22} + A_{13}\varepsilon_{12}\\ \sigma_{22} &= A_{21}\varepsilon_{11} + A_{22}\varepsilon_{22} + A_{23}\varepsilon_{12}\\ \sigma_{12} &= A_{31}\varepsilon_{11} + A_{32}\varepsilon_{22} + A_{33}\varepsilon_{12}\end{aligned} \tag{1.6}$$

where:

$$\begin{aligned}A_{11} &= \overline{E}_1\cos^4\varphi + \overline{E}_2\sin^4\varphi + 2[\overline{E}_1\nu_{12} + 2G_{12}]\sin^2\varphi\cos^2\varphi\\ A_{22} &= \overline{E}_1\sin^4\varphi + \overline{E}_2\cos^4\varphi + 2[\overline{E}_1\nu_{12} + 2G_{12}]\sin^2\varphi\cos^2\varphi\\ A_{12} &= A_{21} = \overline{E}_1\nu_{12} + [\overline{E}_1 + \overline{E}_2 - 2(\overline{E}_1\nu_{12} + 2G_{12})]\sin^2\varphi\cos^2\varphi\\ A_{13} &= A_{31} = \sin\varphi\cos\varphi[\overline{E}_1\cos^2\varphi - \overline{E}_2\sin^2\varphi - (\overline{E}_1\nu_{12} + 2G_{12})\cos 2\varphi]\\ A_{23} &= A_{32} = \sin\varphi\cos\varphi[\overline{E}_1\sin^2\varphi - \overline{E}_2\cos^2\varphi + 2(\overline{E}_1\nu_{12} + 2G_{12})\cos 2\varphi]\\ A_{33} &= [\overline{E}_1 + \overline{E}_2 - 2\overline{E}_1\nu_{12}]\sin^2\varphi\cos^2\varphi + G_{12}\cos^2\varphi\end{aligned} \tag{1.7}$$

The graphical examples of functions $A_{11}(\varphi)$; $A_{22}(\varphi)$,....$A_{33}(\varphi)$ (for different ratios of E_1/E_2 and after using Equation 1.1) are presented below (see Figs. 1.3 and 1.4).

From Figs. 1.3 and 1.4 one can see that they are periodic functions ($\varphi = \pi$ or $\pi/2$).

Next, we write the relations connecting shear stresses and shear strains in the coordinates (11; 22 and 33) and 1, 2, 3 (see Fig. 1.2):

$$\begin{aligned}\sigma_{13} &= \tau_{13}\cos\varphi - \tau_{23}\sin\varphi\\ \sigma_{23} &= \tau_{23}\cos\varphi + \tau_{13}\sin\varphi\end{aligned} \tag{1.8}$$

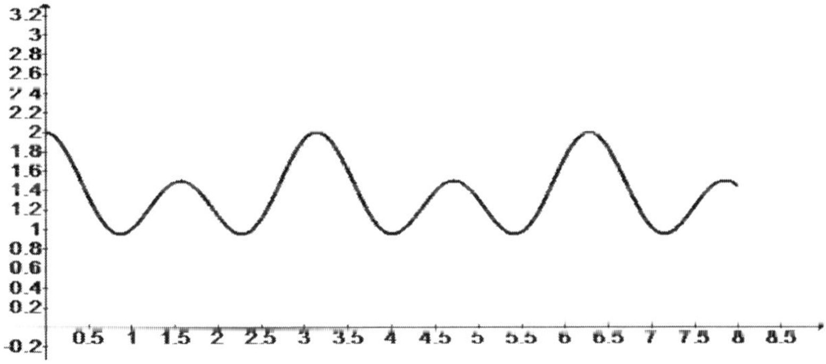

Figure 1.3: Example of functions A_{11}; ....;A_{33}.

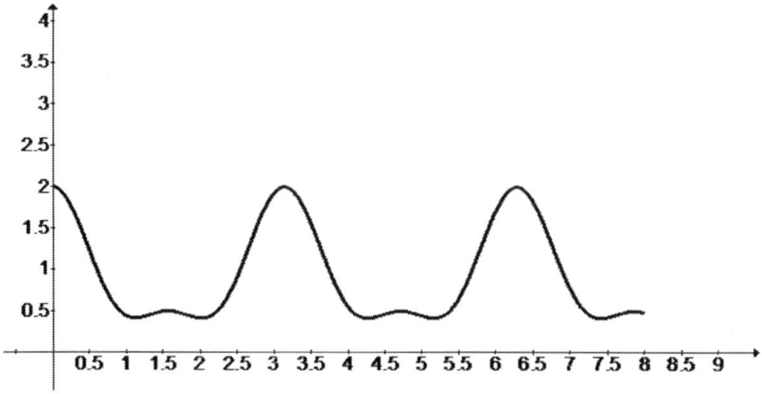

Figure 1.4: Example of functions A_{11}; ....;A_{33} (periodic functions).

$$\begin{aligned} \gamma_{13} &= \varepsilon_{13}\cos\varphi + \varepsilon_{23}\sin\varphi \\ \gamma_{23} &= \varepsilon_{23}\cos\varphi - \varepsilon_{13}\sin\varphi \end{aligned} \tag{1.9}$$

Using Equations (1.3), (1.8), (1.9), we can obtain relations of the type (1.6), that is:

$$\begin{aligned} \sigma_{13} &= A_{44}\varepsilon_{13} + A_{45}\varepsilon_{23} \\ \sigma_{23} &= A_{54}\varepsilon_{13} + A_{55}\varepsilon_{23} \\ A_{44} &= G_{13}\cos^2\varphi + G_{23}\sin^2\varphi \\ A_{55} &= G_{13}\sin^2\varphi + G_{23}\cos^2\varphi \\ A_{45} &= A_{54} = (G_{13} - G_{23})\sin\varphi\cos\varphi \end{aligned} \tag{1.10}$$

In multilayered composites and structural elements, a layer with an angle of reinforcement φ^+, as a rule, corresponds to the same layer with an angle of reinforcement φ^-. Virtually all automated production processes of a multilayer composite package, and in some cases mutual interlacing of adjacent symmetric layers with angles φ ±, are the most economical. In this case it is natural to consider such two layers as one symmetrically reinforced layer in the structural design calculation. This assumption completely corresponds to the real structure of layered composites and greatly simplifies the relationships connecting stresses and strains. If each of the symmetric layers is anisotropic in coordinates (1; 2) and α_i, then, working together, they form an orthotropic layer, Hooke's law for which has a simpler formula than the relations above for an individual layer. To obtain this law, we can rewrite the equations for multilayer with reinforcement angles $\varphi \pm$ as follows:

$$\begin{aligned} \sigma_1^{\pm} &= \overline{E}_1(\varepsilon_1^{\pm} + \nu_{12}\varepsilon_2^{\pm}) \\ \sigma_2^{\pm} &= \overline{E}_1(\varepsilon_2^{\pm} + \nu_{12}\varepsilon_1^{\pm}) \\ \tau_{12} &= G_{12}\gamma_{12}^{\pm} \end{aligned} \tag{1.11}$$

$$\begin{aligned} \sigma_{11}^{\pm} &= \sigma_1^{\pm}\cos^2\varphi + \sigma_2^{\pm}\sin^2\varphi \mp \tau_{12}^{\pm}\sin 2\varphi \\ \sigma_{22}^{\pm} &= \sigma_1^{\pm}\sin^2\varphi + \sigma_2^{\pm}\cos^2\varphi \pm \tau_{12}^{\pm}\sin 2\varphi \\ \sigma_{12}^{\pm} &= \pm[\sigma_1^{\pm} - \sigma_2^{\pm}]\sin\varphi\cos\varphi + \tau_{12}^{\pm}\cos\varphi \\ \varepsilon_1^{\pm} &= \varepsilon_{11}\cos^2\varphi + \varepsilon_{22}\sin^2\varphi \pm \varepsilon_{12}\sin\varphi\cos\varphi \\ \varepsilon_2^{\pm} &= \varepsilon_{11}\sin^2\varphi + \varepsilon_{22}\cos^2\varphi \mp \varepsilon_{12}\sin\varphi\cos\varphi \\ \gamma_{12}^{+} &= \pm[\varepsilon_{22} - \varepsilon_{11}]\sin 2\varphi\cos\varphi + \varepsilon_{12}\cos 2\varphi \end{aligned} \tag{1.12}$$

Here the signs correspond to layers with angles $\pm\ \varphi$. In relations (1.11) it is taken into account, that the properties of the material of both layers are the same, and equalities (1.11) allow for compatibility conditions of deformations of these layers: $\varepsilon_{11}^{\pm} = \varepsilon_1$; $\varepsilon_{22}^{\pm} = \varepsilon_2$; $\varepsilon_{12}^{\pm} = \varepsilon_{12}$. Substituting the deformations into the Hook's law and the stresses in the relations (1.13), we obtain the stresses σ expressed in terms of deformations, by averaging stresses: $\sigma_{11} = \frac{1}{2}(\sigma_{11}^{+} + \sigma_{11}^{-})$; $\sigma_{22} = \frac{1}{2}(\sigma_{22}^{+} + \sigma_{22}^{-})$; $\sigma_{12} = \frac{1}{2}(\tau_{12}^{+} + \tau_{12}^{-})$ according to the formulas:

$$\begin{aligned} \sigma_{11} &= A_{11}\varepsilon_{11} + A_{12}\varepsilon_{22} \\ \sigma_{22} &= A_{21}\varepsilon_{11} + A_{22}\varepsilon_{22} \\ \sigma_{12} &= A_{33}\varepsilon_{12} \\ A_{mn}&(mn = 11;12;21;22;33) \end{aligned} \tag{1.13}$$

We now derive the relations connecting shear stresses σ_{13} and σ_{23} with corresponding deformations of symmetrically reinforced layers. For layers with angles $\varphi \pm$, according to equalities (1.13), we have:

$$\begin{aligned} \sigma_{13}^{\pm} &= A_{44}\varepsilon_{13}^{\pm} + A_{45}\varepsilon_{23}^{\pm} \\ \sigma_{23}^{\pm} &= A_{54}\varepsilon_{13}^{\pm} + A_{55}\varepsilon_{23}^{\pm} \end{aligned} \tag{1.14}$$

Consider a symmetrical pair of layers loaded by stresses σ_{13}. Then $\sigma_{13}^{+} = \sigma_{13}^{-} = \sigma_{13}$, and, since the system is orthotropic, in the relation (1.14) it is necessary to take $\varepsilon^{+}_{13} = \varepsilon^{-}_{13} = \varepsilon_{13}$ and $\varepsilon^{+}_{23} = \varepsilon^{-}_{23} = 0$. Similarly, when the shear stresses σ_{23} are loaded we have: $\varepsilon^{+}_{13} = \varepsilon^{-}_{13} = 0$; and $\varepsilon^{+}_{23} = \varepsilon^{-}_{23} = \varepsilon_{23}$. As a result, we get

$$\sigma_{13} = A_{44}\varepsilon_{13};\ \sigma_{23} = A_{55}\varepsilon_{23} \tag{1.15}$$

We note that the system of coupled symmetrically reinforced layers in general case is more rigid than the asymmetrically reinforced layer of the same thickness. Thus, the physical relationships for the cross-reinforced composite material are defined by equalities (1.14) and (1.15).

1.5 Optimization of multi-layered composite structure parameters

The structure of composites (number of layers n and their thicknesses δ_i, reinforcement angles φ_i and the sequences of layers application for the total thickness $t = n\,\delta_i$) depend on their physical and mechanical characteristics, and hence the efficiency of their application. Therefore, designing the composite structure is a fundamental problem of any structural design development. To date, there is no complete and correct solution of this problem, which in the class of orthotropic composites is formulated as follows: to determine such thickness values of the layers δ_i, their reinforcement angles φ_i and their total number in the packet n, which would deliver a minimum mass $M = ab\sum_{i=1}^{n}\delta_i\rho_i$ while ensuring the strength of each layer in case of creep deformations due to high temperature effect with the assumption that the limit state of a composite structure is in compliance with a given strength criterion, for example, maximum stresses (Mises Hill criteria) or maximum allowable deformation (failure criteria). For example:

$$\frac{\sigma_{1i}^2}{F_{1i}^2} - \frac{\sigma_{1i}\sigma_{2i}}{F_{1i}F_{2i}} + \frac{\sigma_{2i}^2}{F_{2i}^2} + \frac{\tau_{12i}^2}{F_{12i}^2} \le 1 \tag{1.16}$$

F_{1i} and F_{2i}—the corresponding allowable strength limits of the composite material layer based on the solution of creep constitutive integral type equation (see Chapter 3). Stresses in layers σ_{1i}; σ_{2i} and τ_{12i} are defined by formulas

$$\sigma_{1i} = \varepsilon_{xi}\left(\cos^2\varphi_i + \nu_{21}\sin^2\varphi_i\right) + \varepsilon_{yi}\left(\sin^2\varphi_i + \nu_{21}\cos^2\varphi_i\right) + \gamma_{xy}(1-\nu_{12})\sin\varphi_i\cos\varphi_i$$

$$\sigma_{2i} = \varepsilon_{xi}\left(\sin^2\varphi_i + \nu_{21}\cos^2\varphi_i\right) + \varepsilon_{yi}\left(\cos^2\varphi_i + \nu_{21}\sin^2\varphi_i\right) + \gamma_{xy}(1-\nu_{12})\sin\varphi_i\cos\varphi_i$$

$$\tau_{12i} = (\varepsilon_{yi} - \varepsilon_{xi})\sin 2\varphi_i + \gamma_{xy}\cos 2\varphi_i$$

(1.17)

After substituting formulas (1.16) into criterion (1.15), we have inequality:

$$\frac{\sigma_{1i}^2}{F_{1i}^2} - \frac{\sigma_{1i}\sigma_{2i}}{F_{1i}F_{2i}} + \frac{\sigma_{2i}^2}{F_{2i}^2} + \frac{\tau_{12i}^2}{F_{12i}^2} \le 1$$

$$\frac{[\varepsilon_x A_{1i} + \varepsilon_y A_{2i} + \gamma_{xy} A_{31}]^2}{F_{1i}^2} - \frac{[\varepsilon_x A_{1i} + \varepsilon_y A_{2i} + \gamma_{xy} A_{31}][\varepsilon_x A_{2i} + \varepsilon_y A_{1i} - [1-\nu_{21}]\gamma_{xy} A_{31}]}{F_{1i}F_{2i}} +$$

$$+\frac{\varepsilon_x A_{2i} + \varepsilon_y A_{1i} - [1-\nu_{21}]\gamma_{xy} A_{3i}}{F_{2i}^2} + \frac{[\varepsilon_y - \varepsilon_x]A_{4i} + \gamma_{xy} A_{5i}}{F_{12i}^2} < 1$$

$$\Phi(\varepsilon_x;\varepsilon_y;\gamma_{xy};\varphi_i) < 1; \quad i = 1;2 \tag{1.18}$$

For uniaxial loading condition:

$$\frac{\sigma_{11}^2}{F_{11}^2} + \frac{\tau_{12}^2}{F_{12}^2} \le 1$$

$$\frac{[\varepsilon_x A_{11} + \gamma_{xy} A_{31}]^2}{F_{11}^2} + \frac{[-\varepsilon_x A_{41} + \gamma_{xy} A_{51}]^2}{F_{12}^2} \le 1 \tag{1.19}$$

If the strains ε_x and γ_{xy} are given, for example, by restrictions on strength, stability and deflection of limit state of the whole multilayered composite element than inequality (1.18) becomes a function of angle φ_i only, and the optimum design (finding optimum values of φ_i) can be achieved by equating first derivative of $\Phi(\varphi_i)$ to zero.

For uniaxial loading condition without shear deformation:

$$\frac{\sigma_{1i}^2}{F_{1i}^2} \le 1; \quad \frac{[\varepsilon_x A_{1i}]^2}{F_{1i}^2} \le 1$$

$$\varepsilon_x A_{1i} = \varepsilon_x \cos^2\varphi_i \le F_{1i} \tag{1.20}$$

$$i = 1 \to A_{11} = [\bar{E}_1 \cos^4\varphi + \bar{E}_2 \sin^4\varphi + 2[\bar{E}_1\nu_{12} + 2G_{12}]\sin^2\varphi\cos^2\varphi]\varepsilon_x \le F_{11}$$

$$i = 2 \to A_{22} = [\bar{E}_1 \sin^4\varphi + \bar{E}_2 \cos^4\varphi + 2[\bar{E}_1\nu_{12} + 2G_{12}]\sin^2\varphi\cos^2\varphi]\varepsilon_y \le F_{22}$$

The average deformations of the composite element with ε_x; ε_y and γ_{xy} are found from the equilibrium equations of the orthotropic structural element. Consider the formulation of the problem of creep deformation of composite elements of frame and beam structures under high-temperature loading. It is assumed that the structural element has a regular structure in thickness and is formed by gluing together the composite layers. At the same time, the kinematic model of deformation of structural elements is based on hypotheses of the Bernoulli type for the entire package as a whole. The connection between stress and strain tensors in composite layers is established on the basis of Hooke's law for an orthotropic body with effective elastic characteristics. The derivation of the system of equations is based on the principle of possible displacements and is given in Chapter 3.

1.5.1 Influence of fiber length

The critical fiber length is necessary for effective strengthening and stiffening of the composite material. This critical length is dependent on the fiber diameter d; its ultimate tensile stress, and on the fiber–matrix bond strength (shear stress of the matrix, whichever is smaller), and it is calculated by

$$l_c = \frac{d_f^* d}{2\tau_c} \tag{1.21}$$

For glass and carbon fiber–matrix combinations, this critical length is on the order of 1 mm, which ranges between 20 and 150 times the fiber diameter.

Fibers for which $l >> l_c$ (normally $l > 15l_c$) are termed continuous; discontinuous or short fibers have lengths shorter than this. For discontinuous fibers of lengths significantly less virtually no stress transference occurs. To affect a significant improvement in strength of the composite, the fibers must be continuous.

1.5.2 Elastic behavior—longitudinal loading

Let's consider now the elastic behavior of a continuous and oriented fibrous composite that is loaded in the direction of fiber alignment. It is assumed here that the fiber–matrix interfacial bond is very good, such that deformation of both matrix and fibers is the same (an isostrain situation). Under these conditions, the total load sustained by the composite is equal to the sum of the loads carried by the matrix phase and the fiber phase.

1.6 Molecular mechanisms of chemical reactions

1.6.1 Chemical reaction kinetics

The most important kinetic characteristic is the rate of chemical reaction, which is defined as the number of particles (molecules, ions, atoms) of a given species (N) reacting per unit time in a unit of reaction space. With a constant volume for a closed system, the rate of chemical reaction can be expressed by the following Equation [17]:

$$r = \pm \frac{1}{\nu_i} \frac{dC_i}{dt} \tag{1.22}$$

where C_i is the concentration of the reacting i-th substance and νi is the stochiometric coefficient in the reaction equation before the i-th reactant. It is generally considered that the reaction rate is a positive quantity, so if a change in the amount of the initial substance is used in its calculation, then it is necessary to put a minus sign in Equation (1.22), if the measurements are carried out for the reaction products—the plus sign. To calculate the reaction rate, it is sufficient to determine the reaction rate for one of the substances, the changes in the concentrations of the others can be established on the basis of the stochiometry of the reaction. For example, for a one-way reaction occurring in a closed system and written in the form of a stochiometric equation $\nu_1 A + \nu_2 B \rightarrow \nu_3 C + \nu_4 D$ the following expression holds

$$r = -\frac{1}{\nu_1}\frac{dC_A}{dt} = -\frac{1}{\nu_2}\frac{dC_B}{dt} = \frac{1}{\nu_3}\frac{dC_C}{dt} = \frac{1}{\nu_4}\frac{dC_D}{dt} \quad (1.23)$$

The rate of chemical reaction depends on many factors, among which it is necessary to distinguish the dependence on the concentrations of the reacting substances studied in formal kinetics. The basic law on which the quantitative regularities of the course of chemical reactions in time are based is the kinetic law of the acting masses: the rate of the chemical reaction at each moment of time is proportional to the product of the concentrations of the reacting substances. At the heart of the law is a simple physical principle: chemical interaction is possible in the collision of reacting molecules, and the probability of the latter is proportional to the product of concentrations. For the simple reaction the mathematical expression of the law of acting masses has the following form:

$$r = kC_A^{n_A} C_B^{n_B} \quad (1.24)$$

where the exponents at concentrations of reagents n_A and n_B are called partial orders. For elementary reactions occurring in one stage, particular orders are integers equal to the stochiometric coefficients in Equation (1.22); in complex reactions having several stages, the order can be fractional and even negative. The sum of the exponents $n = n_A + n_B$ is called the general order of the reaction. The latter for elementary reactions is the number of molecules participating in one elementary act of chemical transformation. Thus, simple mono-, bi- and tri-molecular reactions are at the same time first-, second- and third-order reactions. Naturally, for complex reactions one can speak only of the molecular nature of the individual stages, and not of the molecular nature of the reaction as a whole. Molecularity of the reaction is a theoretical concept. To determine the molecularness, it is necessary to know the mechanism of the process, its limiting stage, the speed of which determines the speed of the process as a whole. Order, in contrast to molecularness, is an experimental value. In the formula for the effective mass law (1.23), k is called the rate constant (specific rate) of the chemical reaction. It can be seen from

Equation (1.23) that the rate constant is numerically equal to the rate of this reaction at unit concentrations of all the reacting components. The rate constant depends on the same factors as the reaction rate, in addition to the concentration of the reactants and time. The law of acting masses is satisfied under certain conditions, which for a simple reaction consist in the following:

1. The chemical reaction proceeds slowly enough in comparison with the physical processes of energy exchange between the reacting molecules. The violation of this principle is reflected in the kinetic equation. For example, a monomolecular reaction in the gas phase proceeds according to the bimolecular law, when the process is limited by energy transfer in bimolecular collisions.
2. The change in the concentration of reactants does not significantly change the properties of the medium (viscosity, polarity, etc.) and the physical state of the reactants.
3. The chemical reaction is carried out only at the expense of thermal energy. If the reaction occurs under the influence of electromagnetic or radiation, electric current, etc., then the rate depends on these factors, and the law of the acting masses varies.

1.6.2 *Methods for determining the order of reactions*

Formal description of the kinetics of chemical reactions is based on experimental determination of the rate constant and of the partial orders into each of the reacting substances. If several reactants enter the reaction, the following methods are used to determine the order. Assume that all reactants are taken in the stochiometric ratio. In this case, the concentrations of reactants vary in a constant proportion, and the reaction rate is determined by the concentration of any of the starting materials. Thus, for example, if the reaction takes place, then its rate according to the law of mass action can be described by the Equation (1.23). In case of the stochiometric ratio of reactants ($C_A :_B = \nu_1 : \nu_2$), the reaction rate is:

$$r = k(\nu_2 / \nu_1)^{n_B} C_A^n \tag{1.25}$$

where n is the total reaction order determined by one of the methods described below.

1.6.3 *Ostwald's "isolation method"*

One of the reactants is taken in the deficit in such a way that during the experiment it is possible to neglect the expenditure of the others. In this case, the change in the reaction rate is determined only by the concentration of the reactants taken in the deficit: $r = k_{eff.} C_i^{n_i}$, where $k_{eff.}$ includes the product of the rate constant and the concentrations of the reactants. The order of the reaction for the reactants taken in (n_i) is determined by one of the methods described below:

1.6.4 *Graphic method*

The order of the reactant, the concentration of which is practically unchanged during the kinetic experiment, can be determined from the dependences of the reaction rate on the initial concentration of the reactant (C_0). The linear character of the $r - C_0$ dependence indicates the first order for the reactants taken in excess, the linearity in coordinates $r - (C_0)^2$ indicates the second order, etc.

1.6.5 *The differential method of Van't Hoff*

A series of experiments with different initial concentrations of reactants is carried out. The initial sections of the kinetic curves determine the initial reaction rate (r_0) by the graph method or numerical differentiation. If all the reactants, with the exception of one, are taken in excess, a particular order of reaction for the reactants taken in the deficiency is found from the relationship:

$$\ln r_0 = \ln k_{eff.} + n_i \ln C_{0,i} \tag{1.26}$$

Equation (1.26) can be used to determine the order of reactions from the data of one experiment if instead of the initial values of the rates and concentrations of reactants, the corresponding current values determined from the kinetic curve are used. On the graph in the system of coordinate's $\ln r - \ln C$ a linear dependence is obtained, the slope of which to the axis of abscissa is equal to the order of the reaction n, and the initial ordinate corresponds to ln k.

1.6.6 *Dependence of the reaction rate on temperature*

For most chemical reactions, the reaction rate increases with increasing temperature. This is due to an increase in the rate of thermal motion of molecules and the total number of collisions of molecules of reacting substances. There is an empirical Van't Hoff rule, established for homogeneous chemical reactions in solutions, according to which the velocity and the rate constant increase by 2 to 4 times with increasing temperature by 10°C. The exact ratio of the rate constants, called the Van't Hoff temperature coefficient, is determined by the following equation:

$$k_{T_1}/k_{T_2} = \gamma^{\frac{T_1 - T_2}{10}} \tag{1.27}$$

The chemical interaction between molecules occurs when they collide, but the collision frequency is immeasurably higher than the number of molecules reacting per unit time. This is understandable, since each particle (atom, molecule or ion) is a fairly stable structure. Therefore, in order for the chemical interaction to occur, it is necessary that the molecule (atom, ion) have excess energy in comparison with the average particle energy, the so-called activation energy. The latter is not a special kind of energy, it is an increased translational, rotational, vibration, electronic energy of interacting particles. Their activation occurs when they collide with each other and with the walls of the reaction vessel, absorb electromagnetic radiation, and so on. The fraction of particles whose energy exceeds the activation energy (E), according to the Boltzmann law, is exp (–E/RT), so the rate constant can be represented as: $k = A[\exp(-E/RT)]$. This equation is called the integral form of the Arrhenius law. For elementary reactions in a sufficiently wide temperature range, the constants A (the pre-exponential factor) and E are practically independent of temperature. Having determined the numerical values of the activation energy and the pre-exponential factor A, we can calculate the rate constant of the chemical reaction at any given temperature. Constants of the Arrhenius equation are usually determined experimentally from the temperature dependence of the rate constant. In the presence of at least five values of the rate constants for different temperatures, the logarithmic form of the equation is used:

$$\ln k = \ln A - E/RT \tag{1.28}$$

The angular coefficient of the linear dependence ln k – 1/T and the initial ordinate are determined, which are numerically equal to –E/R and ln A, respectively. If the values of the constants are known at two temperatures, then the activation energy can be calculated from the equation:

$$E = [RT_1T_2 \ln(k_2/k_1)]/(T_2 - T_1) \tag{1.29}$$

The main objective of this section is to introduce the reader to the technical terminology and few practical experimental techniques of a very broad discipline of physical chemistry. The practical applications of kinetics of chemical reactions to the subject of creep deformations of composites and nanocomposites are presented in Chapter 5.

References

[1] Composite Materials Handbook. 2002. Department of Defense Handbook.
[2] de Silva, C.W. (ed.). 2008. Mechatronic Systems: Devices, Design, Control, Operation and Monitoring, CRC Press, London, N.Y.

[3] Smith, W.R. 2003. Computational aspects of chemical equilibrium in complex systems. Theoretical Chemistry: Advances and Perspectives, Academic Press, N.Y. 28: 185–259.

[4] Gerstle, F.P. 1991. Composites, Encyclopedia of Polymer Science and Engineering. Wiley, New York.

[5] Hashin, Z. 1972. Theory of Fiber Reinforced Materials. National Aeronautics Space Administration CR-1974.

[6] Christensen, R.M. 1979. Mechanics of Composite Materials, Wiley-Interscience, N.Y.

[7] Hughes, W.J. 2011. Determining the Fatigue Life of Composite Aircraft Structures Using Life and Load-Enhancement Factors, FAA, National Technical Information Services (NTIS), Springfield, V.A.

[8] Yang, Qing-Sheng and Qin, Qing-Hua. 2001. Fiber interactions and effective elastic-plastic properties of short-fiber composites. Composite Structures Elsevier, N.Y. 54: 523–528.

[9] Hyer, M.W. 1998. Stress Analysis of Fiber-reinforced Composite Materials. McGraw-Hill N.Y.

[10] Kelly, A. and Tyson, W.R. 1965. Tensile properties of fiber-reinforced metals: copper/tungsten and copper/molybdenum. J. Mech. Phys. Solids 6(1965): 329.

[11] Kelly, A. 1973. Strong Solids. 2nd Edition, Clarendon Press, Oxford.

[12] Kelly, A. 2009. Engineering triumph of carbon fiber. Composites and Nanostructures 2009(1): 38–49.

[13] Hill, R. 1948. A theory of the yielding and plastic flow of anisotropic metals. Proc. Roy. Soc. London 193: 281–297.

[14] Tsai, S.W. and Wu, E.M. 1971. A general theory of strength for anisotropic materials. Journal of Composite Materials 5: 58–80.

[15] Hoffman, O. 1967. The brittle strength of orthotropic materials. J. Composite Materials 1: 200–206.

[16] Chamis, C.C. 1969. Failure criteria for filamentary composites. Composite Materials: Testing and Design, STP 460, ASTM, Philadelphia, 336–351.

[17] ASM Handbook, Vol. 21. 2001. Composites, ASM International, Materials Park.

[18] Chamis, C.C. and Murthy, P.L.N. 1991. Probabilistic Composite Analysis. First NASA Advanced Composites Technology Conference, Part 2, NASA CP3104-PT-2, 891–900.

[19] Chamis, C.C., Shiao, M.C. and Kan, H.P. 1993. Probabilistic Design and Assessment of Aircraft Composite Structures. Fourth NASA/DoD Advanced Composites Technology Conference, June 7–11, 1993, Salt Lake City, Utah.

[20] Mallick, P.K. 1997. Composites Engineering Handbook. Marcel Dekker Inc., N.Y.

[21] Herakovich, C.T. 1998. Mechanics of Fibrous Composites. John Wiley & Sons, Inc. N.Y.

[22] Reddy, J.N. 1997. Mechanics of Laminated Composite Plates: Theory and Analysis. CRC Press, F.L.

[23] Voigt, W. 1889. Ueber die Beziehung zwischen den beiden Elasticitätsconstanten isotroper Körper. Annalen der Physik 274: 573–587.

[24] Reuss, A. 1929. Berechnung der Fließgrenze von Mischkristallen auf Grund der Plastizitätsbedingung für Einkristalle. Zeitschrift für Angewandte Mathematik und Mechanik 9: 49–58.

2
Creep Laws for Composite Materials

2.1 Introduction

Modern classical methods of investigation in mechanics of deformable rigid bodies are based on three hierarchical levels: mechanics of micro-inhomogeneous media, phenomenological models of a continuous medium and boundary-value problems that are seemingly little related to each other. Within the continuum mechanics, the traditional way of constructing a phenomenological model begins with a specially organized creep test of the material. The results are analyzed and the material model is constructed, which is then applied to the solution of the corresponding boundary-value problem. For non-stationary external loads, the boundary value problem must be solved taking into account the loading history, which in the case of a high temperature effect on the structure means the analytical dependence of the temperature on time.

Undoubtedly, this approach has both advantages and disadvantages. On the one hand, the laws of inelastic deformation in phenomenological theories are formulated for an arbitrary body and can be described as the creep theory of materials of a very diverse nature (metals, polymers, concrete, soils and so on). On the other hand, with the concretization of these general laws for one or another type of materials and environmental loading conditions, it actually describes the phenomenon, but does not explains any specific data of it. Also there is a need for a defining macro-experiment results wherein experiments are carried out in the rigid limits of temperature-force loading and of limit state condition that is used as well as the problem of extrapolation of calculated results for a particular phenomenological theory beyond the boundaries of this framework. Therefore, in order to more adequately reflect the processes of the inelastic creep deformation along with phenomenological theories in parallel, theories based on micro-inhomogeneity development of irreversible deformations are needed. Indeed, from the point of view of continuum mechanics material is a single homogeneous media as a whole; however at the same time it is known, that composites have a very complex structure, and this is how the composite material is treated on the microscopic (mechanics of inhomogeneous media, metallurgy) and submicroscopic (the physics of metals, composites, and nanocomposites based on the theory of dislocations) levels. Within the framework of the mechanics of micro-inhomogeneous media, there is a large number of various structural models built with the involvement of formal considerations for the representation of the material in the form of structured composition based on different levels of complexity. At the level of mechanics of micro-inhomogeneous media (one should note the physical models of creep of the material that are based on the physical nature of the fields of micro-deformations, the theory of dislocations, slip and other structured processes) have been proposed by L.M. Brown and M.F. Askby [1], Shanti V. Nair and Karl Jakus [2] and others. However, when assessing the stress-strain state of creep deformations of real structural element or the whole structural system, the structural analyses are often based on the phenomenological creep theory that treats the materials on the macroscopic level. Therefore, for a complete and adequate description of the deformation of structural elements, experimental studies of materials on a time base commensurate with the time of operation of the structure itself are necessary. It is clear that the main purpose of microstructural theories is not to solve boundary problems on their basis, but in establishing the nature of the creep deformation and the justified qualitative choice of the desired function between the microstructural relationship of deformations and stresses, describing at the phenomenological level the creep process (accurate to the constant parameters of the material and the type of creep functions).

2.2 Phenomenological creep model: single integral type constitutive equation (CE)

2.2.1 Temperature and kinetic energy

The integral representation of creep process is a very appealing theoretical concept, since it is not limited to a particular material or class of materials. Composite elements and heterogeneous materials from which they are made make essential changes to the classical scheme for creating the structural model of composite elements. In the case the phenomenological integral creep model, which is used in this chapter, the specificity of the composite material should manifest itself in the choice of two basic functions of the integral creep equation, namely E (t) and K (t, τ). The function of the instantaneous elastic modulus should reflect the fact that (1) the composite material is a two-phase system with a parameter T_g (very important for practical purposes: transient temperature); and also that (2) the composite material is a composition of two (or more) different materials in its physical parameters of materials, interconnected by internal forces of adhesion, which under the influence of high temperature change with time. The function K is a function of memory and consequently should reflect all changes in the energy balance with changes in temperature and time, such as the change in the free Gibbs energy due to the change in the stress state of the composite material and free chemical energy (in the case of nanocomposites). Thus, in the case of composite the integral creep equation, as noted in the author's work [3], has the following form:

$$E(t)\alpha_0(T)\sigma(t) = \sigma(t) + \int_0^t e^{-[\frac{E}{RT} - A(\frac{E_a}{RT})\sigma(t')^n]} K_1(t-t')\sigma(t')dt' \tag{2.1}$$

$$K(t) = e^{-[\frac{E}{RT} + A(\frac{E_a}{RT})\sigma(t')^n]} K_1(t-t')$$

Creep laws for composites under conditions which favor dislocation processes may be represented by a thermally activated power law:

$$\dot{\varepsilon}_c = A\sigma^n \exp(-H/RT) \tag{2.2}$$

$$H = E_a + PV$$

Here $\dot{\varepsilon}_c$ is the steady-state creep rate; H-activation enthalpy (E_a is the activation energy, V is the activation volume that has a dimension of activation energy, and σ is mean normal stress). Parameter A is obtained from experimental data. Although Equation (2.2) is non-Newtonian by definition, it can be expressed in terms of a simple viscous relation:

$$\dot{\varepsilon}_c = \frac{1}{\eta}\sigma \tag{2.3}$$

Let's define now an effective viscosity as follows:

$$\eta_{eff} = \frac{\exp(H/RT)}{A} \tag{2.4}$$

One can see now that an effective viscosity may vary locally as a function of stress. The final feature we emphasize in the relation (2.4) is its temperature–time dependence. If we invert Equation (2.2), the steady-state stress at fixed strain rate is:

$$\sigma_c = \left(\frac{\dot{\varepsilon}}{A}\right)\exp(H/RT) \tag{2.5}$$

This form of the flow law exhibits a powerful exponential effect of inverse temperature on the steady state strength. In order to analyze the engineering creep deformations of composite materials, one should classify them from the macro-structural point of view, living aside the manufacturing process and dealing with the final composite material product. It has to be mentioned again that the final product of any engineering creep analyses is limited in most cases to obtaining the stress-strain relationship as function of temperature (time) and its application to structural analysis and design. The matrix material can enhance the stiffness and therefore creep performance of the structure and it can be regulated by the mathematical expression of modulus of elasticity E(θ) in Equation (2.1) [4]. It is assumed here that the memory function K (t, t') in this case is affected by high temperature and it might be (or might be not) affected by the rheological composition of the material, for instance, non-Newtonian nonlinear viscosity. A strong dependence on temperature (time) of the properties of soft composites (PMS) exists compared with those of other materials such as metals and ceramics (MMC and CMC). This strong dependence is due to the viscoelastic nature of PMC. Generally, PMC behave in a more elastic fashion in response to a rapidly applied temperature rise (force) and in a more viscous fashion in response to a slowly applied temperature increase. Viscoelasticity means behavior similar to both purely elastic solids in which the deformation is proportional to the applied force and to viscous liquids in which the rate of deformation is proportional to the applied force. The behavior of composites is very complex due to this dual nature. This model (see Equation 2.1) presents the combined effects of thermal gradients and the strong dependence of creep upon temperature; the non-Newtonian exponential law dependence upon stress and the time-dependent nature of strength within the composite material.

When subjected to an applied stress, composites may deform by either or both of two fundamentally different atomistic mechanisms. The lengths and angles of the chemical bonds connecting the atoms may distort, moving the atoms to new positions of greater internal energy. This is a small motion and occurs very quickly, requiring only $\approx 10^{-12}$ seconds. If the composite material has sufficient molecular mobility, larger-scale rearrangements of the atoms may also be possible. For instance, the relatively facile rotation around backbone carbon-carbon single bonds can produce large changes in the conformation of the molecule. Depending on the mobility, a molecule can extend itself in the direction of the applied stress, which decreases its conformational entropy (the molecule is less "disordered"). Elastomers—rubber—respond almost wholly to this entropic mechanism, with little distortion of their covalent bonds or change in their internal energy. However, it is a well-known fact that when the rate of deformation and temperature varies over a wide range, certain difficulties arise in the separation of instantaneous deformation and creep deformation. Therefore, in most of the theories discussed here, the difference in the inelastic deformations of the transient and stationary creep process is often blurred, and both deformations are combined into one-inelastic. When considering the problem of the failure of composites in the case of high-temperature inelastic deformation, it should be taken into account the fact that the micro-mechanisms of viscoelastic failure are different, and therefore it is necessary to divide the creep strains into three main parts in accordance with the energy balance of the Gibbs free energy nonlinear theory. In connection with the foregoing, the energy approach to the description of creep deformation and failure of composites and nanocomposites, are based on the principle of *modified superposition* method of elastic, viscoelastic, and potential chemical energy, as well as the method of separation of the creep deformation as described in [4]. To describe the softening stage and the subsequent stage of material failure, a hypothesis is introduced according to which the instantaneous modulus of elasticity of the composite material is assumed to be proportional to the temperature changes only, which in turn is a predetermined function of time. The basic dimensionless integral creep equation for a uniaxial stress state has the form [5]:

$$
\begin{aligned}
&E(\theta)[\theta] = \sigma(\theta) + \int_0^\theta e^{\frac{\tau}{1+\beta\tau}}\{K_1(\theta,\tau) + A\sigma^s K_2(\theta,\tau)\}\sigma^n(\tau)\,m\mathbf{1}d\tau \\
&K_1(\theta,\tau) = \varphi_1(\theta)f_1(\tau) = m\mathbf{1}(\tau)\sum_{i=1}^{N}\exp(-\alpha_i\, m(\theta))\exp(\alpha_i\, m(\tau)) \\
&K_2(\theta,\tau) = \varphi_2(\theta)f_2(\tau) = m\mathbf{1}(\tau)\sum_{i=1}^{N}\exp(-\beta_i\, m(\theta))\exp(\beta_i\, m(\tau)) \\
&\theta = \frac{E_a}{RT_*^2}[T - T_*];\ \beta = \frac{RT_*}{E_a}
\end{aligned}
\tag{2.6}
$$

T_* - Base Temperature [°K]

$E = E_0(a - b*(\tanh(c*(\theta - \theta_g))))$

$s \geq \mathbf{0}; n \geq \mathbf{0}; A \geq \mathbf{0}$

Here K (θ) is, in comparison with linear theory, the degenerate kernel of time-temperature variant creep function and E denotes the modified dimensionless modulus of elasticity with a; b; c, and θ_g (dimensionless transition temperature)—constants of each component of composite material. The total value of E should comply with the so-called "rule of mixtures" or "volume fraction law". For fiber reinforced composite (FRC) Halpin and Tsai [6] developed a widely used model to predict the mechanical properties of aligned fiber reinforced composite laminates with the notation as follows: f and m represent the fiber and matrix, respectively; L the fiber length; D the fiber diameter; φ the volume fiber fraction, the longitudinal (L) and transverse (T). The basic assumptions used for the development of the rule of mixtures (ROM) and Halpin and Tsai (H-T) equations, as well as most micromechanical approaches are: both matrix and fibers are linearly elastic, isotropic, and homogeneous, fibers are perfectly aligned and spaced, matrix is void free, and bonding between matrix and fibers is perfect.

The longitudinal and transverse modulus, E_1 and E_2 are given by [7]:

$$E_0 = E_c = E_m \varphi_m + E_f \varphi_f \quad \text{Longitudinal direction of the fibers}$$

$$E_0 = E_c = \frac{E_m E_f}{E_m \varphi_m + E_f \varphi_f} \quad \text{Transverse direction of the fibers} \tag{2.7}$$

For dispersed fibers or nanoparticles:

$$E_0 = E = E_m[1 + 2.5\varphi_f + 14.1\varphi_f^2] \quad \text{Dispersed fibers or nanoparticles}$$

$$\varphi_f = \frac{V_f}{V_f + V_m} \tag{2.8}$$

The elastic modulus of the composite estimated with Equations (2.7 and 2.8) is presented in Table 2.1.

The ROM can also be used to evaluate the properties of composites reinforced by hybrid reinforcements that consist of two different fibers. In

Table 2.1: Computed composite modulus with existing relations of φ_m; φ_f.

φ_m	φ_f	Equation (2.7 L), E(GPa)	Equation (2.7 T), E(GPa)	Equation (2.8), E(GPa)
0.95	0.05	3.162	2.0550	2.21191
0.9	0.01	2.024	2.175	2.00928
0.85	0.15	5.566	2.2756	2.7572
0.8	0.2	6.768	2.4049	3.05054
0.75	0.25	7.97	2.5493	3.3577
0.7	0.3	9.1722	2.7123	3.6787
0.65	0.35	10.374	2.8977	4.0135
0.6	0.4	11.576	3.1103	4.3622
0.55	0.45	12.778	3.3566	4.7246
0.5	0.5	13.98	3.6452	5.1009
0.4	0.6	16.384	4.4022	5.8949

this case, another assumption is made in the derivation of the equations, i.e., that the longitudinal modulus of elasticity of the matrix material is very small compared to that of the two types of fibers [8]. Thus, the elastic modulus of the composite is obtained from:

$$E_{hb} = E_{LG}(1 - V_m) + (E_C - E_L)V_{LC}$$

E_{LG} - Longitudinal direction of glass fibers (2.9)

E_{LC} - Longitudinal direction of carbon fibers

Halpin and Tsai [5] developed empirical generalized equations that readily give quite satisfactory approximation of more complicated micromechanics results. These equations are quite accurate at low fiber volume fraction. They are also useful in determining the properties of composites that contain discontinuous fibers oriented in the loading direction. The Halpin-Tsai (H-T) equations can be written as [5, 9]:

$$M/M_m = (1 + \xi\eta V_f)/(1 - \eta V_f) \tag{2.10}$$

where: M represents composite modules, e.g., E_{11}, E_{22}, G_{12}, G_{23}, V_{12}, V_{23}, etc., M_f and M_m represent corresponding fiber and matrix modules, respectively. V_f is the fiber volume fraction, and ξ is a measure of reinforcement of the composite material that depends on the fiber geometry, packing geometry and loading conditions. The term ξ is an empirical factor that is used to make Equation (2.6) conform to the experimental data. The function η in Equation (2.7) is constructed in such a way that when $V_f = 0$, $M = M_m$, and when $V_f = 1$, $M = M_f$. So, for $\xi \rightarrow 0$

$$\frac{1}{M} = \frac{V_m}{M_m} + \frac{V_f}{M_f} \tag{2.11}$$

and for $\xi \rightarrow \infty$:

$$M = M_f V_f + M_m V_m \tag{2.12}$$

These two extremes bound the composite properties. Equation (2.11) gives a lower bound, while Equation (2.12) which is the well-known rule of mixtures equation gives an upper bound.

Whitney [10] suggests $\xi = 1$ or 2 for the transverse modulus E_{22} depending on the fiber array type, e.g., hexagonal, square, etc. Nielsen [11], modified the H-T equations to include the maximum packing factor ϕ_{max} of reinforcement.

2.2.2 *Use of classical relations of composite elastic modulus*

The properties of a composite material are function of the starting materials [11] so that the following relations are also found in literature for estimating the elastic modulus of particle fillers. The modulus of elasticity of the particle filled composite may be predicted using the following equations:

$$E_L = E_m\left(\frac{1+\xi\eta\varphi}{1-\eta\varphi}\right); E_T = E_m\left(\frac{1+\eta\varphi}{1-\eta\varphi}\right); G_{LT} = G_m\left(\frac{1+\lambda\varphi}{1-\lambda\varphi}\right)$$

$$\eta = \frac{\left(\frac{E_f}{E_m}-1\right)}{\left(\frac{E_f}{E_m}+\xi\right)}; \lambda = \frac{\left(\frac{G_f}{G_m}-1\right)}{\left(\frac{G_f}{G_m}+1\right)}; \xi = 2\frac{L}{D} \tag{2.13}$$

Most models accurately predict the longitudinal modulus; however, differences do exist between models when predicting the transverse modulus [12]. The stiffness decreases as one rotates away from the longitudinal axis for an aligned fiber reinforced composite with different volume fraction fiber content [13]. For high volume fraction fiber contents, only a slight misalignment of the fibers from the loading direction results in drastic property reductions. The stiffness in a long fiber reinforced composite with a random planar orientation, such as encountered in sheet molding compound (SMC) charges, can be estimated using:

$$E_{11} = E_{22} = E_{random} = \left(\frac{3}{8}\frac{1}{E_L} + \frac{3}{8}\frac{1}{E_T} - \frac{1}{4}\frac{\nu_{LT}}{E_L} + \frac{1}{8}\frac{1}{G_{LT}}\right)^{-1} \tag{2.14}$$

2.2.3 *Instantaneous creep modulus*

In most cases, the creep modulus, defined as the ratio of applied stress to the time-dependent strain, decreases with increasing temperature [14]. Generally speaking, an increase in temperature correlates to a logarithmic decrease in the time required to impart equal strain under a constant stress. In other words, it takes less work to stretch a viscoelastic material at an equal distance at a higher temperature than it does at a lower temperature. Constitutive modeling of elastic modulus in Equation (2.1) as a function of temperature (time) can be obtained from compression and tension testing data and presented by the hyperbolic tangent [15]:

$$E(\theta) = a - b\,[\tanh(c(\theta - \theta_g))]; \tag{2.15}$$

Below are some examples of modulus of elasticity E (θ) that are used throughout this book.

Example 2.1a

$E_c(\theta) = E_0\{a - b\,[\tanh c\,(\theta - \theta_g)]\}$; $E(\theta) = E_c(\theta)/E_0$

θ_g – dimensionless transition temperature of a composite material.

$$\theta_g = \frac{E_a[T_g - T_*]}{RT_*^2};\quad \beta = \frac{RT_*}{E_a} = 0.067$$

$$T_* \approx 300^{o}K;\quad T_g \approx 400^{o}K;\quad \theta_g \approx 5$$

At small temperatures, the stress is at a high plateau corresponding to a "glassy" modulus E_g, and then falls exponentially to a lower equilibrium "rubbery" modulus E_r as the composite material molecules gradually accommodate the strain by conformational extension rather than bond distortion.

Parameter 'c' defines the width of transition temperatures zone and can be selected by superimposing the relaxation curves by means of the "relaxation modulus" (see Fig. 2.1). Let's assume here c = 5. The value of dimensionless "glassy" modulus $E_g = 1$ and the value of dimensionless "rubbery" modulus E_r is a fraction of E_g and should be obtained from experimental data. Let's assume here that $E_r/E_g = 0.25$. Parameters 'a' and 'b' can be easily computed from the boundary conditions as follows:

E (θ = 0) = 1 and E (θ = ∞) = 0.25. Since tanh (0) = 0 and tanh(θ →∞) ≅ 1, we have in our case:

$$\begin{cases} a + b = 1 \\ a - b = 0.25 \end{cases} \quad \text{or:}\ a = 0.625\ \text{ and }\ b = 0.375$$

Finally, the functions E (θ) are (see Fig. 2.1):

$E(\theta) = 0.625 - 0.375\,[\tanh 5\,(\theta - 5)]$

$E(\theta) = 0.625 - 0.375\,[\tanh 5\,(\theta - 4)]$

$E(\theta) = 0.625 - 0.375\,[\tanh 3\,(\theta - 4)]$

As mentioned above, in order to analyze the engineering creep deformations of composite materials one should consider them from the macro-structural point of view [16], i.e., dealing with the final composite material product. It has to be mentioned again that the final product of any engineering creep analyses is limited in most cases to obtaining the stress-strain relationship as function of temperature (time) and its application to structural analysis and design [17].

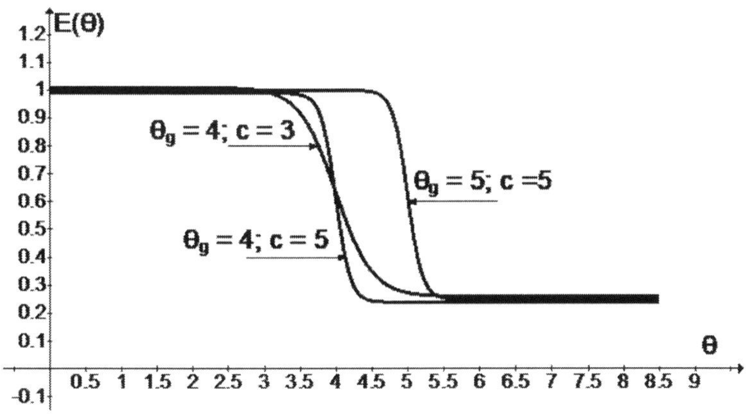

Figure 2.1: Composite effective elastic modulus.

2.2.4 *Effects of θ_g on modulus of elasticity*

Phenomenological models describe the instantaneous modulus of elasticity E (t) in terms of mathematical equations of a "modified Voce hardening law" the tanh rule (where experimental test data fits the hyperbolic tangent function of the modulus of elasticity). These models can provide excellent fits for a given deformation condition. Well-known phenomenological models based on the hardening law with strain rate and temperature effects were described by Van den Bogart et al. [18, 19], Cobden [20] and Abedrabbo et al. [21–23]. These models indirectly consider the microstructure evolution of creep process and account for the effects of micro level processes on the macro level. They usually include the dislocation density-based material theory developed by Bergström [24] and Nes [25]. The concept about dislocation density evolution processes, specifically with regard to storage and dynamic recovery processes is assumed in these studies.

Given the strong coupled effect of temperature and strain rate on creep constitutive response, it was considered important to utilize a material model that captures these effects in simulations of high temperature creep processes. In the current work, the instantaneous modulus of elasticity $E_c(t)$ is described by the physically-based model which takes into account the structured classification of composites: (1) for continuously fiber reinforced and uniformly distributed (or dispersed) in the bulk composites, the hyperbolic tangent function E(t) is the sum of two hyperbolic tangent functions $\varphi_m E_m(t)$ and $\varphi_f E_f(t)$, where each of them represents the matrix and filler material respectfully and φ is the coefficient of "rule of mixtures"; (2) for multilayered and sandwich type composite structures the hyperbolic tangent function represents the modulus of elasticity of each layer separately, but the stress distribution across the whole cross-section of a composite element should obey the linear distribution within each layer of a composite structure. The generic normalized dimensionless modulus of elasticity–temperature relationship is presented below (see Fig. 2.2).

The generic dimensionless modulus-temperature relationship is approximated here as:

$$E = A - B^{*}\tanh[C(T - T_g)] \tag{2.16}$$

Stress and strain data obtained from compression and tension testing should fit to both the hyperbolic tangent functions (filler and matrix). The constants A and B should be determined utilizing a least squares minimization technique. The hyperbolic tangent function has been found to fit load-displacement data well by Murphy, Lockyear and Laws [26, 27 and 28].

The glass transition temperature (T_g) is one of the most important properties of any composite material. It indicates the temperature region where the material transitions from a hard, glassy phase to a soft, rubbery material. The temperature T_g, not to be confused with melting point (T_{melt}), is where the composite material becomes a more pliable, compliant or "rubbery" state. In actuality T_g is not a discrete thermodynamic transition, but a temperature range over which the mobility of the molecules and atoms increase significantly. The ultimate T_g is determined by a number of factors: the chemical structure of the matrix material and the degree of cure. T_g is usually measured using the Differential Scanning Calorimeter (DSC). *The glass transition temperature of a polymer can be measured which is key to understanding mechanical and thermal expansion properties* [29]. The format of the T_g scan is similar to that of a kinetic scan except that it is performed with a cured sample. Temperature is plotted on the X axis and the heat flow

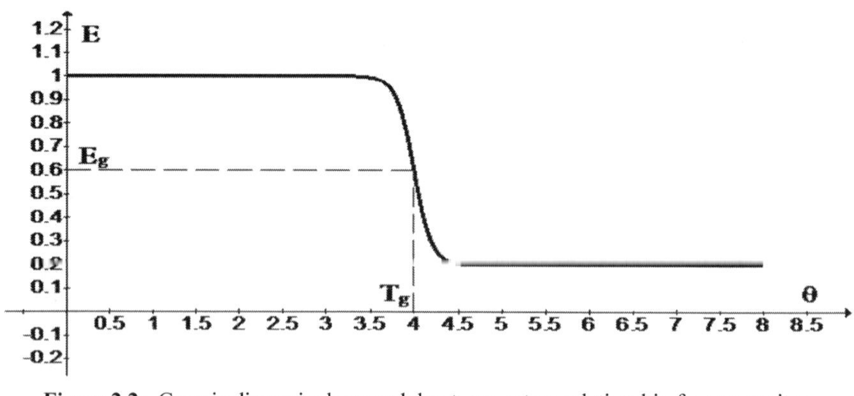

Figure 2.2: Generic dimensionless modulus-temperature relationship for composites.

response on the Y axis. As discussed previously, T_g is actually a temperature range, rather than a specific temperature. The convention, however, is to report a single temperature defined as the midpoint of the temperature range, bounded by the tangents to the two flat regions of the heat flow curve. For the epoxy material T_g spans a temperature range, rather than occurring at a specific temperature, due to the cross-linked polymer chains having multiple degrees of freedom and modes of polymer chain movement in response to any applied thermal energy. The T_g value can also vary depending on its degree of cure. Generally, the reported T_g for a material is based on 100% conversion (full cure). The T_g is strongly dependent on the cure schedule. Low temperature cures such as room temperature (RT) will result in the lowest possible T_g of all for that chemistry. Very high T_g values are not achievable by room temperature curing. If the same material is cured at an elevated temperature, a higher T_g will result. As an example an adhesive could have a T_g between 60°C and 110°C, based on the cure schedule. This is why it is important to maintaining tight temperature control in any production setting. Below are some examples of modulus of elasticity E (0) that are used throughout this book.

Example 2.1

$E_c(\theta) = E_0\{a - b\,[\tanh c\,(\theta - \theta_g)]\}$; $E(\theta) = E_c(\theta)/E_0$

θ_g – dimensionless transition temperature of a composite material.

$$\theta_g = \frac{E_a[T_g - T_*]}{RT_*^2};\quad \beta = \frac{RT_*}{E_a} = 0.067$$

$$T_* \approx 300°K;\quad T_g \approx 400°K;\quad \theta_g \approx 5$$

At small temperatures, the stress is at a high plateau corresponding to a “glassy” modulus E_g, and then falls exponentially to a lower equilibrium “rubbery” modulus E_r as the composite material molecules gradually accommodate the strain by conformational extension rather than bond distortion. Parameter ‘c’ defines the width of transition temperatures zone and can be selected by superimposing the relaxation curves by means of the “relaxation modulus”. Let’s assume here c = 5. The value of dimensionless “glassy” modulus $E_g = 1$ and the value of dimensionless “rubbery” modulus E_r is a fraction of E_g and can be also obtained from experimental data. Let’s assume here that $E_r/E_g = 0.25$. Parameters ‘a’ and ‘b’ can be easily computed from the boundary conditions as follows: E ($\theta = 0$) = 1 and E ($\theta = \infty$) = 0.25. Since tanh (0) = 0 and tanh($\theta \to \infty$) ≅ 1, we have in our case:

$$\begin{cases} a + b = 1 \\ a - b - 0.25 \end{cases} \quad \text{or:}\ a = 0.625\ \text{and}\ b = 0.375$$

Finally, the functions E (θ) are:

$E(\theta) = 0.625 - 0.375\ [\tanh 5(\theta - 5)]$

$E(\theta) = 0.625 - 0.375\ [\tanh 5(\theta - 4)]$

$E(\theta) = 0.625 - 0.375\ [\tanh 3(\theta - 4)]$

2.3 Engineering creep of composites

Generally a mechanical model is used to explain the *engineering creep* behavior of composite materials (not the microscopically observed processes of molecular movements!) which consist of springs and dashpots. The spring complies with Hooke's law as a mechanical model with an elastic element. When a spring is fixed at one end and a load is applied at the other, it becomes instantaneously extended. Then, after the load is removed, it immediately recovers its original length. On the other hand, the dashpot complies with Newton's law (linear or nonlinear) of viscosity as a mechanical model with a viscous element. When a load is applied to a dashpot, it opens gradually, strain being a function of time. Then, after the load is removed, the dashpot remains open and recovery does not occur. The mechanical model which consists of a Hookean spring and a Newtonian dashpot in parallel is a Voigt element, and that in series is a Maxwell element. These are simple models and the behavior of composites can be explained by combining these models. It has been mentioned above that the behavior of *matrix material* can be analyzed by changing the mathematical expression of effective modulus of elasticity as function of temperature (time) in 'simple' mechanical models, for example, the stress relaxation curves of the cross-linked polymers are evaluated by the analogies of a three-element model in which one Maxwell element and one spring are connected in parallel [30]. The equilibrium modulus to be E_e, then the relaxation modulus $E_r(t)$ for the three-element model is defined as:

$$E_r(t) = \sigma(t) / \gamma_0 = \frac{\sigma_1(0)\exp(-t/\tau_1) + \sigma_e}{\gamma_0} = E_1(0)\exp(-t/\tau_1) + E_e \qquad (2.17)$$

The stress relaxation curves of the uncross-linked polymers are evaluated by the analogies of a four-element model in which two Maxwell elements are connected in parallel [31]. $E_r(t)$ for the four-element model is defined as:

$$E_r(t) = \sigma(t)/\gamma_0 = \frac{\sigma_1(0)\exp(-t/\tau_1) + \sigma_2(0)\exp(-t/\tau_2)}{\gamma_0} = \qquad (2.18)$$

$$= E_1(0)\exp(-t/\tau_1) + E_2(0)\exp(-t/\tau_2)$$

where the stress on Maxwell elements is σ_1 and σ_2, elastic modulus are E_1 and E_2, the coefficient of viscosity η_1 and η_2, and the relaxation time τ_1 and $\tau_2 (\tau_1 > \tau_2)$. Therefore the relaxation modulus $E_r(t)$ in Equations (2.17) and (2.18) is a function of time (temperature). Let us analyze first the applicability of general form of the relaxation modulus $E_r(t)$ (before we start analyzing the general mechanical integral type model of *engineering creep* behavior of composite materials) that is approximated by the hyperbolic tangent to a simple Maxwell elements combination [32]. Using these expressions and performing the change of variables in the corresponding differential equation from [32], we obtain:

$$\frac{d\sigma}{d\theta} = -\frac{1}{n}\sigma m1 + E_0 A \tag{2.19}$$

$$m1 = (0.0405 - 0.02252*t^\wedge 1 + 0.004386*t^\wedge 2 - 0.0002747*t^\wedge 3)$$

Let's illustrate the application of the methodology above to the data from Example 2.1.

2.4 Maxwell model

Example 2.2

Data: $\alpha_0 = 10^{-4}$ [1/°K]; E_0 [GPa] = 2.055 = 298 [ksi] = 0.298 (10^3); $\beta T_* =$ 20.1°K; A = 0.067(10^{-4})300 = 20.1(10^{-4}) = 2.01(10^{-3}).

The numerical solution of the Equation (2.19) is given below (using POLYMATH software).

Calculated values of DEQ variables

	Variable	Initial value	Minimal value	Maximal value	Final value
1	m1	0.0405	0.0003976	0.0405	0.0003976
2	t = θ	0	0	8.	8.
3	Z	0	0	0.905932	0.6123323

Differential equations

1 d(Z)/d(t) = –(1/0.01)*Z*m1 + 0.298*2.01*E

Explicit equations

2 m1 = (0.0405 – 0.02252*t^1 + 0.004386*t^2 – 0.0002747*t^3)

3 E = 0.625 – 0.375*tanh(5* (t – 4))

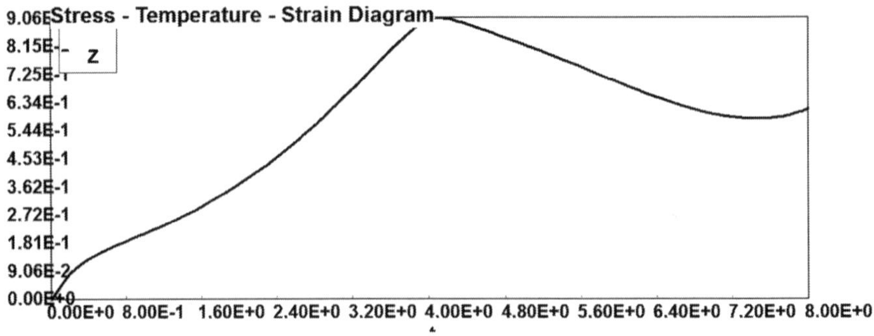

Figure 2.3: Stress-temperature-strain diagram: Z-maxwell model.

Model: σ = Z = a0 + a1*t + a2*t^2 + a3*t^3 + a4*t^4

Variable	Value
a0	0.1524145
a1	–0.1867576
a2	0.248135
a3	–0.0526616
a4	0.0031852

$$\sigma = 0.152 - 0.187\theta + 0.248\theta^2 - 0.0527\theta^3 + 0.00318\theta^4 \tag{2.20}$$

$\sigma_{max} = 0.906(0.625)2.01(0.298) = 0.339$ [GPa] = 49.17ksi

The maximum allowable stress at temperature $\theta = 7.5$; $T = 7.5(20.1) + 300 = 450^\circ$ K ≅ 150°C is:

$\sigma_{all} = 0.581(0.25)2.01(0.298) = 0.087$ [GPa] = 12.62ksi; $\varepsilon_{all} = [7.5(20.1)]10^{-4} = 0.0151$

2.5 Standard linear model

Let's illustrate the application of the methodology above to the *Standards Linear Model.*

Example 2.3

Data: $\alpha_0 = 10^{-4}$ [1/°K]; E_0 [GPa] = 2.055 = 298 [ksi] = 0.298 (10^3); $\beta T_* = 20.1^\circ$ K; $A = 0.067(10^{-4})300 = 20.1(10^{-4}) = 2.01(10^{-3})$; $E_1 = E_2$

The equations for this model are [31]:

$$\varepsilon = \varepsilon_1 + \varepsilon_2;\ \sigma = \sigma_1 + \sigma_2;\ \sigma = E_1\varepsilon_1;$$
$$\sigma_1 = E_2\varepsilon_2;\ \sigma_2 = \eta\dot{\varepsilon}_2 \tag{2.21}$$

We can eliminate the four unknowns from these five equations and after simplifications we have:

$$\sigma + \frac{\eta}{E_1 + E_2}\dot{\sigma} = \frac{E_1E_2}{E_1 + E_2}\varepsilon + \frac{E_1\eta}{E_1 + E_2}\dot{\varepsilon} \tag{2.22}$$

The creep compliance function in this case is:

$$J(t) = \frac{1}{E_1}e^{-(E_2/\eta)t} + \frac{E_1 + E_2}{E_1E_2}\left(1 - e^{-(E_2/\eta)t}\right)$$
$$\text{where: } E = E_1; H = \frac{E_1E_2}{E_1 + E_2}; n = \frac{\eta}{E_1 + E_2} \tag{2.23}$$

The Equation (2.22) will have a standard form now:

$$n\dot{\sigma} + \sigma = En\dot{\varepsilon} + H\varepsilon \tag{2.24}$$

$$\frac{d\sigma}{d\theta} = -\frac{1}{n}\sigma m1 + E\frac{d\varepsilon}{d\theta} + \frac{H}{n}m1(\varepsilon)$$
$$H = \frac{E_1E_2}{E_1 + E_2} \tag{2.25}$$

The numerical solution of the Equation (2.25) is given below (using POLYMATH software).

Calculated values of DEQ variables

	Variable	Initial value	Minimal value	Maximal value	Final value
1	E	1.	0.25	1.	0.25
2	m	0	0	0.0705907	0.0705907
3	m1	0.0405	0.0003976	0.0405	0.0003976
4	t	0	0	8.	8.
5	Y	0	0	1.575479	1.125071
6	Z	0	0	0.9058583	0.6123323

Differential equations

1 d(Z)/d(t) = –(1/0.01)*Z*m1 + 0.298*2.01*E

2 d(Y)/d(t) = –(1/0.01)*Y*m1 + 0.298*2.01*E + 0.29*2.01*0.5*E*(1/0.01)*m1*t

Explicit equations

1 m = (0.0405*t – 0.01126*t^2 + 0.001462*t^3 – 0.00006868*t^4)

2 m1 = (0.0405 – 0.02252*t^1 + 0.004386*t^2 – 0.0002747*t^3)

3 E = 0.625 – 0.375*tanh(5* (t – 4))

Model: σ = a0 + a1*t + a2*t^2 + a3*t^3 + a4*t^4

Variable	Value
a0	0.1317665
a1	–0.0247626
a2	0.3072643
a3	–0.0732268
a4	0.0046558

$$\sigma = 0.132 - 0.0248\theta + 0.307\theta^2 - 0.0732\theta^3 + 0.00466\theta^4 \qquad (2.26)$$

σ_{max} = (1.58)2.01(0.298) = 0.946 [ksi] = 6.53 [MPa]

The maximum allowable stress at temperature θ = 7.5 (T = 7.5(20.1) + 300 = 450° K ≅ 150°C is:

σ_{all} = 1.0925(2.01)(0.298) = 0.654 [ksi] = 4.51[MPa]; ε_{all} = [7.5(20.1)]10^{-4} = 0.0151

The allowable stress is defined here as a stress at maximum allowable strain (before the material enters the tertiary stage of creep).

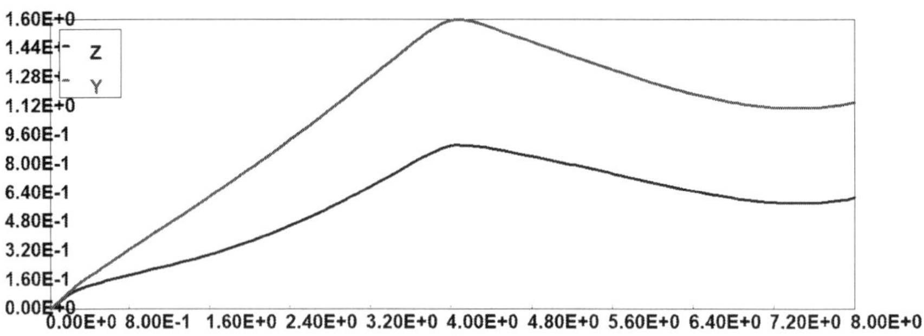

Figure 2.4: Stress-temperature-strain diagram (standard model).

2.6 Effect of variable dimensionless parameters on STS diagram (Standard Linear model)

Consider now the effects of variable dimensionless parameters: $n = \eta/E$; $\beta = RT_*/E_a$; c; θ_g on stress-temperature-strain (STS) diagrams.

Case 1: Small relaxation time 'n' and large maximum value of dimensionless temperature θ_{max}. $n = 0.01$; $\theta_{max} = 10$ ($T = 20.1(10) + 300 = 501$°K. Numerical solution of the Equation (2.25) is given below (using POLYMATH software).

Calculated values of DEQ variables

	Variable	Initial value	Minimal value	Maximal value	Final value
1	E	1.	0.25	1.	0.25
2	m	0	0	0.0706048	0.0542
3	m1	0.0405	–0.0208	0.0405	0.0208
4	n	0.01	0.01	0.01	0.01
5	t	0	0	10.	10.
6	Y	0	0	4.029122	4.029122
7	Z	0	0	4.20561	4.20561

Differential equations

1 d(Y)/d(t) = –(1/n)*Y*m1 + 0.29*2.01*E + 0.29*2.01*0.5*E*(1/n)*m1*t

2 d(Z)/d(t) = –(1/0.01)*Z*m1 + 0.298*2.01*E

Explicit equations

1 m = (0.0405*t – 0.01126*t^2 + 0.001462*t^3 – 0.00006868*t^4)

2 m1 = (0.0405 – 0.02252*t^1 + 0.004386*t^2 – 0.0002747*t^3)

3 n = 0.01

4 E = 0.625 – 0.375*tanh(5* (t – 4))

One can see from Fig. 2.14 that the creep stress decreases steady after dimensionless transitional temperature $\theta_g > 4$ and starting from $\theta_g \cong 8$ composite materials are entering the tertiary stage and fails. As parameter 'n' increases, so does the maximum allowable creep stress and the tertiary stage is less transparent (see Figs. 2.6 and 2.7) below.

Obviously, parameter 'n' plays very important role in qualitative and quantitative evaluation of creep deformation process. Conversely, parameter 'c' has very small effect on maximum value of creep stress and quantitative evaluation of creep.

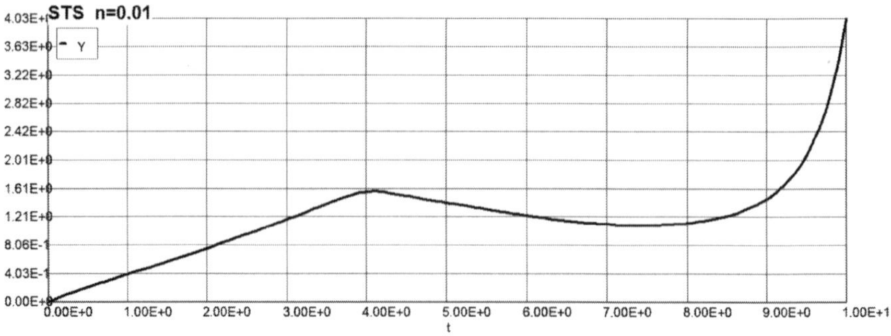

Figure 2.5: Stress-temperature-strain diagram: y-standard model; n = 0.01.

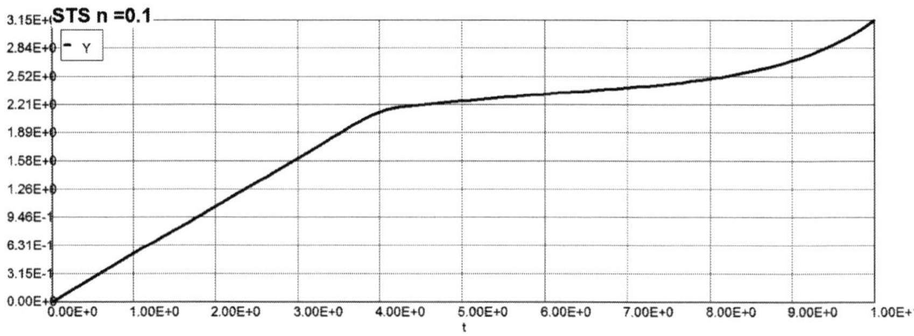

Figure 2.6: Stress-temperature-strain diagram: y-standard model; n = 0.1.

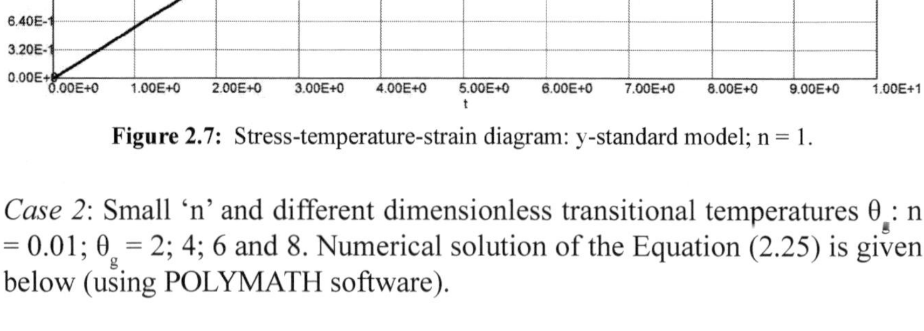

Figure 2.7: Stress-temperature-strain diagram: y-standard model; n = 1.

Case 2: Small 'n' and different dimensionless transitional temperatures θ_g: n = 0.01; θ_g = 2; 4; 6 and 8. Numerical solution of the Equation (2.25) is given below (using POLYMATH software).

Calculated values of DEQ variables $\theta_g = 2$

	Variable	Initial value	Minimal value	Maximal value	Final value
1	E	1.	0.25	1.	0.25
2	m	0	0	0.070605	0.0542
3	m1	0.0405	–0.0208	0.0405	–0.0208
4	n	0.01	0.01	0.01	0.01
5	t	0	0	10.	10.
6	Y	0	0	2.55929	2.55929

Differential equations

1 d(Y)/d(t) = –(1/n)*Y*m1 + 0.29*2.01*E + 0.29*2.01*0.5*E*(1/n)*m1*t

Explicit equations

1 m = (0.0405*t – 0.01126*t^2 + 0.001462*t^3 – 0.00006868*t^4)
2 m1 = (0.0405 – 0.02252*t^1 + 0.004386*t^2 – 0.0002747*t^3)
3 n = 0.01
4 E = 0.625 – 0.375*tanh(5* (t – 2))

Model: Y = a0 + a1*t + a2*t^2 + a3*t^3 + a4*t^4

Variable	Value
a0	0.1380808
a1	0.2476432
a2	–0.0048617
a3	–0.0093566
a4	0.0009298

$$\sigma = 0.138 + 0.248\theta - 0.00486\theta^2 - 0.00936\theta^3 + 0.00093\theta^4 \quad (2.27)$$

The Figs. 2.9 – 2.11 represent the cases $\theta_g = 4$; $\theta_g = 6$; $\theta_g = 8$; accordingly. Computer outputs for these cases are similar to the previous case ($\theta_g = 2$) and therefore are not presented here. However, summary of results are given in Table 2.2 and Fig. 2.12 below.

One can see from Fig. 2.12 that the maximum creep stress linearly increases with the transitional temperature rise, but allowable stress has clearly a nonlinear relationship, since minimum (allowable) stress occurs usually in the rubbery zone (after the composite material passes the

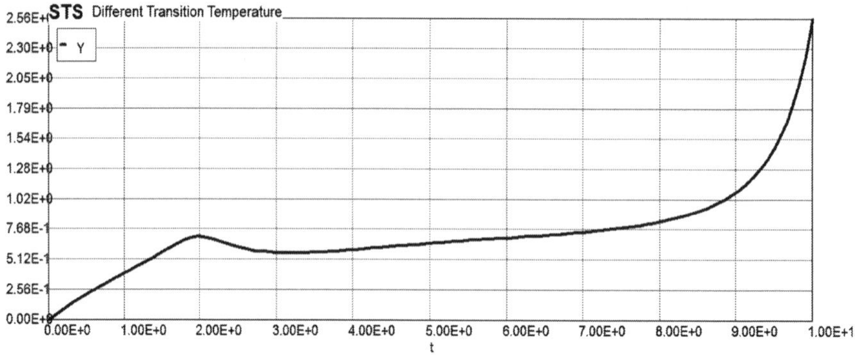

Figure 2.8: Stress-temperature-strain (STS) diagram: $\theta_g = 2$; n = 0.01.

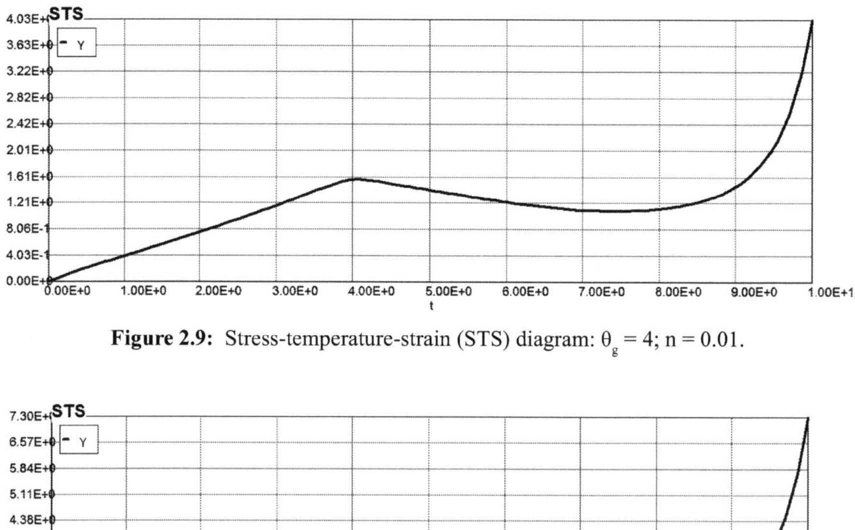

Figure 2.9: Stress-temperature-strain (STS) diagram: $\theta_g = 4$; n = 0.01.

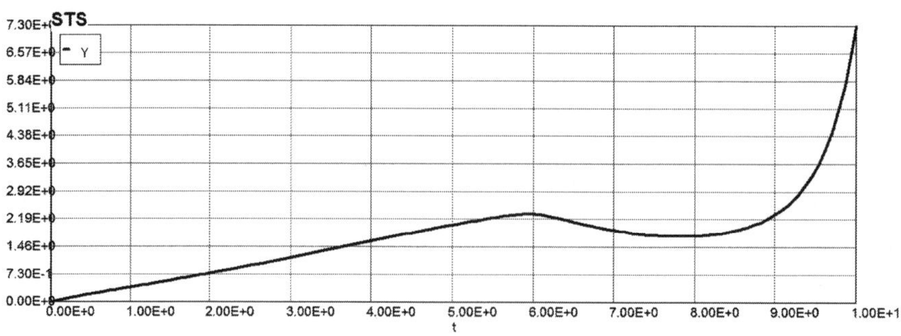

Figure 2.10: Stress-temperature-strain (STS) diagram: $\theta_g = 6$; n = 0.01.

transitional–leathery zone) and therefore part of the energy transforms from storage to the dissipative part. In this respect the energy transformation effect is characterized by dimensionless parameter β that is the average energy over activation energy ratio.

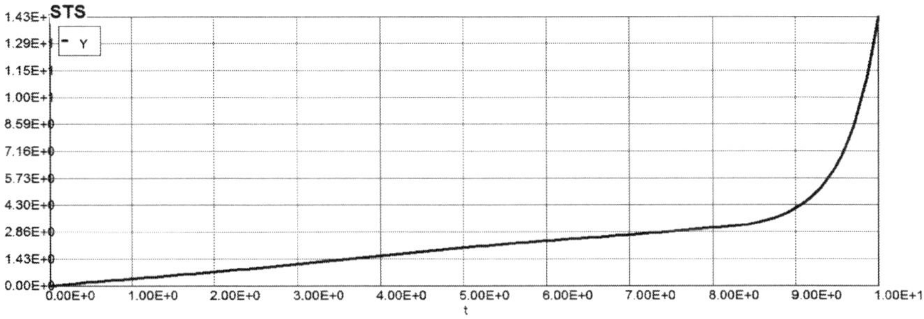

Figure 2.11: Stress-temperature-strain (STS) diagram: $\theta_g = 8$; n = 0.01.

Table 2.2: Summary of results.

	$\theta_g = 2$	$\theta_g = 4$	$\theta_g = 6$	$\theta_g = 8$
σ_{max} [MPa]	0.424 [2.924]	0.941 [6.49]	1.4 [9.67]	1.856 [12.8]
σ_{all} [MPa]	0.34 [2.345]	0.672 [4.64]	1.056 [7.28]	1.856 [12.8]
T_{max} ^{0}K	360	461	461	461

For instance: σ_{max} = 3.1(2.01)(0.298) = 1.856 [ksi] = 12.8 MPa

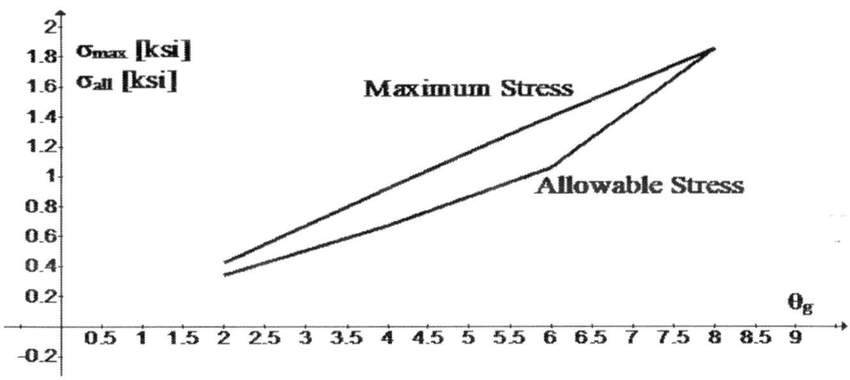

Figure 2.12: Maximum creep stress vs. glass transition temperature.

2.6.1 *Allowable creep stresses vs. parameter β*

Consider now the effect of parameter β on qualitative and quantitative evaluation of the creep deformation process (namely, maximum creep stress and allowable stress); for all practical purposes (in engineering practice) this parameter ranges from 0.01 till 0.1.

Case 3: Parameter β: $0.01 < \beta < 0.1$ and different dimensionless transitional temperatures θ_g: 4; 6 and 8. Numerical solution of the Equation (2.25) is given below (using POLYMATH software).

Consider now: $\theta_g = 4$ and $\beta = 0.01$

Data: $\alpha_0 = 10^{-4}$ [1/° K]; E_0 [GPa] = 2.055 = 298 [ksi] = 0.298 (10^3); A = 0.01 (10^{-4})300 = 3(10^{-4}) = 0.3(10^{-3}); n = 0.01

Calculated values of DEQ variables

	Variable	Initial value	Minimal value	Maximal value	Final value
1	E	1.	0.25	1.	0.25
2	m	0	0	0.0705946	0.0676285
3	m1	0.0405	–0.0071703	0.0405	–0.0071703
4	n	0.01	0.01	0.01	0.01
5	t	0	0	9.	9.
6	Y	0	0	0.2416182	0.2261014

Differential equations

1 d(Y)/d(t) = –(1/n)*Y*m1 + 0.298*0.30*E + 0.298*0.3*0.5*E*(1/n)*m1*t

Explicit equations

1 m = (0.0405*t – 0.01126*t^2 + 0.001462*t^3 – 0.00006868*t^4)

2 m1 = (0.0405 – 0.02252*t^1 + 0.004386*t^2 – 0.0002747*t^3)

3 n = 0.01

4 E = 0.625 – 0.375*tanh(5* (t – 4))

Model: Y = a0 + a1*t + a2*t^2 + a3*t^3 + a4*t^4

Variable	Value
a0	0.011934
a1	0.0177287
a2	0.0342059
a3	–0.0085686
a4	0.0005394

$$\sigma = 0.119 + 0.0177\theta + 0.0342\theta^2 - 0.00857\theta^3 + 0.000539\theta^4 \quad (2.28)$$

The maximum allowable stress at temperature θ = 7.5; T = 7.5(3) + 300 = 322.5° K ≅ 32.5°C is:

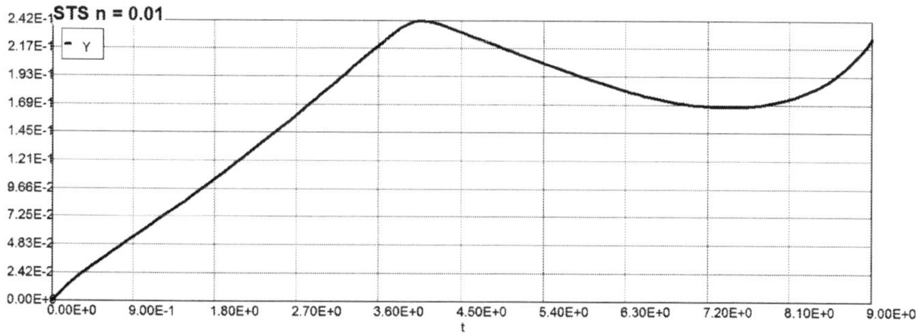

Figure 2.13: Stress-temperature-strain (STS) diagram: $\theta_g = 4$; $\beta = 0.01$.

$\sigma_{all} = 0.1675(0.3)(0.298) = 0.015$ [ksi] $= 0.103$[MPa]; $\varepsilon_{all} = [7.5(3)]10^{-4} = 0.00225$

Consider now: $\theta_g = 4$ and $\beta = 0.02$

$A = 0.02(10^{-4})300 = 6(10^{-4}) = 0.6(10^{-3})$; $n = 0.01$

Calculated values of DEQ variables

	Variable	Initial value	Minimal value	Maximal value	Final value
1	E	1.	0.25	1.	0.25
2	m	0	0	0.0705946	0.0676285
3	m1	0.0405	–0.0071703	0.0405	–0.0071703
4	n	0.01	0.01	0.01	0.01
5	t	0	0	9.	9.
6	Y	0	0	0.4832364	0.4522028

Differential equations

1 d(Y)/d(t) = –(1/n)*Y*m1 + 0.298*0.60*E + 0.298*0.6*0.5*E*(1/n)*m1*t

Explicit equations

1 m = (0.0405*t – 0.01126*t^2 + 0.001462*t^3 – 0.00006868*t^4)

2 m1 – (0.0405 – 0.02252*t^1 + 0.004386*t^2 – 0.0002747*t^3)

3 n = 0.01

4 E = 0.625 – 0.375*tanh(5* (t – 4))

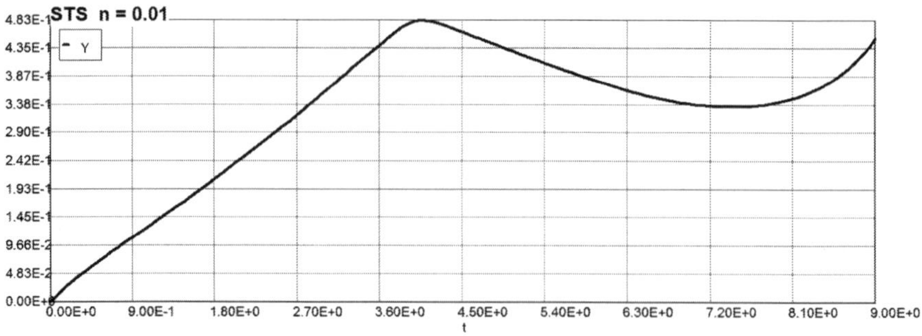

Figure 2.14: Stress-temperature-strain (STS) diagram: $\theta_g = 4$; $\beta = 0.02$.

Model: Y = a0 + a1*t + a2*t^2 + a3*t^3 + a4*t^4

Variable	Value
a0	0.023868
a1	0.0354574
a2	0.0684119
a3	–0.0171373
a4	0.0010789

$$\sigma = 0.0239 + 0.0355\theta + 0.0684\theta^2 - 0.0171\theta^3 + 0.00108\theta^4 \quad (2.29)$$

$\sigma_{min} = 0.335$

The maximum allowable stress at temperature $\theta = 7.5$; $T = 7.5(6) + 300 = 345°$ K $\cong$ 45°C is:

$\sigma_{all} = 0.335(0.6)(0.298) = 0.06$ [ksi] = 0.413[MPa]; $\varepsilon_{all} = [7.5(6)]10^{-4} = 0.0045$

Consider now: $\theta_g = 4$ and $\beta = 0.04$

$A = 0.04(10^{-4})300 = 12(10^{-4}) = 1.2(10^{-3})$; $n = 0.01$

Calculated values of DEQ variables

	Variable	Initial value	Minimal value	Maximal value	Final value
1	E	1.	0.25	1.	0.25
2	m	0	0	0.0705946	0.0676285
3	m1	0.0405	–0.0071703	0.0405	–0.0071703
4	n	0.01	0.01	0.01	0.01
5	t	0	0	9.	9.
6	Y	0	0	0.9664728	0.9044056

Differential equations

1 d(Y)/d(t) = –(1/n)*Y*m1 + 0.298*1.20*E + 0.298*1.2*0.5*E*(1/n)*m1*t

Explicit equations

1 m = (0.0405*t – 0.01126*t^2 + 0.001462*t^3 – 0.00006868*t^4)

2 m1 = (0.0405 – 0.02252*t^1 + 0.004386*t^2 – 0.0002747*t^3)

3 n = 0.01

4 E = 0.625 – 0.375*tanh(5* (t – 4))

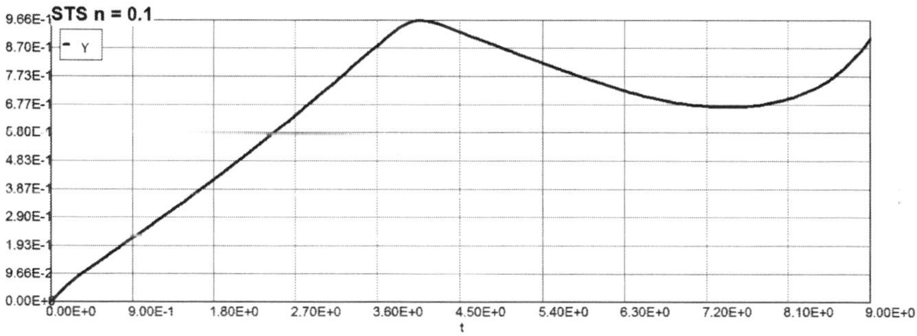

Figure 2.15: Stress-temperature-strain (STS) diagram: $\theta_g = 4$; $\beta = 0.04$.

Model: Y = a0 + a1*t + a2*t^2 + a3*t^3 + a4*t^4

Variable	Value
a0	0.0477361
a1	0.0709147
a2	0.1368238
a3	–0.0342746
a4	0.0021578

$$\sigma = 0.0477 + 0.0709\theta + 0.137\theta^2 - 0.0343\theta^3 + 0.00216\theta^4 \qquad (2.30)$$

$\sigma_{min} = 0.670$

The maximum allowable stress at temperature $\theta = 7.5$; $T = 7.5(12) + 300 = 390°K \cong 90°C$ is:

$\sigma_{all} = 0.670(1.2)(0.298) = 0.24$ [ksi] = 1.652 [MPa]; $\varepsilon_{all} = [7.5(12)]10^{-4} = 0.0090$

Consider now: $\theta_g = 4$ and $\beta = 0.06$

$A = 0.06(10^{-4})300 = 18(10^{-4}) = 1.8(10^{-3})$; n = 0.01

Calculated values of DEQ variables

	Variable	Initial value	Minimal value	Maximal value	Final value
1	E	1.	0.25	1.	0.25
2	m	0	0	0.0706051	0.0676285
3	m1	0.0405	–0.0071703	0.0405	–0.0071703
4	n	0.01	0.01	0.01	0.01
5	t	0	0	9.	9.
6	Y	0	0	1.448809	1.356608

Differential equations

1 d(Y)/d(t) = –(1/n)*Y*m1 + 0.298*1.80*E + 0.298*1.8*0.5*E*(1/n)*m1*t

Explicit equations

1 m = (0.0405*t – 0.01126*t^2 + 0.001462*t^3 – 0.00006868*t^4)

2 m1 = (0.0405 – 0.02252*t^1 + 0.004386*t^2 – 0.0002747*t^3)

3 n = 0.01

4 E = 0.625 – 0.375*tanh(5* (t – 4))

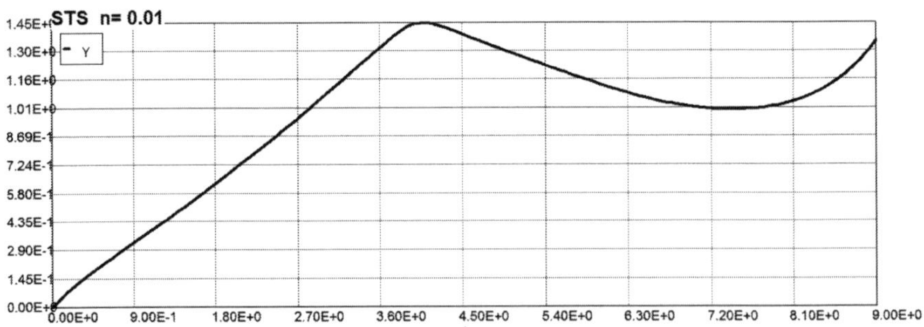

Figure 2.16: Stress-temperature-strain (STS) diagram: $\theta_g = 4$; $\beta = 0.06$.

Model: Y = a0 + a1*t + a2*t^2 + a3*t^3 + a4*t^4

Variable	Value
a0	0.0708292
a1	0.107384
a2	0.2048285
a3	–0.0513484
a4	0.0032333

$$\sigma = 0.0708 + 0.0107\theta + 0.205\theta^2 - 0.0513\theta^3 + 0.003236\theta^4 \quad (2.31)$$

$\sigma_{min} = 1.0054$

The maximum allowable stress at temperature $\theta = 7.5$; $T = 7.5(18) + 300 = 435°\,K \cong 135°C$ is:

$\sigma_{all} = 1.054(1.8)(0.298) = 0.565$ [ksi] $= 3.9$ [MPa]; $\varepsilon_{all} = [7.5(18)]10^{-4} = 0.0135$

Consider now: $\theta_g = 4$ and $\beta = 0.08$

$A = 0.08(10^{-4})300 = 24(10^{-4}) = 2.4(10^{-3})$; $n = 0.01$

Calculated values of DEQ variables

	Variable	Initial value	Minimal value	Maximal value	Final value
1	E	1.	0.25	1.	0.25
2	m	0	0	0.0705946	0.0676285
3	m1	0.0405	–0.0071703	0.0405	–0.0071703
4	n	0.01	0.01	0.01	0.01
5	t	0	0	9.	9.
6	Y	0	0	1.932946	1.808811

Differential equations

1 d(Y)/d(t) = –(1/n)*Y*m1 + 0.298*2.4*E + 0.298*2.4*0.5*E*(1/n)*m1*t

Explicit equations

1 m = (0.0405*t – 0.01126*t^2 + 0.001462*t^3 – 0.00006868*t^4)
2 m1 = (0.0405 – 0.02252*t^1 + 0.004386*t^2 – 0.0002747*t^3)
3 n = 0.01
4 E = 0.625 – 0.375*tanh(5* (t – 4))

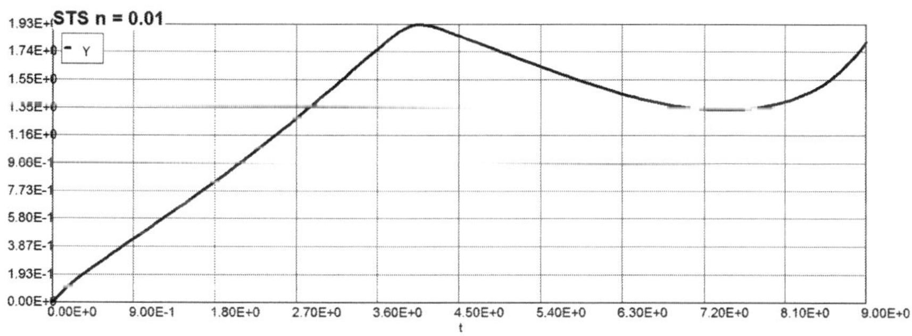

Figure 2.17: Stress-temperature-strain (STS) diagram: $\theta_g = 4$; $\beta = 0.08$.

Model: Y = a0 + a1*t + a2*t^2 + a3*t^3 + a4*t^4

Variable	Value
a0	0.0954721
a1	0.1418294
a2	0.2736475
a3	–0.0685492
a4	0.0043156

$$\sigma = 0.0954 + 0.01420\theta + 0.274\theta^2 - 0.0685\theta^3 + 0.00432\theta^4 \tag{2.32}$$

$\sigma_{min} = 1.340$

The maximum allowable stress at temperature $\theta = 7.5$; $T = 7.5(24) + 300 = 480^{\circ}$ K $\cong 180^{\circ}$C is:

$\sigma_{all} = 1.34(2.4)(0.298) = 0.958$ [ksi] $= 6.61$ [MPa]; $\varepsilon_{all} = [7.5(24)]10^{-4} = 0.018$

Consider now: $\theta_g = 4$ and $\beta = 0.1$

$A = 0.1(10^{-4})300 = 30(10^{-4}) = 3.0(10^{-3})$; $n = 0.01$

Calculated values of DEQ variables

	Variable	Initial value	Minimal value	Maximal value	Final value
1	E	1.	0.25	1.	0.25
2	m	0	0	0.0705946	0.0676285
3	m1	0.0405	–0.0071703	0.0405	–0.0071703
4	n	0.01	0.01	0.01	0.01
5	t	0	0	9.	9.
6	Y	0	0	2.416181	2.261014

Differential equations

1 d(Y)/d(t) = –(1/n)*Y*m1 + 0.298*3.0*E + 0.298*3.0*0.5*E*(1/n)*m1*t

Explicit equations

1 m = (0.0405*t – 0.01126*t^2 + 0.001462*t^3 – 0.00006868*t^4)

2 m1 = (0.0405 – 0.02252*t^1 + 0.004386*t^2 – 0.0002747*t^3)

3 n = 0.01

4 E = 0.625 – 0.375*tanh(5* (t – 4))

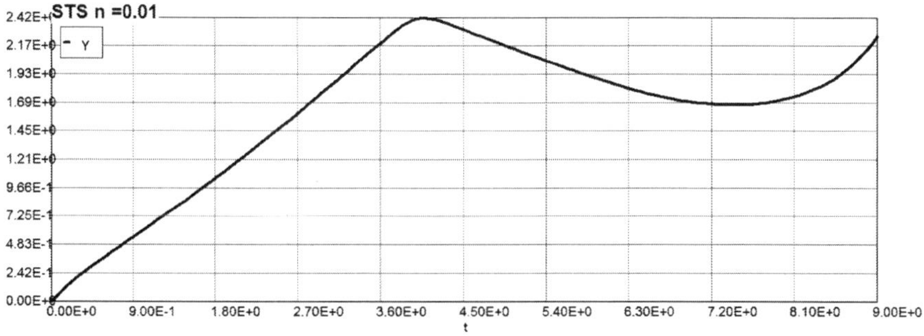

Figure 2.18: Stress-temperature-strain (STS) diagram: $\theta_g = 4$; $\beta = 0.1$.

Model: Y = a0 + a1*t + a2*t^2 + a3*t^3 + a4*t^4

Variable	Value
a0	0.1193406
a1	0.1772851
a2	0.3420604
a3	–0.0856867
a4	0.0053945

$$\sigma = 0.119 + 0.177\theta + 0.342\theta^2 - 0.0857\theta^3 + 0.00539\theta^4 \tag{2.33}$$

$\sigma_{min} = 1.676$

The maximum allowable stress at temperature $\theta = 7.5$; $T = 7.5(30) + 300 = 525° K \cong 225°C$ is:

$\sigma_{all} = 1.676(3)(0.298) = 1.498$ [ksi] = 10.33 [MPa]; $\varepsilon_{all} = [7.5(30)]10^{-4} = 0.0225$

The summary of results is presented in Table 2.3 and Fig. 2.19 below.

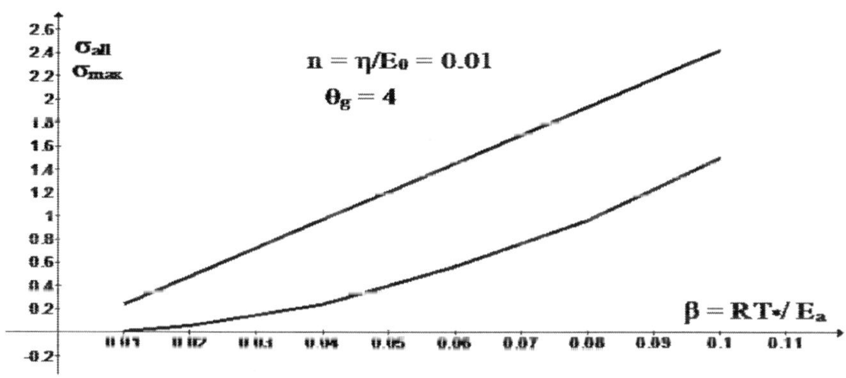

Figure 2.19: Allowable creep stresses vs. parameter β.

Table 2.3: Summary of results (allowable creep stress vs. parameter β).

β	0.01	0.02	0.04	0.06	0.08	0.1
σ_{all}	0.015 [0.103]	0.06 [0.413	0.24 [1.652]	0.565 [3.9]	0.958 [6.61]	1.498 [10.33]
ε_{all}	0.00225	0.045	0.0090	0.0135	0.018	0.0225

2.6.2 *Allowable creep stresses vs. parameter β and n*

β = 0.01; n = 0.03

Calculated values of DEQ variables

	Variable	Initial value	Minimal value	Maximal value	Final value
1	E	1.	0.25	1.	0.25
2	m	0	0	0.0706044	0.0676285
3	m1	0.0405	–0.0071703	0.0405	–0.0071703
4	n	0.03	0.03	0.03	0.03
5	t	0	0	9.	9.
6	Y	0	0	0.332149	0.332149

Differential equations

1 d(Y)/d(t) = –(1/n)*Y*m1 + 0.298*0.30*E + 0.298*0.30*0.5*E*(1/n)*m1*t

Explicit equations

1 m = (0.0405*t – 0.01126*t^2 + 0.001462*t^3 – 0.00006868*t^4)

2 m1 = (0.0405 – 0.02252*t^1 + 0.004386*t^2 – 0.0002747*t^3)

3 n = 0.03

4 E = 0.625 – 0.375*tanh(5* (t – 4))

Model: Y = a0 + a1*t + a2*t^2 + a3*t^3 + a4*t^4

Variable	Value
a0	0.0121276
a1	0.0343844
a2	0.0292593
a3	–0.0071469
a4	0.0004354

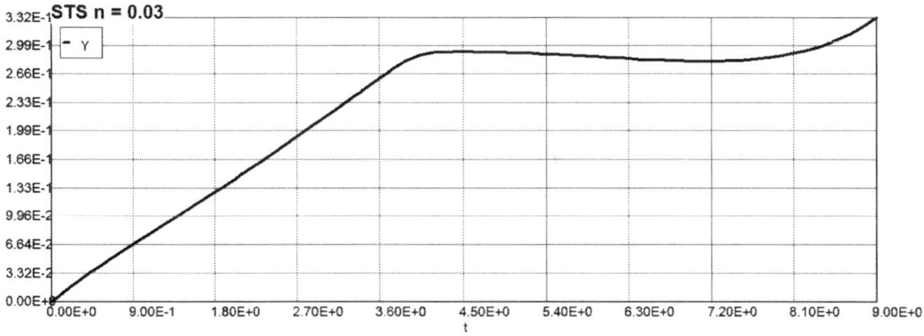

Figure 2.20: Stress-temperature-strain (STS) diagram: $\theta_g = 4$; $\beta = 0.01$; n = 0.03.

$$\sigma = 0.0121 + 0.0344\theta + 0.0292\theta^2 - 0.0715\theta^3 + 0.00043\theta^4 \tag{2.34}$$

$\sigma_{min} = 0.288$

The maximum allowable stress at temperature $\theta = 8.0$ is:

$\sigma_{all} = 0.288(0.3)(0.298) = 0.0258$ [ksi] = 0.1776 [MPa]; $\varepsilon_{all} = [8(3)]10^{-4} = 0.0024$

$\beta = 0.02$; n = 0.03

Calculated values of DEQ variables

	Variable	Initial value	Minimal value	Maximal value	Final value
1	E	1.	0.25	1.	0.25
2	m	0	0	0.0706044	0.0676285
3	m1	0.0405	–0.0071703	0.0405	–0.0071703
4	n	0.03	0.03	0.03	0.03
5	t	0	0	9.	9.
6	Y	0	0	0.664298	0.664298

Differential equations

1 d(Y)/d(t) = –(1/n)*Y*m1 + 0.298*0.60*E + 0.298*0.60*0.5*E*(1/n)*m1*t

Explicit equations

1 m = (0.0405*t – 0.01126*t^2 + 0.001462*t^3 – 0.00006868*t^4)

2 m1 = (0.0405 – 0.02252*t^1 + 0.004386*t^2 – 0.0002747*t^3)

3 n = 0.03

4 E = 0.625 – 0.375*tanh(5* (t – 4))

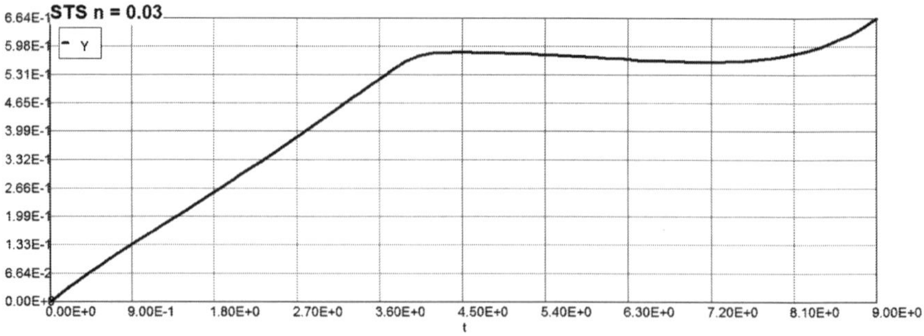

Figure 2.21: Stress-temperature-strain (STS) diagram: $\theta_g = 4$; $\beta = 0.02$; $n = 0.03$.

Model: Y = a0 + a1*t + a2*t^2 + a3*t^3 + a4*t^4

Variable	Value
a0	0.0242554
a1	0.0687685
a2	0.0585187
a3	–0.0142939
a4	0.0008708

$$\sigma = 0.0242 + 0.0688\theta + 0.0585\theta^2 - 0.0143\theta^3 + 0.00087\theta^4 \quad (2.35)$$

$\sigma_{min} = 0.577$

The maximum allowable stress at temperature $\theta = 8.0$ is:

$\sigma_{all} = 0.577(0.6)(0.298) = 0.103$ [ksi] $= 0.712$ [MPa]; $\varepsilon_{all} = [8(6)]10^{-4} = 0.0048$

$\beta = 0.04$; $n = 0.03$

Calculated values of DEQ variables

	Variable	Initial value	Minimal value	Maximal value	Final value
1	E	1.	0.25	1.	0.25
2	m	0	0	0.0706028	0.0676285
3	m1	0.0405	–0.0071703	0.0405	–0.0071703
4	n	0.03	0.03	0.03	0.03
5	t	0	0	9.	9.
6	Y	0	0	1.328596	1.328596

Differential equations

1 d(Y)/d(t) = –(1/n)*Y*m1 + 0.298*1.20*E + 0.298*1.20*0.5*E*(1/n)*m1*t

Explicit equations

1 m = (0.0405*t – 0.01126*t^2 + 0.001462*t^3 – 0.00006868*t^4)

2 m1 = (0.0405 – 0.02252*t^1 + 0.004386*t^2 – 0.0002747*t^3)

3 n = 0.03

4 E = 0.625 – 0.375*tanh(5* (t – 4))

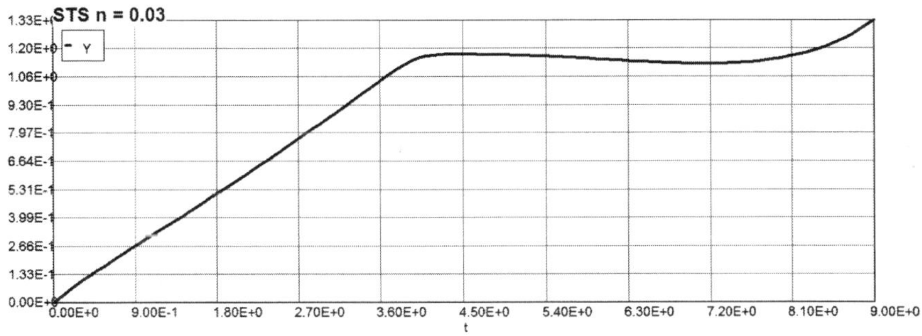

Figure 2.22: Stress-temperature-strain (STS) diagram: $\theta_g = 4$; $\beta = 0.04$; n = 0.03.

Model: Y = a0 + a1*t + a2*t^2 + a3*t^3 + a4*t^4

Variable	Value
a0	0.0481914
a1	0.1380466
a2	0.1168141
a3	–0.0285514
a4	0.0017398

$$\sigma = 0.0482 + 0.138\theta + 0.117\theta^2 - 0.0286\theta^3 + 0.00174\theta^4 \quad (2.36)$$

$\sigma_{min} - 1.154$

The maximum allowable stress at temperature $\theta = 8.0$ is:

$\sigma_{all} = 1.154(1.2)(0.298) = 0.4127$ [ksi] = 2.846 [MPa]; $\varepsilon_{all} = [8(12)]10^{-4} = 0.0096$

$\beta - 0.06$; n – 0.03

Calculated values of DEQ variables

	Variable	Initial value	Minimal value	Maximal value	Final value
1	E	1.	0.25	1.	0.25
2	m	0	0	0.070605	0.0676285
3	m1	0.0405	–0.0071703	0.0405	–0.0071703
4	n	0.03	0.03	0.03	0.03
5	t	0	0	9.	9.
6	Y	0	0	1.992894	1.992894

Differential equations

1 d(Y)/d(t) = –(1/n)*Y*m1 + 0.298*1.80*E + 0.298*1.80*0.5*E*(1/n)*m1*t

Explicit equations

1 m = (0.0405*t – 0.01126*t^2 + 0.001462*t^3 – 0.00006868*t^4)

2 m1 = (0.0405 – 0.02252*t^1 + 0.004386*t^2 – 0.0002747*t^3)

3 n = 0.03

4 E = 0.625 – 0.375*tanh(5* (t – 4))

Model: Y = a0 + a1*t + a2*t^2 + a3*t^3 + a4*t^4

Variable	Value
a0	0.0735132
a1	0.2057197
a2	0.1756378
a3	–0.0428732
a4	0.0026111

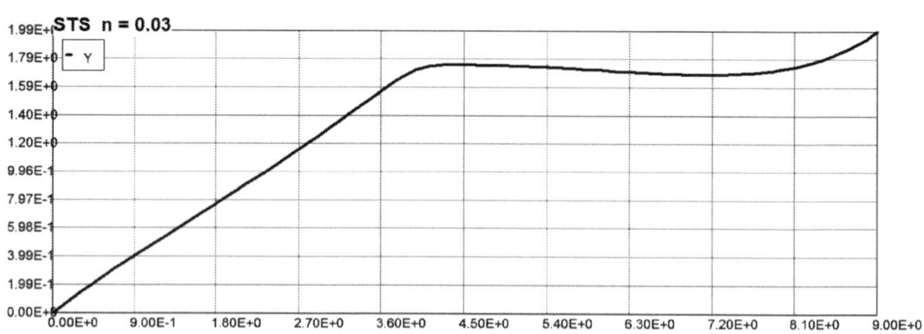

Figure 2.23: Stress-temperature-strain (STS) diagram: $\theta_g = 4$; $\beta = 0.06$; n = 0.03.

$$\sigma = 0.0735 + 0.206\theta + 0.176\theta^2 - 0.0429\theta^3 + 0.00261\theta^4 \quad (2.37)$$

$\sigma_{min} = 1.730$

The maximum allowable stress at temperature $\theta = 8.0$ is:

$\sigma_{all} = 1.73(1.8)(0.298) = 0.928$ [ksi] $= 6.4$ [MPa]; $\varepsilon_{all} = [8(18)]10^{-4} = 0.0144$

$\beta = 0.08$; $n = 0.03$

Calculated values of DEQ variables

	Variable	Initial value	Minimal value	Maximal value	Final value
1	E	1.	0.25	1.	0.25
2	m	0	0	0.0705984	0.0676285
3	m1	0.0405	–0.0071703	0.0405	–0.0071703
4	n	0.03	0.03	0.03	0.03
5	t	0	0	9.	9.
6	Y	0	0	2.657192	2.657192

Differential equations

1 d(Y)/d(t) = –(1/n)*Y*m1 + 0.298*2.40*E + 0.298*2.40*0.5*E*(1/n)*m1*t

Explicit equations

1 m = (0.0405*t – 0.01126*t^2 + 0.001462*t^3 – 0.00006868*t^4)

2 m1 = (0.0405 – 0.02252*t^1 + 0.004386*t^2 – 0.0002747*t^3)

3 n = 0.03

4 E = 0.625 – 0.375*tanh(5* (t – 4))

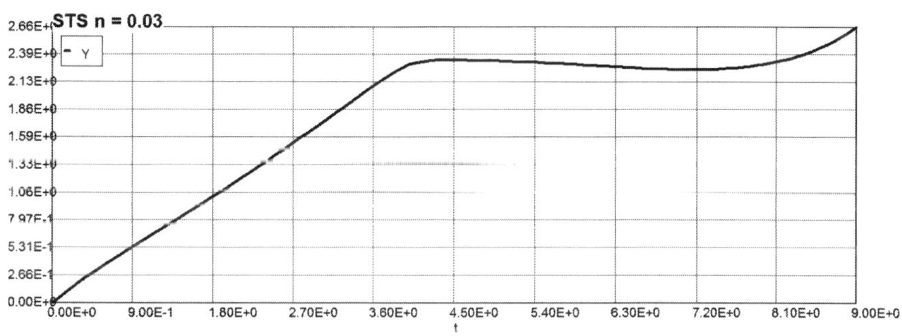

Figure 2.24: Stress-temperature-strain (STS) diagram: $\theta_\rho = 4$; $\beta = 0.08$; $n = 0.03$.

Model: Y = a0 + a1*t + a2*t^2 + a3*t^3 + a4*t^4

Variable	Value
a0	0.0993823
a1	0.2726974
a2	0.2347532
a3	–0.057243
a4	0.0034852

$$\sigma = 0.0994 + 0.273\theta + 0.235\theta^2 - 0.0572\theta^3 + 0.00348\theta^4 \quad (2.38)$$

$\sigma_{min} = 2.307$

The maximum allowable stress at temperature $\theta = 8.0$ is:

$\sigma_{all} = 2.307(2.4)(0.298) = 1.65$ [ksi] $= 11.38$ [MPa]; $\varepsilon_{all} = [8(24)]10^{-4} = 0.0192$

$\beta = 0.1$; $n = 0.03$

Calculated values of DEQ variables

	Variable	Initial value	Minimal value	Maximal value	Final value
1	E	1.	0.25	1.	0.25
2	m	0	0	0.0706051	0.0676285
3	m1	0.0405	–0.0071703	0.0405	–0.0071703
4	n	0.03	0.03	0.03	0.03
5	t	0	0	9.	9.
6	Y	0	0	3.32149	3.32149

Differential equations

1 d(Y)/d(t) = –(1/n)*Y*m1 + 0.298*3.0*E + 0.298*3.0*0.5*E*(1/n)*m1*t

Explicit equations

1 m = (0.0405*t – 0.01126*t^2 + 0.001462*t^3 – 0.00006868*t^4)

2 m1 = (0.0405 – 0.02252*t^1 + 0.004386*t^2 – 0.0002747*t^3)

3 n = 0.03

4 E = 0.625 – 0.375*tanh(5* (t – 4))

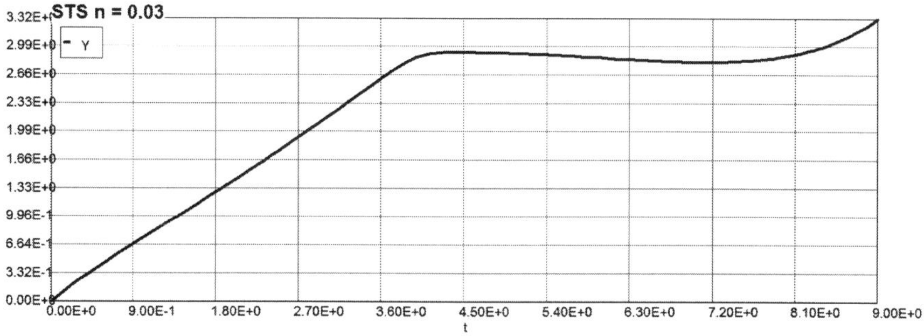

Figure 2.25: Stress-temperature-strain (STS) diagram: $\theta_g = 4$; $\beta = 0.1$; $n = 0.03$.

Model: Y = a0 + a1*t + a2*t^2 + a3*t^3 + a4*t^4

Variable	Value
a0	0.1224176
a1	0.3429371
a2	0.2927251
a3	–0.0714578
a4	0.0043522

$$\sigma = 0.122 + 0.343\theta + 0.293\theta^2 - 0.0715\theta^3 + 0.00435\theta^4 \qquad (2.39)$$

$\sigma_{min} = 2.884$

The maximum allowable stress at temperature $\theta = 8.0$ is:

$\sigma_{all} = 2.884(3.0)(0.298) = 2.578$ [ksi] = 17.78[MPa]; $\varepsilon_{all} = [8(30)]10^{-4} = 0.024$

The summary of results is presented in Table 2.4 and Fig. 2.26 below.

As can be seen from the Fig. 2.26, the increase in allowable stress occurs as a consequence of the increase of two main parameters: the relaxation time 'n' and the dimensionless parameter β. From a physical point of view as seen from an analysis of solutions of the differential Equation (2.25), the increase in relaxation time (at temperature monotonically increasing in time) increases the stress [33, 34], and the increase of the parameter β decreases the activation energy E_a, therefore, also increases the allowable stress [35]. Computer calculations for the different parameters 'n' and 'β' are similar to the above and the results are presented in Table 2.5 below.

The allowable stresses from Table 2.5 can be used in a preliminary structural engineering design of composite materials and structural systems.

Table 2.4: Summary of results (allowable creep stress vs. parameter β, but n = 0.03).

β	0.01	0.02	0.04	0.06	0.08	0.1
σ_{all}	0.0258 [0.1776]	0.103 [0.712]	0.413 [2.846]	0.928 [6.4]	1.65 [11.38]	2.578 [17.78]
ε_{all}	0.0024	0.0048	0.0096	0.0144	0.0192	0.024

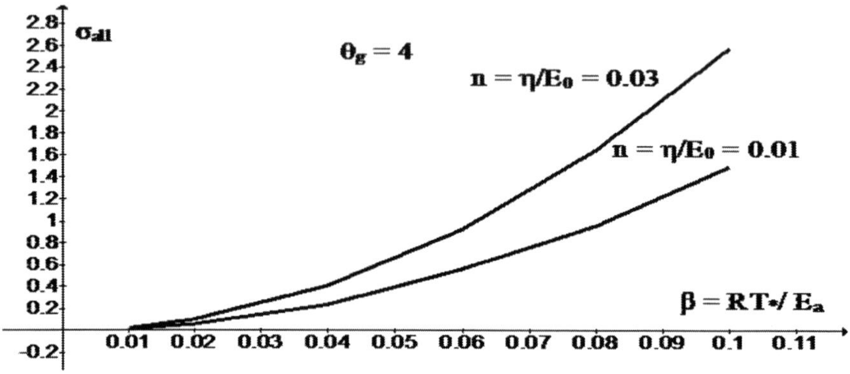

Figure 2.26: Allowable stresses vs. relaxation time.

Table 2.5: Summary of results (allowable creep stress vs. parameters β and n).

β / n	0.01	0.02	0.04	0.06	0.08	0.1
0.01 [MPa]	0.015 [0.103]	0.06 [0.413	0.24 [1.652]	0.565 [3.9]	0.958 [6.61]	1.498 [10.33]
0.03 [MPa]	0.0258 [0.1776]	0.103 [0.712]	0.413 [2.846]	0.928 [6.4]	1.65 [11.38]	2.578 [17.78]
0.05 [MPa]	0.03 [0.207]	0.12 [0.829]	0.48 [3.32]	1.08 [7.46]	1.92 [13.3]	3.0 [20.7]
0.075 [MPa]	0.0327 [0.226]	0.131 [0.904]	0.524 [3.615]	1.18 [8.138]	2.097 [14.46]	3.276 [22.6]
0.1 [MPa]	0.0355 [0.245]	0.142 [0.978]	0.568 [3.92]	1.277 [8.8]	2.269 [15.65]	3.545 [24.45]

References

[1] Ashby, M.F. and Brown, L.M. 2014. Perspectives in Creep Fracture. Elsevier, N.Y.

[2] Nair, S.V. and Jakus, K. 1995. High Temperature Mechanical Behavior of Ceramic Composites. Elsevier, N.Y.

[3] Razdolsky, L. 2016. Reliability index and structural fire resistance of spacecraft and aircraft framing systems. AIAA SPACE 2016, AIAA SPACE Forum (AIAA 2016–5413).

[4] Razdolsky, L. 2017. Phenomenological high temperature creep models of composites and nanomaterials. Proceedings of the AIAA SPACE 2017, Orlando, FL, USA.
[5] Razdolsky, L. 2014. Probability-Based Structural Fire Load. Cambridge University Press, UK.
[6] Halpin, J.C. and Tsai, S.W. 1969. Effects of Environmental Factors on Composite Materials. AFML-TR-67–423.
[7] Jones, R.M. 1999. Mechanics of Composite Materials. 2nd ed., Taylor & Francis, USA, 127 p.
[8] Daken, H.H. and Ismail, A.A. 1995. Effect of carbon on the mechanical behavior of carbon/glass woven hybrid composites. 6th ASAT conf., V. 2 Cairo, Egypt 1995: 423–442.
[9] Chawla, K.K. 1987. Composite Materials Science and Engineering. Springer-Verlag, N.Y.
[10] Whitney, J.M. 1973. J. Structural Div. Amer. Soc. Civil Eng. 113, Jan 1973, N.Y.
[11] Nielsen, L.E. 1974. Mech. Prop. of polymers and composites. Vol. 2, Marcel Dekker, N.Y.
[12] Ferry, J.D. 1980. Viscoelastic Properties of Polymers. John Wiley & Sons Inc., N.Y.
[13] Tobolosky, A. and Eyring, H. 1943. Mechanical properties of polymeric materials. J. Chem. Phys. 11: 125–134.
[14] Alger, M.S.M. 1997. Polymer Science Dictionary (2nd ed.). Springer Publishing, N.Y.
[15] Campbell, F.C. 2010. Stiffness of long fiber composites. Advanced Composite Materials Technology Research Centre, Hong Kong University of Science and Technology. Hong Kong.
[16] Odqvist, F.K.G. 1974. Mathematical Theory of Creep and Creep Rupture (2nd ed.). Clarendon Press, UK.
[17] Skrzypek, J. and Ganczarski, A. 1998. Modeling of Material Damage and Failure of Structures. Foundation of Engineering Mechanics, Springer, Berlin.
[18] Van den Bogart, A.H. and Huetink, J. 2006. Simulation of aluminum sheet forming at elevated temperatures. Computer Methods in Applied Mechanics and Engineering 195(48-49): 6691–6709.
[19] Kurukuri, S., Van den Bogart, A.H., Miroux, A. and Holmdel, B. 2009. Warm forming simulation of Al-Mg sheet. Journal of Materials Processing Technology 209: 5636–5645.
[20] Cobden, R. 1994. Aluminum: Physical properties, characteristics and alloys. European Aluminum Association.
[21] Abedrabbo, N., Pourboghrat, F. and Carsley, J. 2007. Forming of AA5182-O and AA5754-O at elevated temperatures using coupled thermo-mechanical finite element models. Int. J. Plast. 23: 841–875.
[22] Carsley, J., Krajewski, P., Schroth, J. and Lee, T. 2006. Aluminum forming technologies: Status and research opportunities. Proceedings of the Conference New Developments in Sheet Metal Forming, IFU, Stuttgart.
[23] Abedrabbo, N., Pourboghrat, F. and Carsley, J. 2006. Forming of aluminum alloys at elevated temperatures—Part 1. Material characterization. International Journal of Plasticity 22: 314–341.
[24] Bergström, Y. 1983. The plastic deformation of metals—A dislocation model and its applicability. Reviews on Powder Metallurgy and Physical Ceramics 2/3: 79–265.
[25] Nes, E. 1997. Modeling of work hardening and stress saturation in FCC metals. Progress in Materials Science 41(3): 129–193.
[26] Lockyear, S.A. 1999. Mechanical Analysis of Transversely Loaded Wood/Plastic Sections. Master Thesis, Washington State University.
[27] Murphy, J.F. 2010. Characterization of Nonlinear Materials. USDA Forest Products Laboratory Technical Note, Madison, WI.

[28] Laws, V. 1982. The relationship between tensile and bending properties of non-linear composite materials. Journal of Materials Science 17: 2919–2924.
[29] ASTM E1356. Standard Test Method for Assignment of the Glass Transition Temperature by Differential Scanning Calorimetric.
[30] Murphy, J.F. 2003. Characterization of Nonlinear Materials. USDA Forest Products Laboratory Technical Note, Madison, WI.
[31] Nielsen, L.E. and Landel, R.F. 1994. Mechanical Properties of Polymers and Composites. 2nd edition. Marcel Dekker Inc., N.Y., 63–232.
[32] Razdolsky, L. 2017. Probability Based High Temperature Engineering. Springer Nature Publishing Co., AG Switzerland.
[33] Callister, W.D. 1991. Material Science and Engineering (Second Edition). John Wiley & Sons, Inc., N.Y.
[34] ASTM. 1994. Plastics (II), 08.02, ASTM , Philadelphia, PA.
[35] Erhard, G. 2006. Designing with Plastics. Hansen Publishers, Munich.

3

Creep Models of Fibrous and Dispersed Composites

Deterministic Approach

3.1 Micromechanics and macromechanics

Modeling of the properties of multi-structured materials at various levels of its behavior has become the subject of considerable interest in the scientific and engineering community in recent years, which is explained by the development of modeling methods that predict the behavior of materials. The integrated model of knowledge (data and models) includes the information necessary to conclude a prediction about the behavior of the material (including non-critical defects) during operation or in the technological process. The focus of modern research in the field of integrated material data is the discovery of new materials, the development of analytical information, the implementation of models of the information infrastructure of materials, as well as the development of the best methods for supporting data storage based on materials type. However, at the present stage, there are a number of problems that need to be resolved to achieve this integration and link the presentation of the structured materials at different length and time scales. The latter requires the development of fundamental models of data harmonization, as well as the implementation of computational codes describing the evolution of the structured material on various dimensional levels. Computational models of structured materials can be conditionally divided into 3 dimensional categories: molecular and atomic level, microstructure level and macro level (see Fig. 3.1 below).

However, when modeling composite materials by standard methods, it is necessary to take into account the fact that their macroscopic and mechanical properties obtained from testing samples do not reflect the actual properties, since they do not take into account the influence of the orientation of the reinforcing fibers (micro-level), which in turn depends on the parameters

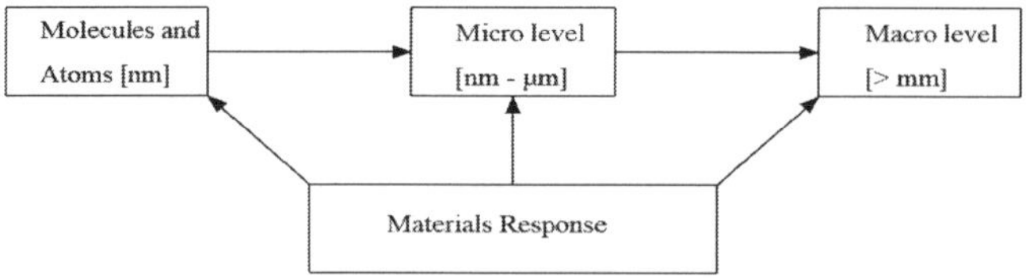

Figure 3.1: Dimensional categories.

characterizing the technological process. In general, the multistage modeling technique [1] for calculating the creep of composites and nanocomposites is reduced to the following steps:

1. All variables and parameters (stress, strain, temperature, etc.) are presented in dimensionless form. This has a threefold effect: first, this reduces the number of parameters that should be determined based on the experimental data to a minimum; second, there is no need to include the aging function (so-called "shift factor") in the integral creep constitutive equation (the data of material creep tests results can be used to analyze the behavior of the material in the future), and finally—the dimensionless form of the integral type constitutive creep equation can be solved in numerical (tabulated) form and be approximated by the analytical function afterwards for further use in structural engineering analysis and design.
2. At any point (time and space) of the composite, the total elongation is the elongation of the material at the initial instant time, minus the elongations due to the external temperature effect, the joint action of temperature and stress, and the elongations due to heat released from the chemical reaction (for example, nanoparticles growth process in the case of nanocomposites).
3. In order to simplify the formulation of creep deformation problem of a composite material, the latter will be divided conditionally into two groups: multilayered composites and dispersed composites. The modulus of elasticity of the material entering the integral type creep equation is the average weighted value of the components (the fiber and the matrix) and obeys so called "rule of mixtures".
4. The memory function (the kernel) should include the Arrhenius law's exponential function, since the high temperature effect on creep deformation is the main topic of this work.
5. Composite material is presented as a viscoelastic nonlinear (non-Newtonian) material.

Brief review of such non-Newtonian viscosity of materials and special forms of it that are used in the analysis of composites creep deformations is the next topic.

3.2 Brief classification of non-newtonian fluids

In a large number of problems of classical mechanics of viscous incompressible liquids, the linear law of the connection between the viscous stress tensor τ_{ij} and the shear rate tensor γ_{ij} (shear strain rates) is applied.

$$\tau_{ij} = \eta\gamma_{ij} \quad \gamma_{ij} = \left(\frac{\partial u_i}{\partial x_j} + \frac{\partial u_j}{\partial x_i}\right) \tag{3.1}$$

η is the dynamic viscosity. Such liquids are usually called Newtonian [2, 3].

However, many liquids that are involved in technological processes possess the nonlinearity of viscosity, as well as other properties, such as ductility, elasticity, are also manifested. Real liquids have a wide range of properties, structured and mechanical characteristics and compositions. Any classification and any approach to their study are to a certain extent idealization and simplification of the actual behavior of the substance. There are many different methods for studying non-Newtonian liquids. One of the most effective approaches is theoretical and rheological, when for a class of liquids with similar properties a certain rheological model is written and on its basis the behavior of liquids of this class is explained. At the present time, a large number of different rheological models are known, or state equations, as they are also called, describing real fluids with a certain degree of accuracy. The simplest and most reasonable classification of these models was proposed by Dodge [4] and described by Wilkinson W.L. in [5]. According to this classification, non-Newtonian fluids depend on the nature of the flow curve, i.e., type of rheological equation $\tau = f(\gamma')$, where τ is the shear stress, and $\gamma' = \partial u/\partial z$ is the shear strain rate.

The non-Newtonian behavior of structured composite systems manifests itself in a change in viscosity with the change in shear flow velocity. Rheological model of structured systems explains the phenomenon of non-Newtonian flow in suspensions, emulsions, solutions and melts of polymers, as well as in liquid crystals. Structured approach is an alternative to classical theory viscoelasticity and relates the change in viscosity with the change in the structure of matter.

To take into account the nonlinear part on the flow curve of a nonlinear-viscoelastic fluids, various rheological laws were proposed, the most common of which is Kaye's phenomenological equation [6]

$$\tau = \left(\tau_0^{1/n} + [\mu_p\dot{\gamma}]^{1/m}\right)^n \tag{3.2}$$

Most known rheological equations can be obtained from (3.2), by choosing the values of the exponents n and m. For example, the well-known Casson equation is obtained [7] if m = n = 2, the modification of the Ostwald equation (the so-called Herschel-Balkley equation [8]) is obtained if n = 1. Equation (3.2) reduces to the Newton equation at $\tau_0 = 0$ and m = n. In all of these equations, there is static yield strength. In [9], a generalized Kasson equation was proposed to describe the rheological properties of a wide class of non-Newtonian fluids, including nonlinear viscoelastic fluids.

$$\tau^{1/2} = \left(\tau_0^{1/n} \frac{\tau_c^{1/n}}{\chi + \dot{\gamma}^{1/2}} + [\mu_\infty^{1/2}] \right) \dot{\gamma}^{1/2} \tag{3.3}$$

In the present work, to describe the rheological properties of viscoelastic medium, the Equations (3.2) and (3.3) are used. The position of the interface between the rigid (quasi-hard) and the liquid zones is determined during the problem solving, and the formulation of additional conditions is not required on the boundary itself. The Williamson equation, like Equation (3.3), satisfactorily describes the experimentally observed shape of the viscous-plastic fluid flow curve over the whole range shear rates. In contrast to the model (3.2), in the Williamson rheological equation there is no ultimate static viscous shear stress. In the framework of the model (3.2) it is not expected to restore the rigid internal structure of the material if the viscous stresses are less than a certain limiting value. Under a rigid (quasi-solid) zone, within the framework of the Williamson model, it means an area in a non-stationary fluid flow where the shear rates are negligibly small in comparison with the surrounding liquid. When choosing rheological models for different types of generalized Newtonian fluids, the guidance was done by some formal, but fairly logical rules. Rheological equations should:

1. Give a good approximation of the experimental data over a wide range of shear rates;
2. Include a minimum number of independent constants that must be amenable to experiment and have a physical meaning;
3. In the limit—the model, should give the rheological Newton's equation for the ordinary liquid.

Let's discuss specific rheological laws of generalized Newtonian liquids, which are used in this work.

There are many cases where in the process of processing the same product passes from one rheological state to another, often the opposite in properties to the first. These changes in raw materials to the finished product are due to a change in the internal energy state, which are characterized by work accomplished through the flow of various processes, in this case due to the processes of fine grinding, mixing and forming. The work carried out during the course of the processes can be represented in the form.

$$A = \int f(F)dx \tag{3.4}$$

f(F)—the function of forces affecting the raw material; X is the parameter determining the effect of the force. The viscosity of non-Newtonian materials depends on the rate of deformation, is related to the structure and its change in flow. In turn, the flow of material depends on its physic-chemical characteristics—from the shape and arrangement of molecules, concentration, temperature, etc. By adding ingredients to a pure solvent, it is possible to increase its viscosity and thereby change the nature of its flow. To describe the equilibrium state of a two-phase system in a steady flow, an effective viscosity is used, which in turn depends on changes in the velocity and shear stress.

3.3 Composite design process

When developing new composites, the reliable mathematical models allow an engineer to predict their properties over a wide range of concentrations and physical parameters of constituent components. The use of adequate phenomenological models of material properties allow purposeful search for promising composites for solving many applied problems. In this section, let's analyze the mathematical modeling of two-phase multilayer composites, one of the components of which is filler, the second can be a matrix with different physical properties (for instance, material with different orientation with respect to applied load). For calculations purposes, the effective medium method is introduced in which the composite is considered as a homogeneous material. It is also assumed that at the boundary section of the layers, the mechanical connection may not be ideal. Using a joint solution of the equations of linear creep deformation of the whole system of multilayer composites, and non-linear creep constitutive single integral type equation for each component together with the conditions of continuity of deformation on the boundary of the matrix and the filler, an effective (approximate) stress-strain diagram (as function of time/temperature) for a multilayer composite structural element is obtained. The model presented below can be used to calculate the effective (optimal) parameters of composite structure in case where the characteristic dimensions of the structured scale units of the fillers (reinforcement) are much smaller than the characteristic scale units of the composite element itself.

3.3.1 *Artificial composite materials*

In engineering practice, among all composites, fibrous composite materials that combine ease with the strength and rigidity in given directions are used. Fibrous composite materials based on glass, carbon, boron, basalt, organic fibers are widely used in critical structures in engineering, aerospace engineering, industrial and civil construction, etc. The right choice of components, taking into account their properties, relationships and structure, allows obtaining composite materials with specified properties. A distinctive feature of composite materials is that such materials show the advantages of components, and not their shortcomings. At the same time composite materials have properties that are not possessed by the individual components that make up their composition. As an example of such product is a multi-layered panel consisting of thin very strong layers and aggregate (honeycomb, cellular, box, folded type z-gofers), with complex geometry and a large-sized structure. Determination of the *effective mechanical properties* of such products is an important problem in view of the inadequacy of the regulatory framework in this area and, in most cases, the solution of which is limited to mathematical modeling without experimental confirmation. In this case, the development of so-called design solutions and modeling of technological processes are very important for the production of critical parts from composite materials.

3.4 Mechanical testing of composites

In the field of testing of composite materials there is a large experience, but the development of materials science in the creation of new structural composite materials is significantly ahead of the processes of standardization in the development and creation of scientifically sound methods for testing composites. In this regard, it is often necessary to refine existing standard methods for testing composites, or to develop new methods for experimental study of the properties of composites with the design and creation of appropriate specialized equipment. Methodical issues of experimental studies of the composites strength are presented in [9]. The most relevant and practical methods of determination of the mechanical properties of fibrous composites are uniaxial tensile, compression, bending and shear tests. Analysis of the regulatory framework for mechanical testing of composites is presented in [10]. Static tests on uniaxial tension and compression of reinforced composite materials are regulated in [11]. However, standard test methods have limited application, in particular, when studying the mechanical properties of high volumetric percentage of fiber unidirectional composite materials. In this case a number of testing problems arise. The main problem is the correspondence of the boundary conditions to the practical method of fixing the samples at the ends. Another problem is that with an increased percentage of fiber filling, the material has a high strength in the longitudinal direction and a much lower strength in the transverse direction. In order to resolve these problems some researchers tend to reject the use of a direct tensile test method and use alternative methods to determine the mechanical properties of unidirectional composites. One such method is the bending test method, followed by recalculation and processing of test results. However, it should be noted that despite the large volume of results obtained (elasticity modulus, interlayer shear modulus, normal stress strength and interlayer shear strength), flexural tests are considered secondary in view of the characteristics of the materials tested. Principal difficulties encountered in testing composite materials, were analyzed in [12, 13]. Micromechanical models of composites allow revealing the analytical dependences showing the influence of the properties of fibers, matrix, and their adhesion interaction, the structure of the material and the mechanics of failure on the macroscopic elastic-resistance characteristics of a single-directional layer. Most successfully they describe the ultimate modulus of elasticity and the strength of the composite under tension. The elongation of the composite under tension in the transverse direction is made up of deformation of fibers and binder. It should be borne in mind that the modulus of elasticity of the fibers themselves in the transverse direction coincides with the modulus of elasticity in the longitudinal direction only for isotropic glass and boron fibers. For carbon and the organic fiber cross module is substantially lower than the longitudinal one. A similar dependence holds for the shear modulus of a unidirectional composite "in plane" of fibers. Strength of composites in transverse tension-compression and shear depends on many factors, primarily on the properties of the matrix, adhesion interaction, material structured presence of pores and other defects.

3.5 Resilient properties of composites

Creation of hybrid composite materials combining two or more types of fibers —glass, organic, carbon and boric, is a promising direction in the development of modern technology, because it increases the possibility of creating materials with the required properties. The most significant factor affecting the nature of the mechanical behavior of composites, especially when in tension, is the magnitude of the ultimate deformation of the fiber reinforcing materials. The mechanical behavior of such materials in tension, compression, bending and shear basically corresponds to the principle of additivity, that is, to the "rule of mixtures". A different pattern of regularities is observed in the investigation of composites, combining fibers with different deformation parameters. When the carbon fiber is stretched (in carbon-organic combination), the fibers failure occurs not simultaneously. The ultimate deformation of the composite is determined in this case mainly by the deformation of those fibers, the volume content of which prevails. Let's denote the index "1"—high-modulus fibers, the index "2"—low-modulus. With high fiber content the (large modulus elasticity and small value of the limiting deformation ε_1) strength of composite is calculated by the formula:

$$\sigma_{k1} = \varepsilon_1(E_m \varphi_m + E_1 \mu_1 + E_2 \mu_2) \tag{3.5}$$

With high content of fibers (low modulus of elasticity), composite stress σ_k is calculated by the formula:

$$\sigma_{k2} = \varepsilon_2(E_m \mu_m + E_2 \mu_2) \quad (3.6)$$

The mechanism of failure of three-component materials varies according to reaching a certain critical ratio of the multi-modulus fibers μ_{cr}, in which the breaking of the fibers with different tensile elongations equal-probable, i.e., $\sigma_{k1} = \sigma_{k2}$. Neglecting the strength of the matrix, we obtain the relations:

$$\varepsilon_1 E_1 \mu_1 + \varepsilon_1 E_2 \mu_2 = \varepsilon_2 E_2 \mu_2 \quad (3.7)$$

after the transformation of which we have:

$$\mu_1/\mu_2 = k = E_2(\varepsilon_2 - \varepsilon_1)/\varepsilon_1 E_1; \mu_{cr2} = k/(1 + k). \quad (3.8)$$

The concept of a critical volume also holds for composites based on one type fibers. It characterizes the transition from the failure of the binder to failure of fibers. Because of the great difference in their module of elasticity μ_{cr} is very small and amounts to 0.1–0.5% of the fibers.

3.6 The phenomenological approach in mechanics of heterogeneous media

In no way rejecting the direction in the study of micro-stresses but it should nevertheless be noted that such theories on the level of mechanics of micro-inhomogeneous media are quite complex even in the case of a uniaxial stress state, and therefore are of little use for solution, for example, of boundary-value problems in the mechanics of continuous media. In these cases the phenomenological creep theory at the macro level is more preferable. The phenomenological approach in mechanics of heterogeneous media is based on the concept of the existence of a representative element of a deformable volume. According to this concept, the material point of the continuum is identified with a region whose dimensions are small in comparison with the dimensions deformable sample, but large in comparison with individual material particles. The properties of the representative element are assumed to be identical to those of the body. For each element and material in general, the basic laws of the mechanics and thermodynamics of the continuum are fulfilled. The advantage of the continuous representation of composite or nanomaterials is the possibility of rigorous use of the apparatus of mathematical analysis of integral equations Volterra of the second kind and calculus of variations, oriented to the study of continuous functions. The permissibility of the continuum approach is evidenced by experimental data. At the same time, according to many scientists, the phenomenological approach allows to take into account the influence of the technological processes of composites. The phenomenological integral type representation of creep process is a very appealing theoretical concept, since it is not limited to a particular material or class of materials. The Volterra equation for one-dimensional case is written as [14]:

$$\varepsilon(t) = \int_{-\infty}^{t} K_1[t-\tau, \sigma(\tau)]\dot{\sigma}(\tau)d\tau + \int_{-\infty}^{t}\int_{-\infty}^{t} K_2[t-\tau_1, \sigma(\tau_1), t-\tau_2, \sigma(\tau_2)]\dot{\sigma}(\tau_1)\dot{\sigma}(\tau_2)d\tau_1 d\tau_2 + \ldots \quad (3.9)$$

(It is often referred to as nonlinear superposition theory).

Choosing a sufficiently large number of members of this series, and having determined the kernels of the integral equation, one can describe any deformation process with any accuracy. Many works have been devoted to the development of this direction, and the main efforts are aimed at simplifying the concept of the state of the material (incompressibility, the same behavior when stretching and compressing, etc.). This makes it possible to simplify the kernels somewhat and reduce the number of terms in the series. It is obvious that the use of multiple integrals and the determination of a large number of kernels is very difficult, therefore all work in this direction are, in the main, only theoretical developments. For a single simple uniaxial tension creep test, the Equation (3.9) reduces to:

$$\varepsilon(t) = \int_{-\infty}^{t} K_1[t-\tau, \sigma(\tau)]\dot{\sigma}(\tau)d\tau \quad (3.10)$$

Application of linear theory to the description of creep deformation processes of different materials is rather limited. Linear theory gives good results at not too high stresses, but for a number of materials, type of polymers, the linearity region cannot be distinguished at all.

Note that, depending on the temperature-time rates (temperature-time function is assumed to be given), the behavior of the material can be described by Equations state (3.9) and (3.10).

Another way to generalize the Equation (3.9) is to have the nonlinear part of creep constitutive law on the left hand of Equation 3.9 (the Rabotnov equation [15]) and it is as follows:

$$F[\varepsilon(t)] = \sigma(t) + \int_{-\infty}^{t} K(t-\tau)\sigma(\tau)d\tau \quad (3.11)$$

To develop constitutive creep equations for inelastic materials, the concept of a rational thermodynamic approach with state variables (also called irreversible thermodynamics [16]) is used. The state variable approach includes certain internal variables (such as stress, strain, temperature, and possibly external pressure) in order to represent the internal state of material.

Constitutive equations which describe the evolution of the internal state are included as part of the Gibbs theory.

For creep deformation analysis of composites and nanocomposites the generalized and modified Schapery theory [17–19] is used in this book. The final integral equation Volterra of the second kind is as follows:

$$E(t)\varepsilon(t) = J_0(\sigma, t) + \sum_{i=1}^{n} f_i[\sigma(t)] \int_0^t F_i(t-\tau) \frac{\partial g_i[\sigma(\tau)]}{\partial \tau} d\tau \tag{3.12}$$

It is generally accepted that the Schapery equation is more general with respect to the Rabotnov equation because it contains four unknown functions that take into account the nonlinearity, and the Rabotnov equation has only one. *The advantage of Equation (3.11) in comparison with others is that in it all nonlinearity is carried to the left side of the equation in the concept of the so-called curve of instantaneous deformation. Any process of time dependant deformation, as mentioned above, is "slipping" from an instantaneous curve. With an increase in the inversed values of the relaxation time the maximum stress values are also increased and the stress-strain diagrams become more condensed, tending in the limit to the curve of instantaneous stress-strain diagram.*

3.6.1 Structured—phenomenological approach to the solution of boundary value mechanics of composites

This section is concerned with conditions of geometric linearity (small strains and rotations) of the whole multilayered structure and material physical nonlinearity. In addition, attention is restricted to cases in which a stress component, $\sigma_i(t)$ of each individual layer, is expressed only in terms of the corresponding strain component history $\varepsilon_i(\theta)$, $0 < \theta(\tau) < \theta_{max}$, where $\theta(\tau)$ is a given dimensionless temperature-time function (primary load on structural system). The uniaxial creep constitutive equation under consideration has the form:

$$E_i(\theta)[\theta] = \sigma_i(\theta) + \int_0^{\theta} e^{\overline{\frac{\tau}{1+\beta\tau}}} K_i(\theta,\tau)\sigma^n(\tau)\, m\mathbf{1} d\tau$$

$$K_i(\theta,\tau) = \varphi_i(\theta) f_i(\tau) = m\mathbf{1}(\tau)\sum_{j=1}^{N} \exp(-\alpha_j\, m(\theta)) \exp(\alpha_j\, m(\tau)) \tag{3.13}$$

$$i = \mathbf{1}, \mathbf{2}, \mathbf{3} \ldots, M;\ \beta = \frac{RT_*}{E_a};\ \ T = \beta T_* \theta + T_*[^0 K]$$

M - number of layers

Here $K_i(t)$ denotes a relaxation function associated with nonlinear viscoelastic response; $m(\theta)$ and $m1(\theta)$ is a time-temperature function and its first derivative respectfully. It is given in this chapter by the relation [20]:

$$m = (0.0405*\theta - 0.01126*\theta^2 + 0.001462*\theta^3 - 0.00006868*\theta^4) \quad (3.14)$$

$$m1 = (0.0405 - 0.02252*\theta + 0.004386*\theta^2 - 0.0002747*\theta^3) \quad (3.15)$$

This constitutive Equation (3.13) is a one-dimensional version of that used by Shay and Caruthers [21–24] or Knauss and Emri [25]. In both papers, the authors began with a tensorial version of Equation (3.13) which is appropriate for the three dimensional response of isotropic materials, and then applied it to uniaxial tension. It is assumed here that the modulus of elasticity E_i is the function of temperature only, i.e., it is constant in each layer plane, but could differ from one layer to another. The peculiarity of the structured-phenomenological approach in solving problems of the mechanics of deformation of structurally inhomogeneous bodies lies in the fact that the mechanical properties of structured elements are specified by using the phenomenological single integral type equations generally accepted in creep mechanics, and macroscopic deformation (elastic or inelastic) and the strength properties are calculated by averaging over the elementary macro-volume. It is assumed that the structured elements are homogeneous and are connected along the interface, so that the structural properties are piecewise-homogeneous functions of the spatial coordinate. Geometry and arrangement of the elements of the structure is assumed to be given. The environment has the property of macroscopic homogeneity. At each layer of the structured elements phenomenological equations and relations of mechanics remain valid. System of static equilibrium equations of all layers are constructed, as well as the physical creep constitutive equations for each homogeneous medium. Then, using the classical force (or displacements) method, equations of continuity of displacements on the boundaries of two adjacent structural layers are written. The boundary and initial creep conditions of the structural system as a whole are also provided using usual methods of solid-state mechanics.

One of the generally accepted approximate methods for analyzing the creep time variant process of structured composites is the introduction of an 'effective' value of the instantaneous modulus of elasticity for each layer of a multilayered structure. After determining the average creep strains (or stresses) as a function of time (or temperature) for each individual composite component, the stress distributions along the cross section of the whole multilayered structure based on the piece-wise linear Bernoulli theory (see Fig. 3.1) is determined for a set of discrete time values [26]. Despite the fact that this approach is just an approximation to the very complex problem of multilayered structure analysis and design, it provides an opportunity to apply a well-developed simple mathematical tool to the construction of a future probabilistic model of creep deformation of such structures [27, 28].

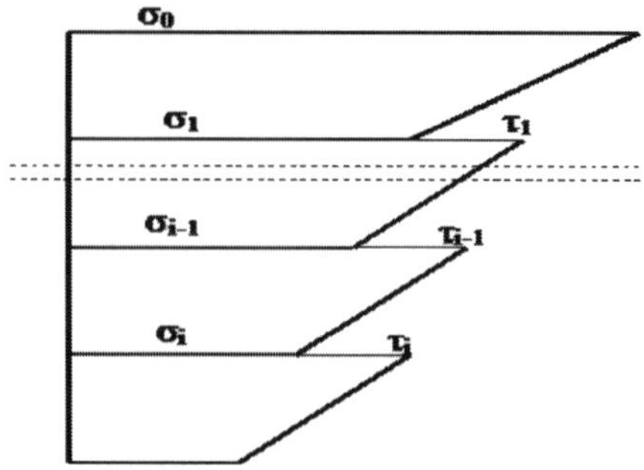

Figure 3.2: Stress distributions (multilayered structure).

In this case it is assumed that the deterministic approximate solutions of the boundary value creep problems of a composite with different discrete values of the dimensionless relaxation (or retardation) time parameter are used as realizations of a random process [29]. The applicability of this method to the analysis of creep strains for a homogeneous isotropic material is shown in [30]. When creating a mathematical model of a structured inhomogeneous material, the problem arises of determining the periodicity of its structure, such as multilayered beams, plates and shells. A detailed analysis of such a structured-phenomenological model is illustrated below in the form of examples.

3.6.2 Equilibrium equations of multilayered composites

Consider a structured composite element consisting of a finite number of (n + 1) layers of materials of different thickness and mechanical properties. It will be assumed that each layer(s) is connected to the previous and subsequent layers by means of longitudinal and transverse rods uniformly distributed along the length of this element and working on transverse (shearing off) and longitudinal forces (tearing off), respectively. It will be assumed that each layer (i) is connected to the previous and subsequent layers by means of diagonal and transverse rods uniformly distributed along the length of this element and working on transverse (shearing off) and longitudinal forces (tearing off), respectively.

During the loading of a multilayer composite structure, in the diagonal shear elements there are forces that are functions of the coordinate x for each fixed time t [31] (see Fig. 3.2).

A composite structural element with its shearing and transverse fictitious elements presents a statically indeterminate system. One can choose a composite beam without these fictitious elements, as a basic statically

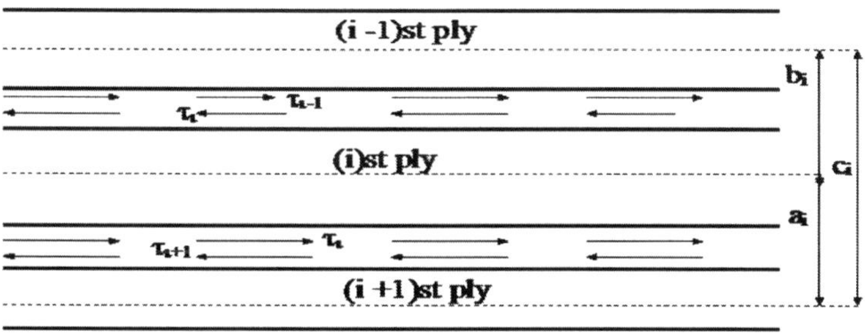

Figure 3.3: Multilayered composite.

determinate system. The removed constraints are replaced by statically indeterminate forces (the transverse elements are assumed to be absolutely rigid, which corresponds to the classical hypothesis of incompressibility of fibers in the transverse direction of the beam). The total rigidity of such a basic system has the form:

$$\sum_{i=1}^{n+1} E_i I_i = \sum EI \tag{3.16}$$

Let M be the total bending moment equal to the sum of the bending moments in the cross sections of each layer of the multilayer composite main system:

$$M^0 = \sum_{i=1}^{n+1} M_i \tag{3.17}$$

In the case of a laminated beam, the stiffness coefficient in the shear stress direction is introduced $\xi = b_{gl} G_{gl}/h_{gl}$. Where b_{gl}; h_{gl}; G_{gl} width; height; and shear modulus of laminated area respectfully.

The stiffness coefficient in the direction perpendicular to the shear stress is $\eta = b_{gl} E_{gl}/h_{gl}$. In the fibers of the constituent elements, longitudinal deformations ε and displacements u arise. The concentrated shift "Γ" along the separating layers of the plane is equal to the difference in the displacement of the upper and lower fibers u^{up} and u^{dw} arise respectfully: $\Gamma = u_{bot} - u_{top}$. Internal shear forces causes bending moments in the ith layer, equal to:

$$\begin{gathered} M_i^1 = -T_{i-1} b_{i-1} - T_i a_i \\ T_i = \int_0^x \tau_i dx; \quad T_{i-1} = \int_0^x \tau_{i-1} dx \end{gathered} \tag{3.18}$$

Summing up, the moment M_i give an addition to the total bending moment and axial forces $N_i = -N_{i-1} + N_i$ of the basic system M^0 and N^0, in connection with which the total value of the bending moment and axial force in the system becomes equal to:

$$M = M^0 + \sum_{i=1}^{n+1} M_i^T$$
$$N = N^0 - T_i + T_{i-1} \tag{3.19}$$

The unknown forces T_i for each layer of a multilayer composite structure can be calculated by solving the following system of linear differential equations given in [31]:

$$\frac{T_1''}{\xi_1} = \Delta_{11}T_1 + \Delta_{12}T_2 + \ldots + \Delta_{1n}T_n + \Delta_{10}$$
$$\frac{T_2''}{\xi_2} = \Delta_{21}T_1 + \Delta_{22}T_2 + \ldots + \Delta_{2n}T_n + \Delta_{20} \tag{3.20}$$
$$\frac{T_n''}{\xi_n} = \Delta_{n1}T_1 + \Delta_{n2}T_2 + \ldots + \Delta_{nn}T_n + \Delta_{n0}$$

where:

$$\Delta_{i0} = \frac{N_{i+1}^0}{E_{i+1}F_{i+1}} - \frac{N_i^0}{E_iF_i} - \frac{M^0 c_i}{\sum_{i=1}^{n} E_i I_i}; \quad \Delta_{ii} = \frac{1}{E_{i+1}F_{i+1}} + \frac{1}{E_iF_i} - \frac{c_i^2}{\sum_{i=1}^{n} E_i I_i}$$
$$\Delta_{i,i+1} = -\frac{1}{E_{i-1}F_{i-1}} + \frac{c_{i+1}c_i}{\sum_{i=1}^{n} E_i I_i}; \quad \Delta_{i,i-1} = -\frac{1}{E_iF_i} + \frac{c_{i-1}c_i}{\sum_{i=1}^{n} E_i I_i}; \quad \Delta_{ij} = \frac{c_j c_i}{\sum_{i=1}^{n} E_i I_i} \tag{3.21}$$

$$\Gamma = u_{bot} - u_{top}; \quad \varepsilon = du/dx = u'; \quad \Gamma' = \varepsilon_{bot} - \varepsilon_{top}$$
$$\Gamma' = \varepsilon_{bot} - \varepsilon_{top} = \frac{N_{i+1}}{E_{i+1}F_{i+1}} - \frac{N_i}{E_iF_i} - \frac{Mc_i}{\sum_{i=1}^{n} E_i I_i}$$

The stiffness coefficient ξ_i is assumed to be constant along the beam length. The boundary conditions are as follows: $T_i(0) = T_i(L) = 0$. We seek a solution of the system of differential Equations (3.20) by expanding the unknown functions T_i and terms N^0 and M^0 into Fourier series. After substituting these expressions into Equation (3.20) and carrying out the corresponding operations, we arrive at a system of linear algebraic equations with respect to the unknowns b_{ik}. Such calculations should theoretically be repeated for each value of k = 1, 2, 3.... However, the series (3.22) is, as a rule, rapidly convergent; therefore, practically two or three terms in the expansion of the series (3.22) are sufficient. The solution of the system of linear algebraic equations at the present time presents no difficulty.

$$\Delta_{i0} = \sum_{k=1}^{\infty} a_{ik} \sin(\frac{k\pi x}{L}); \quad T_{ik} = \sum_{k=1}^{\infty} b_{ik} \sin(\frac{k\pi x}{L})$$

$$i = 1, 2, \ldots, n$$

$$\begin{cases} \left(\Delta_{11} + \dfrac{k^2\pi^2}{L^2\xi_1}\right) b_{1k} + \Delta_{12} b_{2k} + \ldots + \Delta_{1n} b_{nk} + a_{1k} = 0 \\ \Delta_{21} b_{1k} + \left(\Delta_{22} + \dfrac{k^2\pi^2}{L^2\xi_2}\right) b_{2k} + \ldots + \Delta_{2n} b_{nk} + a_{2k} = 0 \\ \Delta_{n1} b_{1k} + \Delta_{n2} b_{2k} + \ldots + \left(\Delta_{nn} + \dfrac{k^2\pi^2}{L^2\xi_n}\right) b_{nk} + a_{nk} = 0 \end{cases} \tag{3.22}$$

$$\text{if } N_i^0 \equiv 0 \text{ and } M^0 \neq \text{const.} \rightarrow a_{ik} = \frac{c_i}{\sum_{i=1}^{n} E_i I_i} \int_0^L M^0(x) \sin\frac{k\pi x}{L} dx$$

Solving Equation (3.20) for each "k", we have:

$$T_i - \sum_{k=1}^{\infty} T_{ik} = \sum_{k=1}^{\infty} b_{ik} \sin(k\pi x/L) \tag{3.23}$$

The equations for determining the transverse (tear off) forces in the joints have the following form:

$$\begin{cases} -s_1 = q_1 + \tau_1' b_1 - \dfrac{E_1 I_1}{\sum EI} \{\sum_{j=1}^{n} \tau_j' c_j + q\} \\ s_1 - s_2 = q_2 + \tau_1' a_1 + \tau_2' b_2 - \dfrac{E_2 I_2}{\sum EI} \{\sum_{j=1}^{n} \tau_j' c_j + q\} \\ s_2 - s_3 = q_3 + \tau_2' a_2 + \tau_3' b_3 - \dfrac{E_3 I_3}{\sum EI} \{\sum_{j=1}^{n} \tau_j' c_j + q\} \\ s_{n-1} - s_n = q_n + \tau_{n-1}' a_{n-1} + \tau_n' b_n - \dfrac{E_2 I_2}{\sum EI} \{\sum_{j=1}^{n} \tau_j' c_j + q\} \\ s_n = q_{n+1} + \tau_n' a_n - \dfrac{E_{n+1} I_{n+1}}{\sum EI} \{\sum_{j=1}^{n} \tau_j' c_j + q\} \end{cases} \tag{3.24}$$

After determining the unknown forces T, the internal forces M and N are calculated without difficulty.

$$\begin{aligned} M_i &= M^0 \frac{E_i I_i}{\sum EI} - \frac{E_i I_i}{\sum EI} \sum_{i=1}^{n+1} T_j c_j \\ N_i &= N_i^0 - T_i + T_{i-1} \\ \sigma_{ix} &= N_i / F_i + M_i z_i / I_i \end{aligned} \tag{3.25}$$

Is the distance from the center of gravity of the cross section of the i-th layer of the multilayer composite element to the one considered [in].

The shear stresses in the ith layer of the multilayer composite element are calculated from:

$$\tau_{xy} = \frac{1}{b(z_i)}\left[-\frac{N_i'}{F_i}F(z_i) - \frac{M_i'}{I_i}S(z_i) + \tau_{i-1}\right] \tag{3.26}$$

$F(z_i)$ and $S(z_i)$ are the area of the part of the cross section of the ith layer located above the level z_i, and the static moment of this area relative to the central axis of the section; $b(z_i)$ is the width of the section at level z_i. The diagram of axial stresses in a multilayer composite element is stepped, but with the same slope to the vertical in all layers. The jumps in the diagram of axial stresses on the boundary of two adjoining layers are calculated as follows: τ'_i/ξ_i.

3.6.3 *Hooke's law for each layer*

Suppose that each element of the multilayered structure is in a plane stressed state. Hooke's law for a layer can be written in the form:

$$\begin{aligned}
&\sigma_1 = \bar{E}_1(\varepsilon_1 + \nu_{12}\varepsilon_2)\\
&\sigma_2 = \bar{E}_2(\varepsilon_2 + \nu_{21}\varepsilon_1)\\
&\tau_{12} = G_{12}\gamma_{12}\\
&\bar{E}_1 = \frac{E_1}{1-\nu_{12}\nu_{21}};\quad \bar{E}_2 = \frac{E_2}{1-\nu_{12}\nu_{21}}
\end{aligned} \tag{3.27}$$

where ε_1 ε_2 γ_{12}—deformations in the directions 1, 2 and the plane 12, respectively, E_1, E_2 and G_{12} are the elastic module in directions 1, 2 and the shear modulus in the plane of the layer, and ν_{12} and ν_{21} are the Poisson coefficients.

Elastic constants in (3.27) are determined either experimentally, or are calculated within the framework of the theory of reinforcement. Since the actual distribution of stresses and strains in the composite body, where the main material has more rigid inclusions, it is very difficult, then in order to obtain any practically useful dependence for determination of the elastic constants E_1, E_2, ν_{12}, G_{12} and it is necessary to make some assumptions:

1. One direction of the reinforced material is a continuous macroscopically homogeneous (transversely isotropic) body.
2. The main material (hereinafter referred to as the matrix) and the reinforcement material (filler) are linearly elastic, isotropic and

homogeneous: the connection between deformations and the stresses in the matrix and reinforcement follow Hooke's law.

3. There is an ideal bond between the matrix and the reinforcement.
4. Additional stresses directed laterally to the reinforcement, which in the general case, in view of the different values of the Poisson coefficients for reinforcement and matrix can occur under the action of stresses, are small and negligible.
5. When the sample is loaded across the direction of reinforcement the stresses in matrix and reinforcement are the same, and they are in proportion of deformations of the components. Materials calculated with such an assumption, in a general deformation composite material is proportional to the volume content of each component.
6. The composite material is reinforced with straight-line fibers.

With sufficiently thick uniform saturation of the main material (matrix) with fibers, the first premise is completely acceptable. To such conclusion come all researchers who study the mechanical properties of dispersed reinforced media. In order to be able to use this premise with sufficient accuracy for practical purposes, it is necessary to know what should be the smallest number of fibers in unit cross-sectional area unidirectional reinforced material. The second premise is introduced because of the need to simplify solution of the problem, limited only to the elastic stage of operation, which is true for many composite materials. From the third premise it follows that the component materials in the direction of the reinforcement are deformed together. The use of this assumption almost excludes errors, if the stresses acting along the reinforcing fibers are not varying along its length. The fourth premise is very close to the real situation and with equal Poisson's ratios of the matrix and fibers generally disappear. When discussing the fifth premise, it should be remembered that actual stresses in the main material near the reinforcing bar increase, because more rigid material for reinforcement is used, but this increase has a local character and it gradually disappears as a result of elastic redistribution. Actual stress distribution is a very complex issue requiring special consideration. When determining the averaged elastic of the deformation characteristics of the composite material, we will operate with the average stresses for the entire volume of the considered element.

Assuming that the distribution of fibers in an elementary volume is uniform, and the fiber diameter is small in comparison with the distances over which the averaged the stress and strain fields vary markedly, with sufficient accuracy of effective stiffness characteristics unidirectional composite material can be calculated from the following formulas [31, 32]:

$$E_1 = \mu E_f + (1-\mu)E_m$$
$$E_2 = \left[\frac{\mu}{E_f} + \frac{1-\mu}{E_m}\right]^{-1} \tag{3.28}$$
$$G_{12} = \frac{(1+\mu)G_f + (1-\mu)G_m}{(1+\mu)G_m + (1-\mu)G_f} G_m$$

where the shear module of the isotropic fiber and the matrix are respectively equal:

$$G_f = \frac{E_f}{2(1+\nu_f)}; \quad G_m = \frac{E_m}{2(1+\nu_m)}$$
$$\nu_{12} = \mu\nu_f + (1-\mu)\nu_m; \quad \nu_{21} = \frac{\nu_{12}E_2}{E_1} \tag{3.29}$$

Formulas for the elastic modulus E_1 and Poisson's ratio ν_{12} are sufficiently accurate for use in engineering calculations. The relations (3.28) for modules E_2 and G_{12} can be recommended for approximate calculations. It is expedient to refine them by the results of experiments using given material. Graphical presentation of Equations (3.28) and (3.29) is as follows:

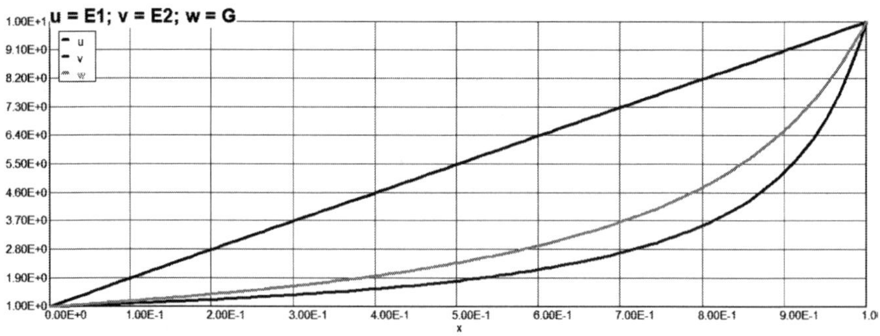

Figure 3.4: Effective modulus of elasticity.

3.6.4 *Elastic deformation model of cross-reinforced composite material*

Let's introduce the orthogonal coordinates (1, 2) and assume that the axis 1 (the reinforced layer) makes an angle φ with the axis σ_{11} (see Fig. 3.5).

Static relationships, the connecting stresses in the coordinate systems σ_{11}, σ_{22} and 1, 2 are as follows:

$$\begin{aligned}\sigma_{11} &= \sigma_1 \cos^2\varphi + \sigma_2 \sin^2\varphi - \tau_{12}\sin 2\varphi \\ \sigma_{22} &= \sigma_1 \sin^2\varphi + \sigma_2 \cos^2\varphi - \tau_{12}\sin 2\varphi \\ \sigma_{12} &= (\sigma_1 - \sigma_2)\sin\varphi\cos\varphi + \tau_{12}\cos 2\varphi\end{aligned} \tag{3.30}$$

Geometric relationships that allow expressing deformations in the system coordinate 1, 2 through deformations in the system coordinate σ_{11}. σ_{22}, can be written as follows:

$$\begin{aligned}\varepsilon_1 &= \varepsilon_{11}\cos^2\varphi + \varepsilon_{22}\sin^2\varphi + \varepsilon_{12}\sin\varphi\cos\varphi \\ \varepsilon_2 &= \varepsilon_{11}\sin^2\varphi + \varepsilon_{22}\cos^2\varphi - \varepsilon_{12}\sin\varphi\cos\varphi \\ \gamma_{12} &= (\varepsilon_{22} - \varepsilon_{11})\sin 2\varphi + \varepsilon_{12}\cos 2\varphi\end{aligned} \tag{3.31}$$

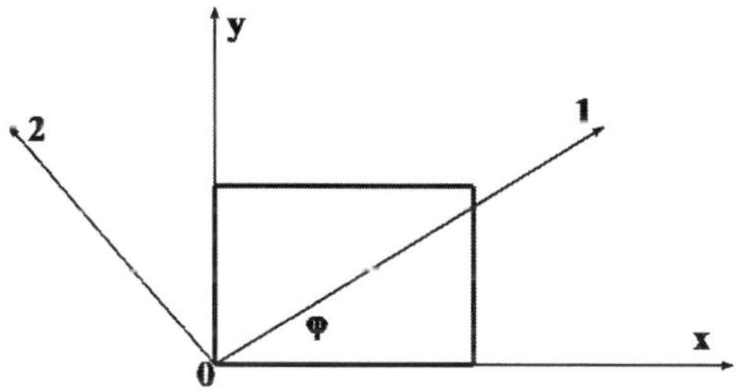

Figure 3.5: Axes rotation.

Now we obtain the relations connecting the stresses σ_{11} σ_{22} σ_{12} with deformations ε_{11} ε_{22} ε_{12}. For this purpose we substitute the deformations ε_1; ε_2 and γ_{12} into Hooke's law, and the stresses obtained as a result of this substitution in the relation (3.30). After some simplification and by using symmetry conditions, we write the physical relationships for the layer reinforced at an angle φ to the axis α_1.

$$\begin{aligned}\sigma_{11} &= A_{11}\varepsilon_{11} + A_{12}\varepsilon_{22} + A_{13}\varepsilon_{12} \\ \sigma_{22} &= A_{21}\varepsilon_{11} + A_{22}\varepsilon_{22} + A_{23}\varepsilon_{12} \\ \sigma_{12} &= A_{31}\varepsilon_{11} + A_{32}\varepsilon_{22} + A_{33}\varepsilon_{12}\end{aligned} \tag{3.32}$$

where:

$$\begin{aligned}A_{11} &= \bar{E}_1\cos^4\varphi + \bar{E}_2\sin^4\varphi + 2[\bar{E}_1\nu_{12} + 2G_{12}]\sin^2\varphi\cos^2\varphi \\ A_{22} &= \bar{E}_1\sin^4\varphi + \bar{E}_2\cos^4\varphi + 2[\bar{E}_1\nu_{12} + 2G_{12}]\sin^2\varphi\cos^2\varphi \\ A_{12} &= A_{21} = \bar{E}_1\nu_{12} + [\bar{E}_1 + \bar{E}_2 - 2(\bar{E}_1\nu_{12} + 2G_{12})]\sin^2\varphi\cos^2\varphi \\ A_{13} &= A_{31} = \sin\varphi\cos\varphi[\bar{E}_1\cos^2\varphi - \bar{E}_2\sin^2\varphi - (E_1\nu_{12} + 2G_{12})\cos 2\varphi] \\ A_{23} &= A_{32} - \sin\varphi\cos\varphi[\bar{E}_1\sin^2\varphi - \bar{E}_2\cos^2\varphi + 2(E_1\nu_{12} + 2G_{12})\cos 2\varphi] \\ A_{33} &= [\bar{E}_1 + \bar{E}_2 - 2\bar{E}_1\nu_{12}]\sin^2\varphi\cos^2\varphi + G_{12}\cos^2\varphi\end{aligned} \tag{3.32a}$$

The graphical examples of functions A_{11}; A_{22},....A_{33} are presented below:
Example:

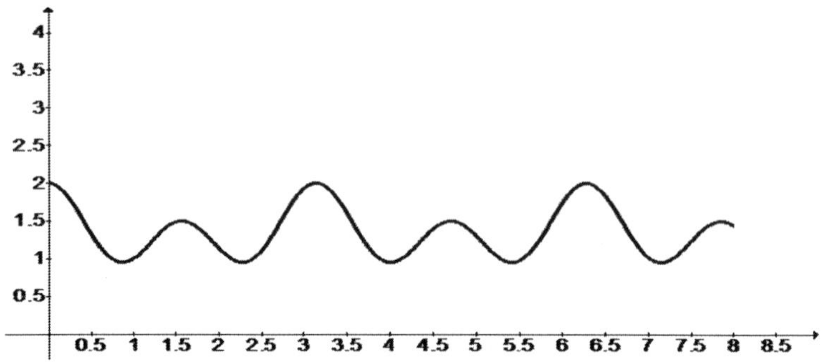

Figure 3.6: Functions type: A_{11}; A_{22},....A_{33}.

From Fig. 3.6 one can see that these functions are periodic functions ($\varphi = \pi$ or $\pi/2$).

Next, we write the relations connecting shear stresses and shear strains in the coordinates (11; 22 and 33) and 1, 2, 3 (see Fig. 3.2):

$$\sigma_{13} = \tau_{13}\cos\varphi - \tau_{23}\sin\varphi$$
$$\sigma_{23} = \tau_{23}\cos\varphi + \tau_{13}\sin\varphi \tag{3.33}$$

$$\gamma_{13} = \varepsilon_{13}\cos\varphi + \varepsilon_{23}\sin\varphi$$
$$\gamma_{23} = \varepsilon_{23}\cos\varphi - \varepsilon_{13}\sin\varphi \tag{3.34}$$

Using Equations (3.27) (3.33) (3.34), we can obtain relations of the type (3.31), that is,

$$\sigma_{13} = A_{44}\varepsilon_{13} + A_{45}\varepsilon_{23}$$
$$\sigma_{23} = A_{54}\varepsilon_{13} + A_{55}\varepsilon_{23}$$
$$A_{44} = G_{13}\cos^2\varphi + G_{23}\sin^2\varphi \tag{3.35}$$
$$A_{55} = G_{13}\sin^2\varphi + G_{23}\cos^2\varphi$$
$$A_{45} = A_{54} = (G_{13} - G_{23})\sin\varphi\cos\varphi$$

In multilayered composites and structural elements, a layer with an angle of reinforcement φ^{+}, as a rule, corresponds to the same layer with an angle of reinforcement φ^{-}. Virtually all automated production processes of a multilayer composite package, and in some cases mutual interlacing of adjacent symmetric layers with angles $\varphi \pm$, are the most economical. In this case such two layers are natural to consider as one symmetrically reinforced layer in the structural design calculation. This assumption completely corresponds to the real structure of layered composites and greatly simplifies the relationships connecting stresses and strains. If each of the symmetric layers is anisotropic in coordinates (1; 2) and α_i, then, working together, they form an orthotropic layer, Hooke's law for which has a simpler formula than the relations above for an individual layer. To obtain this law, we can rewrite the equations for multilayer with reinforcement angles $\varphi \pm$ as follows:

$$\begin{aligned} \sigma_1^{\pm} &= \bar{E}_1(\varepsilon_1^{\pm} + \nu_{12}\varepsilon_2^{\pm}) \\ \sigma_2^{\pm} &= \bar{E}_1(\varepsilon_2^{\pm} + \nu_{12}\varepsilon_1^{\pm}) \\ \tau_{12} &= G_{12}\gamma_{12}^{\pm} \end{aligned} \tag{3.36}$$

$$\begin{aligned} \sigma_{11}^{\pm} &= \sigma_1^{\pm}\cos^2\varphi + \sigma_2^{\pm}\sin^2\varphi \mp \tau_{12}^{\pm}\sin 2\varphi \\ \sigma_{22}^{\pm} &= \sigma_1^{\pm}\sin^2\varphi + \sigma_2^{\pm}\cos^2\varphi \pm \tau_{12}^{\pm}\sin 2\varphi \\ \sigma_{12}^{\pm} &= \pm[\sigma_1^{\pm} - \sigma_2^{\pm}]\sin\varphi\cos\varphi + \tau_{12}^{\pm}\cos\varphi \\ \varepsilon_1^{\pm} &= \varepsilon_{11}\cos^2\varphi + \varepsilon_{22}\sin^2\varphi \pm \varepsilon_{12}\sin\varphi\cos\varphi \\ \varepsilon_2^{\pm} &= \varepsilon_{11}\sin^2\varphi + \varepsilon_{22}\cos^2\varphi \mp \varepsilon_{12}\sin\varphi\cos\varphi \\ \gamma_{12}^{\pm} &= \pm[\varepsilon_{22} - \varepsilon_{11}]\sin 2\varphi\cos\varphi + \varepsilon_{12}\cos 2\varphi \end{aligned} \tag{3.37}$$

Here the signs correspond to layers with angles $\pm\,\varphi$. In relations (3.37) it is taken into account, that the properties of the material of both layers are the same, and equalities (3.37) allow for compatibility conditions of deformations of these layers: $\varepsilon_{11}^{\pm} = \varepsilon_1$; $\varepsilon_{22}^{\pm} = \varepsilon_2$; $\varepsilon_{12}^{\pm} = \varepsilon_{12}$. Substituting the deformations into the Hook's law and the stresses in the relations (3.27), we obtain the stresses σ expressed in terms of deformations (by averaging stresses): $\sigma_{11} = \frac{1}{2}(\sigma_{11}^{+} + \sigma_{11}^{-})$; $\sigma_{22} = \frac{1}{2}(\sigma_{22}^{+} + \sigma_{22}^{-})$; $\sigma_{12} = \frac{1}{2}(\tau_{12}^{+} + \tau_{12}^{-})$ according to the formulas:

$$\begin{aligned} \sigma_{11} &= A_{11}\varepsilon_{11} + A_{12}\varepsilon_{22} \\ \sigma_{22} &= A_{21}\varepsilon_{11} + A_{22}\varepsilon_{22} \\ \sigma_{12} &- A_{33}\varepsilon_{12} \\ A_{mn}&(mn = 11;12;21;22;33) \end{aligned} \tag{3.38}$$

We now derive the relations connecting shear stresses σ_{13} and σ_{23} with corresponding deformations of symmetrically reinforced layers. For layers with angles $\varphi \pm$, according to equalities (3.34), we have:

$$\begin{aligned} \sigma_{13}^{\pm} &= A_{44}\varepsilon_{13}^{\pm} + A_{45}\varepsilon_{23}^{\pm} \\ \sigma_{23}^{\pm} &= A_{54}\varepsilon_{13}^{\pm} + A_{55}\varepsilon_{23}^{\pm} \end{aligned} \tag{3.39}$$

Consider a symmetrical pair of layers loaded by stresses σ_{13}. Then $\sigma_{13}^{+} = \sigma_{13}^{-} = \sigma_{13}$, and, since the system is orthotropic, in the relation (3.27) it is necessary to take $\varepsilon_{13}^{+} = \varepsilon_{13}^{-} = \varepsilon_{13}$ and $\varepsilon_{23}^{+} = \varepsilon_{23}^{-} = 0$. Similarly, when the shear stresses σ_{23} are loaded we have: $\varepsilon_{13}^{+} = \varepsilon_{13}^{-} = 0$; and $\varepsilon_{23}^{+} = \varepsilon_{23}^{-} = \varepsilon_{23}$. As a result, we get:

$$\sigma_{13} = A_{44}\varepsilon_{13};\ \sigma_{23} = A_{55}\varepsilon_{23} \qquad (3.40)$$

We note that the system of coupled symmetrically reinforced layers in the general case is more rigid than the asymmetrically reinforced layer of the same thickness. Thus, the physical relationships for the cross-reinforced composite material are defined by equalities (3.39) and (3.40).

Example 3.1

$$\text{Data}: N^0 \equiv 0;\ n = 6;\ M^0 = \frac{qL^2}{2}(x/L)(x/L-1);$$

$$h = 10'';\ L = 100'';\ M^0_{max} = (0.125)10^4\,k-in;\ q = 1k/in$$

$$E_iI_i = E_{i+1}I_{i+1};\ E_iF_i = E_{i+1}F_{i+1};$$

$$E_i = 1ksi;\ I_i = 10^4 in^4;\ G_i = E_m$$

$$\xi_i = 0.2G_i = 0.1;\ c = 10'' = h$$

$$E_1I_1 = E_2I_2 = \ldots E_6I_6 = 10^4 \Leftrightarrow \varphi = \pm 0;\ \mu \approx 0$$

$$\frac{1}{E_2} = \frac{\mu}{E_f} + \frac{1-\mu}{E_m} \Leftrightarrow \text{assume}: \frac{E_f}{E_m} = 10;\ \Leftrightarrow E_2 \approx E_m$$

$$G_{12} = \frac{(1+\mu)G_f + (1-\mu)G_m}{(1+\mu)G_m + (1-\mu)G_f}G_m \Leftrightarrow \mu \to 0 \Leftrightarrow G_{12} = G_m$$

$$E_1 = \mu E_f + (1-\mu)E_m;\ E_1 \approx E_m$$

$$\Delta_{i0} = -\frac{M^0 c}{\sum_{i=1}^{n} E_iI_i} = \frac{10^5(x/L)(x/L-1)}{6(10^4)} = 1.67(x/L)(x/L-1)$$

$$\Delta_{ii} = \frac{2}{E_iF_i} - \frac{c^2}{\sum_{i=1}^{n} E_iI_i} = \frac{2}{100} - \frac{100}{6(10^4)} = 0.0183$$

$$\Delta_{i,i+1} = -\frac{1}{E_iF_i} + \frac{c^2}{\sum_{i=1}^{n} E_iI_i} = \Delta_{i,i-1} = -\frac{1}{100} + \frac{100}{6(10^4)} = -0.0083$$

$$\Delta_{ij} = \frac{c_jc_i}{\sum_{i=1}^{n} E_iI_i} = \frac{100}{6(10^4)} = 0.00167\ \ |i-j| > 1$$

$$\{\Delta_{ii} + \frac{\pi^2 m^2}{L^2[\xi_1]}\} = 0.0183 + \frac{\pi^2 1^2}{[10^4(0.1)]} = 0.0282$$

$$A_{im} = 1.67\int_0^1 \Delta_{i0}(x)\sin\pi x dx = 1.67\int_0^1 x(x-1)\sin\pi x dx = 0.215$$

0.0282	–0.0083	0.00167	0.00167	0.00167	0.215
–0.0083	0.0282	–0.0083	0.00167	0.00167	0.215
0.00167	–0.0083	0.0282	–0.0083	0.00167	0.215
0.00167	0.00167	–0.0083	0.0282	–0.0083	0.215
0.00167	0.00167	0.00167	–0.0083	0.0282	0.215

Using POLYMATH software we have

Linear Equations Solution

	Variable	Value
1	$x1 = T_1$	9.338148
2	$x2 = T_2$	13.24887
3	$x3 = T_3$	14.31709
4	$x4 = T_4$	13.24887
5	$x5 = T_5$	9.338148

The equations

[1] $0.0282 \cdot x1 - 0.0083 \cdot x2 + 0.00167 \cdot x3 + 0.00167 \cdot x4 + 0.00167 \cdot x5 = 0.215$

[2] $-0.0083 \cdot x1 + 0.0282 \cdot x2 - 0.0083 \cdot x3 + 0.00167 \cdot x4 + 0.00167 \cdot x5 = 0.215$

[3] $0.00167 \cdot x1 - 0.0083 \cdot x2 + 0.0282 \cdot x3 - 0.0083 \cdot x4 + 0.00167 \cdot x5 = 0.215$

[4] $0.00167 \cdot x1 + 0.00167 \cdot x2 - 0.0083 \cdot x3 + 0.0282 \cdot x4 - 0.0083 \cdot x5 = 0.215$

[5] $0.00167 \cdot x1 + 0.00167 \cdot x2 + 0.00167 \cdot x3 - 0.0083 \cdot x4 + 0.0282 \cdot x5 = 0.215$

For the sake of simplicity of calculations in this chapter, we confine ourselves to the case of a high-temperature load uniformly distributed along the length of the composite beam, which varies only with the time; and the elastic module for each layer are constant along the length and depend only on the difference of the temperatures. Moreover, the temperature-time dependence is assumed to be given, and it can be represented as a deterministic function. The high-temperature load is considered to be the main load, and the remaining types of loads are considered secondary when choosing the combination of loads for analysis and calculation of the structural composite element.

Example 3.2

$$\text{Data: } N^0 \equiv 0;\ n = 7;\ M^0 = \frac{qL^2}{2}(x/L)(x/L-1);$$

$$h = 10'';\ L = 100'';\ M^0_{max} = (0.125)10^4 k - in;\ q = 1k/in$$

$$E_iI_i = 1.5(10^4); E_iF_i = 1.5(10^2); \Leftrightarrow \text{If}: \varphi = \mp 0.25\pi;\ \varphi = \mp 1.21\pi$$

$$E_1 = \mu E_f + (1-\mu)E_m;\ E_1 \approx E_m = 1.5$$

$$E_2 = [\mu/E_f + (1-\mu)/E_m]^{-1}; E_2 \approx E_m = 1.5$$

$$G_{12} = \frac{(1+\mu)G_f + (1-\mu)G_m}{(1+\mu)G_m + (1-\mu)G_f}G_m;\ G_{12} \approx G_m = 1.5$$

$$\sum EI = 10.5(10^4);\ \sum EF = 10.5(10^2);$$

$$E_{i,eff} = 1.5\text{ksi};\ \xi_i = 0.07G_{i,eff} = 0.1;\ c = 10'' = h$$

$$\Delta_{i0} = -\frac{M^0 c}{\sum_{i=1}^{n} E_iI_i} = \frac{(0.5)10^5(x/L)(x/L-1)}{10.5(10^4)} = 0.476(x/L)(x/L-1)$$

$$N_i = N_i^0 - T_i + T_{i-1};\ N_1 = 0 - T_1 + 0;\ N_2 = 0 - T_2 + T_1$$

$$T_i(x) = T_i \sin(\pi x/L)$$

$$\Delta_{ii} = \frac{2}{E_iF_i} - \frac{c^2}{\sum_{i=1}^{n} E_iI_i} = \frac{2}{150} - \frac{100}{10.5(10^4)} = 0.0124$$

$$\Delta_{i,i+1} = -\frac{1}{E_iF_i} + \frac{c^2}{\sum_{i=1}^{n} E_iI_i} = \Delta_{i,i-1} = -\frac{1}{150} + \frac{100}{10.5(10^4)} = -0.0057$$

$$\Delta_{ij} = \frac{c_jc_i}{\sum_{i=1}^{n} E_iI_i} = \frac{100}{10.5(10^4)} = 0.001$$

$$\{\Delta_{ii} + \frac{\pi^2 m^2}{L^2[\xi_1]}\} = 0.0124 + \frac{\pi^2 1^2}{[10^4(0.1)]} = 0.0223$$

$$A_{im} = 0.476\int_0^1 \Delta_{i0}(x)\sin\pi x dx = 0.476\int_0^1 x(x-1)\sin\pi x dx = 0.129(0.476) = 0.0614$$

$$\sigma_{11} = A_{11}\varepsilon_{11} \Leftrightarrow A_{11} = E_{eff}$$

$$= A_{11} = \overline{E}_1\cos^4\varphi + \overline{E}_2\sin^4\varphi + 2[\overline{E}_1\nu_{12} + 2G_{12}]\sin^2\varphi\cos^2\varphi \approx$$

$$\approx E_m\cos^4\varphi + E_m\sin^4\varphi + 2[E_m\nu_{12} + 2G_m]\sin^2\varphi\cos^2\varphi \approx$$

$$\approx E_m\left(\cos^4\varphi + \sin^4\varphi + 4\sin^2\varphi\cos^2\varphi\right)$$

$$\frac{E_f}{E_m} = 10;\ \ \nu \to 0;\ E_m = 1$$

The graph of effective modulus of elasticity is presented below.

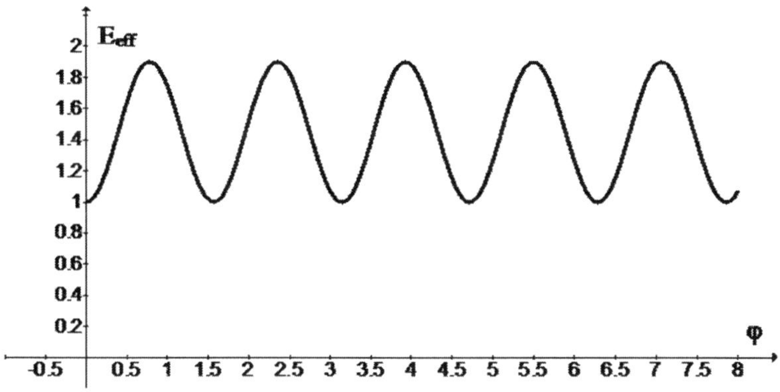

Figure 3.7: Effective modulus of elasticity E_{eff}.

Max E_{eff}: $\varphi = \mp 0.25\pi$ $\varphi = \mp 1.21\pi$ $\varphi = \mp 1.6\pi \rightarrow E_{eff.} = 1.5$

Min E_{eff}: $\varphi = 0$; $\varphi = \mp 0.5\pi \rightarrow E_{eff.} = 1.0$

$E_{1,eff.} = 1.5$ $E_m = 1.5$; $I_1 = 10^4 \rightarrow E_{1,eff.}$ $I_1 = 1.5(10^4)$

0.0124	–0.0057	0.001	0.001	0.001	0.001	0.0614
–0.0057	0.0124	–0.0057	0.111	0.111	0.111	0.0614
0.001	–0.0057	0.0124	–0.0057	0.111	0.111	0.0614
0.001	0.001	–0.0057	0.0124	–0.0257	0.111	0.0614
0.001	0.001	0.001	–0.0057	0.0124	–0.0057	0.0614
0.001	0.001	0.001	0.001	–0.0057	0.0124	0.0614

Linear Equations Solution

	Variable	Value
1	x1	17.34957
2	x2	28.17652
3	x3	8.156276
4	x4	–2.109035
5	x5	0.0218037
6	x6	0.8024961

The equations

[1] $0.0124 \cdot x1 - 0.0057 \cdot x2 + 0.001 \cdot x3 + 0.001 \cdot x4 + 0.001 \cdot x5 + 0.001 \cdot x6 = 0.0614$

[2] $-0.0057 \cdot x1 + 0.0124 \cdot x2 - 0.0057 \cdot x3 + 0.111 \cdot x4 + 0.111 \cdot x5 + 0.111 \cdot x6 = 0.0614$

[3] $0.001 \cdot x1 - 0.0057 \cdot x2 + 0.0124 \cdot x3 - 0.0057 \cdot x4 + 0.111 \cdot x5 + 0.111 \cdot x6 = 0.0614$

[4] $0.001 \cdot x1 + 0.001 \cdot x2 - 0.0057 \cdot x3 + 0.0124 \cdot x4 - 0.0257 \cdot x5 + 0.111 \cdot x6 = 0.0614$

[5] $0.001 \cdot x1 + 0.001 \cdot x2 + 0.001 \cdot x3 - 0.0057 \cdot x4 + 0.0124 \cdot x5 - 0.0057 \cdot x6 = 0.0614$

[6] $0.001 \cdot x1 + 0.001 \cdot x2 + 0.001 \cdot x3 + 0.001 \cdot x4 - 0.0057 \cdot x5 + 0.0124 \cdot x6 = 0.0614$

The stresses in each layer of the seven (7) layered beam in this case are as follows:

$$M_i = M^0 \frac{E_i I_i}{\sum EI} - \frac{E_i I_i}{\sum EI} \sum_{j=1}^{n} T_j c_j$$

$$N_i = N_i^0 - T_i + T_{i-1}$$

$$\sigma_{ix} = N_i / F_i + M_i z_i / I_i$$

$$\sum_{j=1}^{n} T_j c_j = 10(17.35 + 28.18 + 8.16 - 2.11 + 0.022 + 0.802) = 524.0\text{k} - \text{in}$$

$$M_i = \frac{1}{7}(M^0_{max} - 524) = \frac{1}{7}[1250 - 524] = 103.7\text{k} - \text{in} = 8.54\text{ft} - \text{k}$$

$$N_1 = -17.35\text{k};\ \sigma_1^{max} = -17.35/100 - 5(103.7)/1000 = -0.174 - 0.518\text{ksi} = -0.692\text{ksi}$$

$$\sigma_1^{min} = -17.35/100 + 5(103.7)/1000 = +0.344\text{ksi}$$

$$N_2 = 17.35 - 28.17 = -10.83\text{k};\ \sigma_2^{max} = -10.83/100 - 5(103.7)/1000 = -0.626\text{ksi}$$

$$\sigma_2^{min} = -10.83/100 + 5(103.7)/1000 = +0.41\text{ksi}$$

$$N_7 = 0.2 - 0 = 0.2\text{k};\ \sigma_7^{min} = 0.2/100 - 5(103.7)/1000 = -0.52\text{ksi}$$

$$\sigma_7^{max} = 0.2/100 + 5(726)/1000 = +0.52\text{ksi}$$

$$\tau_{xyi} = \frac{1}{b(z_i)}\left[-\frac{N_i'}{F_i}F(z_i) - \frac{M_i' S(z_i)}{I_i} + \tau_{i-1}\right] =$$

$$= \frac{1}{b}\left[-\frac{17.35\pi}{LF_i}F(z_i) - \frac{Q\pi S(z_i)}{LI_i} + \tau_{i-1}\right]_{x=0}$$

$$\tau_{xy1} = \frac{1}{100}\left[-\frac{\pi(17.35)}{1000(100)}500 - \frac{103.7\pi(500)5}{1000(100)} + 0\right] =$$

$$= -0.00272 - 0.0814 = 0.0841\text{ksi}$$

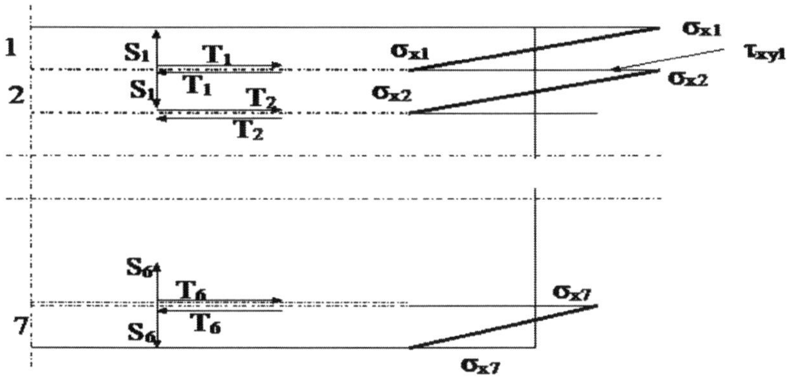

Figure 3.8: Stress distribution.

Example 3.3

$$E_1 = \mu E_f + (1-\mu)E_m \Leftrightarrow \mu \to 0.5 \Leftrightarrow E_1 = 2E_m$$

$$\frac{1}{E_2} = \frac{\mu}{E_f} + \frac{1-\mu}{E_m} \Leftrightarrow \text{assume}: \frac{E_f}{E_m} = 3;\ \mu = 0.5 \Leftrightarrow E_2 = 1.5E_m$$

$$G_{12} = \frac{(1+\mu)G_f + (1-\mu)G_m}{(1+\mu)G_m + (1-\mu)G_f} G_m \Leftrightarrow \mu \to 0 \Leftrightarrow G_{12} = 1.67G_m$$

$$\sigma_{11} = A_{11}\varepsilon_{11} \Leftrightarrow A_{11} = E_{eff}; \nu \to 0;\ E_m = 1$$

$$= A_{11} = \bar{E}_1 \cos^4\varphi + \bar{E}_2 \sin^4\varphi + 2[\bar{E}_1 \nu_{12} + 2G_{12}]\sin^2\varphi\cos^2\varphi \approx$$

$$\approx E_m \cos^4\varphi + E_m \sin^4\varphi + 2[E_m \nu_{12} + 2G_m]\sin^2\varphi\cos^2\varphi$$

$$E_{eff} = E_m\left(2\cos^4\varphi + 1.5\sin^4\varphi + 6.67\sin^2\varphi\cos^2\varphi\right)$$

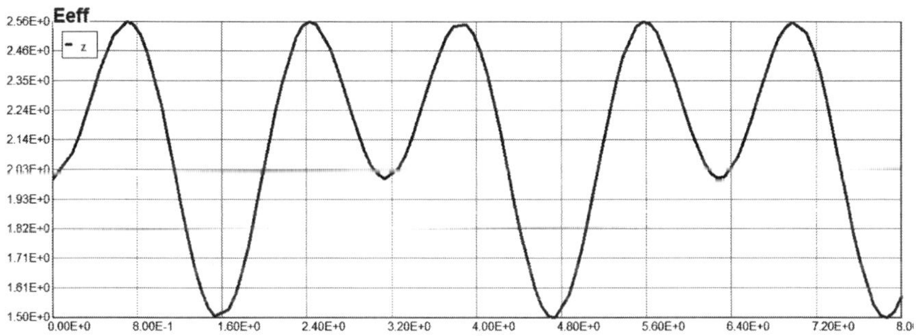

Figure 3.9: Effective modulus of elasticity (Example 3.3).

$\varphi = 0.7 = \pm 0.223 \approx \pm 40^o$; $\varphi = 2.4 = \pm 0.764\pi \approx \pm\ 138^o$; $\varphi = 3.88 = 1.23\pi \rightarrow$
$E_{eff.} = 2.56\ E_m = 2.56$

Data: $N^0 \equiv 0;\ n = 7;\ M^0 = \frac{qL^2}{2}(x/L)(x/L-1);$

$h = 2";\ b = 12";\ H = 14";\ L = 100";\ M^0_{max} = (0.125)10^4 k - in;\ q = 1k/ft$

$E_i I_i = 2.56(8) = 20.48; E_i F_i = 2.56(24); \Leftrightarrow$ If: $\varphi = \mp 0.223\pi;\ \varphi = \mp 0.7641\pi$

$E_1 = \mu E_f + (1-\mu)E_m;\ E_1 = 2E_m;$

$E_2 = [\mu/E_f + (1-\mu)/E_m]^{-1}; E_2 = 1.5E_m$

$$G_{12} = \frac{(1+\mu)G_f + (1-\mu)G_m}{(1+\mu)G_m + (1-\mu)G_f}G_m;\ G_{12} = 1.67G_m$$

$\sum EI = 2.56(7)(2^3) = 143.4;\ \sum EF = 2.56(7)24 = 430;$

$E_{im} = 1ksi;\ \xi_i = 0.1G_{im} = 0.1;\ c = 2" = h$

$$\Delta_{i0} = -\frac{M^0 c}{\sum_{i=1}^{n} E_i I_i} = -\frac{(0.5)10^4(x/L)(x/L-1)}{(12)0.143(10^3)} = -2.92(x/L)(x/L-1)$$

$$\Delta_{ii} = \frac{2}{E_i F_i} - \frac{c^2}{\sum_{i=1}^{n} E_i I_i} = \frac{2}{(2.56)24} - \frac{4}{0.143(10^3)} = 0.0326 - 0.028 = 0.00455$$

$$\Delta_{i,i+1} = -\frac{1}{E_i F_i} + \frac{c^2}{\sum_{i=1}^{n} E_i I_i} = \Delta_{i,i-1} = -\frac{1}{(2.56)24} + \frac{4}{0.143(10^3)} = -0.0163 + 0.028 = 0.0117$$

$$\Delta_{ij} = \frac{c_j c_i}{\sum_{i=1}^{n} E_i I_i} = \frac{4}{0.143(10^3)} = 0.028 \quad |i-j|>1$$

$$\{\Delta_{ii} + \frac{\pi^2 m^2}{L^2[\xi_1]}\} = 0.00455 + \frac{\pi^2 1^2}{[10^4(0.1)]} = 0.00455 + 0.0099 = 0.0145$$

$$A_{im} = 2.92\int_0^1 \Delta_{i0}(x)\sin \pi x dx = 8.74\int_0^1 x(x-1)\sin \pi x dx = 0.129(2.92) = 0.376$$

0.0145	0.0117	0.028	0.028	0.028	0.028	0.376
0.0117	0.0145	0.0117	0.028	0.028	0.028	0.376
0.028	0.0117	0.0145	0.0117	0.028	0.028	0.376
0.028	0.028	0.0117	0.0145	0.0117	0.028	0.376
0.028	0.028	0.028	0.0117	0.0145	0.0117	0.376
0.028	0.028	0.028	0.028	0.0257	0.0145	0.376

Linear Equations Solution

	Variable	Value
1	$x1 = T_1$	–6.608056
2	$x2 = T_2$	12.21993
3	$x3 = T_3$	3.234256
4	$x4 = T_4$	–8.151608
5	$x5 = T_5$	10.26408
6	$x6 = T_6$	6.397687

The equations

[1] $0.0145 \cdot x1 + 0.0117 \cdot x2 + 0.028 \cdot x3 + 0.028 \cdot x4 + 0.028 \cdot x5 + 0.028 \cdot x6 = 0.376$

[2] $0.0117 \cdot x1 + 0.0145 \cdot x2 + 0.0117 \cdot x3 + 0.028 \cdot x4 + 0.028 \cdot x5 + 0.028 \cdot x6 = 0.376$

[3] $0.028 \cdot x1 + 0.0117 \cdot x2 + 0.0145 \cdot x3 + 0.0117 \cdot x4 + 0.028 \cdot x5 + 0.028 \cdot x6 = 0.376$

[4] $0.028 \cdot x1 + 0.028 \cdot x2 + 0.0117 \cdot x3 + 0.0145 \cdot x4 + 0.0117 \cdot x5 + 0.028 \cdot x6 = 0.376$

[5] $0.028 \cdot x1 + 0.028 \cdot x2 + 0.028 \cdot x3 + 0.0117 \cdot x4 + 0.0145 \cdot x5 + 0.0117 \cdot x6 = 0.376$

[6] $0.028 \cdot x1 + 0.028 \cdot x2 + 0.028 \cdot x3 + 0.028 \cdot x4 + 0.0257 \cdot x5 + 0.0145 \cdot x6 = 0.376$

The stresses in each layer of the seven (7) layered beam in this case are as follows:

$$M_i = M^0 \frac{E_i I_i}{\sum EI} - \frac{E_i I_i}{\sum EI} \sum_{j=1}^{n} T_j c_j$$

$$N_i = N_i^0 - T_i + T_{i-1}$$

$$\sigma_{ix} = N_i / F_i + M_i z_i / I_i$$

$$\sum_{j=1}^{n} T_j c_j = 10(-6.61 + 12.22 + 3.23 - 8.15 + 10.26 + 6.4) = 173.5\text{k} - \text{in}$$

$$M_i = \frac{1}{7}(M_{max}^0 - 173.5) = \frac{1}{7}[1250/12 - 173.5] = -9.9\text{k} - \text{in}$$

$$N_1 = -6.61\text{k}; \ \sigma_1^{max} = -6.61/24 - 5(9.9)/8 = -0.275 - 6.19 = -6.46\text{ksi}$$

$$\sigma_1^{min} = -6.61/24 + 5(9.9)/8 = 5.92\text{ksi}$$

$$N_2 = 12.22 + 6.61 = 18.83\text{k}; \ \sigma_2^{max} = 18.83/24 - 6.19 = -5.4\text{ksi}$$

$$\sigma_2^{min} = 18.83/24 + 6.19 = 6.97\text{ksi}$$

$$N_7 = 6.4 - 0 = 6.4\text{k}; \ \sigma_7^{max} = 6.4/24 + 6.19 = 6.46\text{ksi}$$

3.7 Phenomenological creep models of composite structures

Let us consider a multilayered composite beam, obeying the general integral linear creep law. In this case, the linear Volterra integral equation has the form:

$$E_0E(t)I\frac{\partial^4 y(x,t)}{\partial x^4} = q(x,t) + \int_0^t q(x,t)K(t,\tau)d\tau \tag{3.41}$$

We expand the functions of deflections and loads in Fourier series with respect to trigonometric functions of the form (or fundamental beam functions), satisfying the given boundary conditions:

$$q(x,t) = \sum_{k=1}^{\infty} X_k(x)Q_k(t);\quad y(x,t) = \sum_{k=1}^{\infty} X_k(x)y_k(t)$$
$$\text{where: } X_k(x) = \sin(k\pi x/L) \tag{3.42}$$

Substituting the series (3.42) term by term in Equation (3.41), we obtain after contraction of $X_k(x) = \sin(k\pi x/L)$.

$$\lambda_k^4 E(t)y_k(t) = Q_k(t) + \int_0^t Q_k(t)K(t,\tau)d\tau$$
$$\lambda_k^4 = E_0I[\frac{k\pi}{L}]^4 \tag{3.43}$$

Consider, for example, the case of axial tension. Taking into account the fact that the temperature load causes the axial displacement $y = \alpha(T - T_*)L = \varepsilon L$, we have

$$A[\bar{\sigma}] = AE(t)\varepsilon = \sigma_k(t) + \int_0^t \sigma_k(t)K(t,\tau)d\tau$$
$$A = \lambda_k^4 \beta T_* L\alpha_0 \tag{3.44}$$

As mentioned above, the integral representation of creep process is a very appealing theoretical concept, since it is not limited to a particular material or class of materials. Composite elements and heterogeneous materials, from which they are made, make essential changes to the classical scheme for constructing the phenomenological integral creep model of composite elements. In this case the specificity of the composite material should manifest itself in the choice of two basic functions of the integral creep constitutive equation, namely E (t) and K (t, τ). The function of the elastic modulus should reflect the fact that (1) the composite material is a two-phase system with a parameter T_g *very important for practical purposes (transient temperature); and also that (2) the composite material is a composition of two (or more) different in its physical parameters of materials, interconnected by internal forces of adhesion, which under the influence of high temperature change with time. The function K is a function of memory and consequently should reflect all changes in the energy balance with changes in temperature and time, such as the change in the free Gibbs energy due to the change in the stress state of the composite material and free chemical energy (in the case of nanocomposites) [34]. Thus, in the case of composite structures, the integral creep equation, as noted in the author's work [35], has the following form:*

$$E(t)[\varepsilon(t)] = \sigma(t) + \int_{-\infty}^{t} K_1[t-\tau, \sigma(\tau)]d\tau + \int_{-\infty}^{t} K_2[t-\tau, \sigma(\tau)]d\tau \tag{3.45}$$

It is assumed here that the classical concept of entropy is extendable into non-equilibrium situations. First integral on the right hand side of the Equation (3.15) represents the entropy flows into the volume element; second—represents the entropy source due to creep dependence on stress and temperature that are lumped together. Based on the simplifications introduced in author's previous work, and a system of dimensionless parameters and variables (stress and temperature) integral equation (3.46) with time variant kernels K_1 and K_2 is solvable for creep stress-temperature functions.

$$\begin{gathered}
E(\theta)[\theta] = \sigma(\theta) + \int_0^{\theta} e^{\frac{\tau}{1+\beta\tau}} K_1(\theta,\tau)\sigma^n(\tau)\, m\mathbf{1} d\tau + \int_0^{\theta} e^{\frac{\tau}{1+\beta\tau}} f_1[\sigma(\tau)] K_2(\theta,\tau)\sigma^n(\tau)\, m\mathbf{1} d\tau + \\
+ \int_0^{\theta} \frac{E_0}{E_2} e^{\frac{\tau}{1+\beta\tau}} f_2[d^{-1}(\tau)] K_3(\theta,\tau)\sigma(\tau)\, m\mathbf{21} d\tau \\
K_1(\theta,\tau) = m\mathbf{1}(\tau)\sum_{i=1}^{N} \exp(-\alpha_i\, m(\theta))\exp(\alpha_i\, m(\tau)) \\
K_2(\theta,\tau) = m\mathbf{1}(\tau)\sum_{i=1}^{N} e^{\frac{\tau}{1+\beta\tau}} \exp(-\beta_{i1}\, m(\theta))\exp(\beta_{i1}\, m(\tau)) \\
f_1(\sigma) = A_2\sigma^s; \quad s = \mathbf{1,2,3}\ldots, M; \quad \beta = \frac{RT_*}{E_a}; \quad T = \beta T_*\theta + T_*[^\circ K]
\end{gathered} \tag{3.46}$$

It should be noted that the composites in this chapter are conventionally divided into two categories: dispersion composites (the matrix and the filler are ideally mixed and represent a homogeneous mixture with a modulus of elasticity $E = E_1\mu + (1-\mu) E_2$, where μ is the volume fraction coefficient, E_2 and E_1—modulus of elasticity of the matrix and filler, respectively); And multi-layer reinforced structures (where the modulus of elasticity of each layer is equal to E_i and in general can differ from each other or periodically repeat. The actual forms of integrants and parameters E_0; E_2; s; T_*; E_a and θ_g in Equation (3.46) obviously should be based on simple well known uniaxial creep tests data and corresponding temperature-time functions of external heat transfer to the system. The solutions of Equation (3.27) for some particular expressions of integrants are presented via Examples (see below). For the seven ply composite structure:

Example 3.2 (con't)

Data: 7 ply Beam (see Example 3.2); $E_0 = 1$ [GPa]; $0.001 < \alpha_i < 1000$

Calculated values of DEQ variables

	Variable	Initial value	Minimal value	Maximal value	Final value
1	A2	1.	1.	1.	1.
2	E	1.	0.25	1.	0.25
3	m	0	0	0.0705846	0.0705846
4	m1	0.0405	0.0004702	0.0405	0.0004702
5	n	1.	1.	1.	1.
6	s	1.	1.	1.	1.
7	t	0	0	8.	8.
8	Y1	0	0	3.382307	1.996408
9	z	0	0	2.48795	0.4445576
10	z1	0	0	0.6998575	0.6998575
11	z2	0	0	0.8528368	0.8528368

Differential equations

1 d(z1)/d(t) = (exp(t/(1+0.067*t)))*((exp(0.001*m)))*m1*z^n

2 d(z2)/d(t) = (exp(t/(1+0.067*t)))*(A2*(z^s))*((exp(0.001*m)))*m1*z^n

Explicit equations

1 m = (0.0405*t – 0.01126*t^2 + 0.001462*t^3 – 0.00006868*t^4)

2 m1 = (0.0405 – 0.02252*t^1 + 0.004386*t^2 – 0.0002747*t^3)

3 E = (0.625 – 0.375*(tanh(3*(t – 4))))

4 z = t*E–(z1*1+z2)*((exp(–0.001*m)))

5 n = 1.0

6 s = 1.0

7 Y1 = t*E

8 A2 = 1

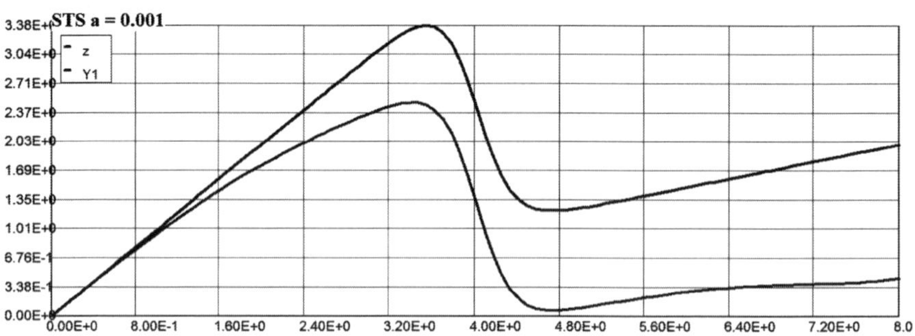

Figure 3.10: Stress-temperature-strain diagram α = 0.001.

$z = \sigma = -0.25*\theta + 1.8*\theta^2 - 0.806^3 + 0.122*\theta^4 - 0.0061*\theta^5$

Variable	Value
a1	–0.2512931
a2	1.808933
a3	–0.8063904
a4	0.1221665
a5	–0.0061373

Finally, maximum stress of a 7-ply beam is:

$$M_i = M^0 \frac{E_i I_i}{\sum EI} - \frac{E_i I_i}{\sum EI} \sum_{j=1}^{n} T_j c_j$$

$$N_i = N_i^0 - T_i + T_{i-1}$$

$$\sigma_{ix} = N_i / F_i + M_i z_i / I_i$$

$$\sum_{j=1}^{n} T_j c_j = 10(-6.61 + 12.22 + 3.23 - 8.15 + 10.26 + 6.4) = 173.5\text{k} - \text{in}$$

$$M_i = \frac{1}{7}(M_{max}^0 - 173.5) = \frac{1}{7}[1250/12 - 173.5] = -9.9\text{k} - \text{in}$$

$$N_1 = -6.61\text{k};\ \sigma_1^{max} = -6.61/24 - 5(9.9)/8 = -0.275 - 6.19 = -6.46\text{ksi}$$

$$\sigma_1^{min} = -6.61/24 + 5(9.9)/8 = 5.92\text{ksi}$$

$$N_2 = 12.22 + 6.61 = 18.83\text{k};\ \sigma_2^{max} = 18.83/24 - 6.19 = -5.4\text{ksi}$$

$$\sigma_2^{min} = 18.83/24 + 6.19 = 6.97\text{ksi}$$

$$N_7 = 6.4 - 0 = 6.4\text{k};\ \sigma_7^{max} = 6.4/24 + 6.19 = 6.46\text{ksi}$$

$$\sigma_{max} = 6.46(2.45)A = 41.94\ [\text{MPa}] = 6.08\text{ksi}$$

$$A = \lambda_k^4 \beta T_* L \alpha_0 = 9.89(0.067)40000(10^{-4}) = 2.65$$

$$\lambda_k^4 = E_0 I[\frac{k\pi}{L}]^4 = 1000(7)10^4(\frac{\pi^4}{10^8}) = 68.2\ [\text{MPa}] = 9.89[\text{ksi}]$$

3.8 Temperature-time dependent structured heterogeneous composites

The recent experiments with temperature-time dependent composite materials revealed that the rate of stress relaxation is strain dependent and the rate of creep is stress dependent. This nonlinear behavior of composites requires a more general description than the separable quasi-linear viscoelasticity theory commonly used in engineering creep analysis and structural design. Hence, the modified superposition model that is suggested here describes the application of principles of irreversible thermodynamics in order to account for both elastic nonlinearity and strain-dependent relaxation rate behavior. Therefore, the creep laws derived for the constant one dimensional stress condition cannot be applied to variable stress conditions and hence must be modified. Time hardening and strain hardening are two popular theories, which take into account the variable stress and variable time-temperature process. In case of strain hardening theory it is assumed that the creep rate is a function of stress and accumulated strain $\dot{\varepsilon}_c = f(\sigma, \varepsilon)$. The particular forms of these laws can be obtained by assuming that the creep curve can be represented by the Bailey-Norton law, which is a common representation of creep in the primary and secondary creep ranges. Creep laws for composites under conditions which favor dislocation processes may be represented by a thermally activated power law [36].

$$\dot{\varepsilon}_c = A\sigma^n \exp(-H/RT)$$
$$H = E + PV \tag{3.47}$$

where $\dot{\varepsilon}_c$ is the steady-state creep rate; H-activation enthalpy (E is the activation energy, V is the activation volume, and P is the pressure or mean normal engineering stress). Although Equation (3.33) is non-Newtonian by definition, it can be expressed in terms of a simple viscous relation

$$\dot{\varepsilon}_c = \frac{1}{2\eta}\sigma \tag{3.48}$$

by defining an effective viscosity

$$\eta_{eff} = \frac{\sigma^{1-n} \exp(H/RT)}{2A} \tag{3.49}$$

which may vary locally as a function of stress. The final feature that is emphasized in the relation (3.49) is its time dependence. If we invert Equation (3.49), the steady-state stress at fixed strain rate is

$$\sigma_c = \left(\frac{\dot{\varepsilon}}{A}\right)^{1/n} \exp(H/nRT) \tag{3.50}$$

It is time dependent for a given strain increment; σ_c is the strain-independent counterpart of $\dot{\varepsilon}$ in constant stress tests. In addition, this form of the flow law exhibits a powerful exponential effect of inverse temperature on the steady state strength. Of greatest impact to flexural models of the composites are the combined effects of thermal gradients and the strong dependence of creep upon temperature, the non-Newtonian power-law dependence upon stress, and the time-dependent nature of strength. The non-linear dependence of strain rate, characteristic of dislocation creep, also has an important mechanical impact, affecting the distribution and pattern of strains resulting from various thermal loading sources. Analysis will commence with the modified form of Equation (3.36) proposed by author [27]. The single integral formulation of the modified superposition method allows the creep function to be depended on stress level. Therefore the Equation (3.32) can be written in dimensionless form as follows:

$$E(\theta)[\theta] = \sigma(\theta) + \int_0^\theta e^{\frac{\tau}{1+\beta\tau}+f(\sigma)} K(\theta,\tau)\sigma^n(\tau)\,m\mathbf{1}d\tau$$

$$K(\theta,\tau) = \varphi(\theta)f(\tau) = m\mathbf{1}(\tau)\sum_{i=1}^{N}\exp(-\alpha_i\, m(\theta))\exp(\alpha_i\, m(\tau)) \qquad (3.51)$$

$$\Phi = e^{\frac{\tau}{1+\beta\tau}+f(\sigma)};\quad m = t = \varphi(\theta);\quad m\mathbf{1} = \dot{\varphi}(\theta)$$

Here K (θ) is, in comparison with linear theory, the degenerate kernel of time-temperature variant creep function and Φ denotes the modified dimensionless Arrhenius law.

This Volterra integral of second kind identity is the natural generalization of the simple one-dimensional relationship proposed by Fung [37], which preserves *objectivity*. As mentioned, thanks to the relative simplicity of Fung's approach over more general nonlinear viscoelastic models, it has proved extremely popular. In the paragraphs (examples) below, a number of practical applications and interpretations of the model is discussed.

Consider now that $f(\sigma) = A\sigma^s$ in Equation (3.51) and A; s; β and n are the kernels constants that must be determined by experiment. Deterministic temperature-time function 'm' and first derivative 'm1' assumed to be given.

$$E(\theta)[\theta] = \sigma(\theta) + \int_0^\theta e^{\frac{\tau}{1+\beta\tau}+A\sigma^s} K(\theta,\tau)\sigma^n(\tau)\,m\mathbf{1}d\tau$$

$$K(\theta,\tau) = \varphi(\theta)f(\tau) = m\mathbf{1}(\tau)\sum_{i=1}^{N}\exp(-\alpha_i\, m(\theta))\exp(\alpha_i\, m(\tau)) \qquad (3.52)$$

$$\theta = \frac{E_a}{RT_*^2}[T - T_*];\ \beta = \frac{RT_*}{E_a}$$

T_* - Base Temperature [°K]

The solution of the creep constitutive Equation (3.52) and numerical analysis of the effect of dimensionless parameters A; s; β and n is illustrated via examples below.

Example 3.3

$\Phi = \exp[\tau/(1+\beta\tau) + A(\sigma)]$

Data: $0 < A < 1$; $\beta = 2(400)/12000 = 0.067$; $T_* = 400°\,K = 127°C$;

$T_{max} = 26.8(8) + 400 = 614°$ K ($\theta_{max} = 8$); $\theta_g = 4$; s =1; n = 1. It is assumed here that $\mu << 1$, therefore $E \approx E_m = E_0[1-(\tanh(3*(\theta-4))]*0.5$

Calculated values of DEQ variables

	Variable	Initial value	Minimal value	Maximal value	Final value
1	A	0.1	0.1	0.1	0.1
2	E	1.	3.775E-11	1.	3.775E-11
3	E1	1.	0.25	1.	0.25
4	m	0	0	0.0705907	0.0705907
5	m1	0.0405	0.0003976	0.0405	0.0003976
6	n	1.	1.	1.	1.
7	s	1.	1.	1.	1.
8	t	0	0	8.	8.
9	Y1	0	0	3.33385	3.02E-10
10	z	0	–0.5001526	2.913951	–0.2176501
11	z1	0	0	0.5439919	0.219192

Differential equations

1 d(z1)/d(t) = (exp(t/(1+0.067*t)))*(exp(A*(z^s)))*((exp(0.1*m)))*m1*z^n

Explicit equations

1 m = (0.0405*t – 0.01126*t^2 + 0.001462*t^3 – 0.00006868*t^4)
2 m1 = (0.0405 – 0.02252*t^1 + 0.004386*t^2 – 0.0002747*t^3)
3 E = (1–(tanh(3*(t – 4))))*0.5
4 E1 = 0.625 – 0.375*tanh(5* (t – 4))
5 z = t*E–z1*((exp(–0.1*m)))
6 n = 1.0
7 s = 1.0
8 A = 0.1
9 Y1 = t*E

Model: z = a* x*(1–(tanh(3*(x – 4))))*0.5

Variable	Initial guess	Value
a	1.	0.884

Model: $z = \sigma = 0.442*\theta*(1-(\tanh(3*(\theta - 4))))$ (3.53)

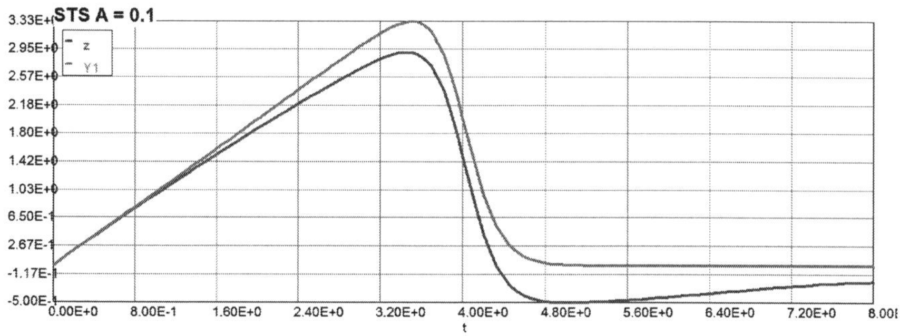

Figure 3.11: Stress-temperature-strain diagram A = 0.1.

Example 3.4

Calculated values of DEQ variables [$A = 0.2$]; E = (a–b*(tanh(c*(θ – θ_g))))

	Variable	Initial value	Minimal value	Maximal value	Final value
1	A	0.2	0.2	0.2	0.2
2	E1	1.	0.25	1.	0.25
3	m	0	0	0.0705907	0.0705907
4	m1	0.0405	0.0003976	0.0405	0.0003976
5	n	1.	1.	1.	1.
6	s	1.	1.	1.	1.
7	t	0	0	8.	8.
8	Y1	0	0	3.33389	3.02E-10
9	z	0	–0.6049704	2.823963	–0.2731613
10	z1	0	0	0.6544733	0.2750964

Differential equations

1 d(z1)/d(t) = (exp(t/(1 + 0.067*t)))*(exp(A*(z^s)))*((exp(0.1*m)))*m1*z^n

Explicit equations

1 m = (0.0405*t – 0.01126*t^2 + 0.001462*t^3 – 0.00006868*t^4)

2 m1 = (0.0405 – 0.02252*t^1 + 0.004386*t^2 – 0.0002747*t^3)

3 E1 = 0.625 – 0.375*tanh(5* (t – 4))

4 z = t*E–z1*((exp(–0.1*m)))

5 n = 1.0

6 s = 1.0

7 A = 0.2

8 Y1 = t*E

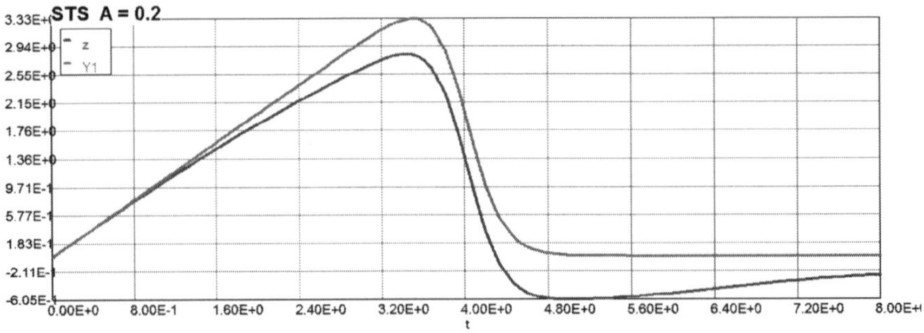

Figure 3.12: Stress-temperature-strain diagram A = 0.2.

Model: z = a* x*(1–(tanh(3*(x – 4))))*0.5

Variable	Initial guess	Value
a	1.	0.862

Model: $z = \sigma = 0.431*\theta*(1-(\tanh(3*(\theta - 4))))$ (3.54)

Example 3.5

Calculated values of DEQ variables [*A = 0.4*]; $E = (a-b*(\tanh(c*(\theta - \theta_g))))$

Calculated values of DEQ variables

	Variable	Initial value	Minimal value	Maximal value	Final value
1	A	0.4	0.4	0.4	0.4
2	E	1.	3.775E-11	1.	3.775E-11
3	E1	1.	0.25	1.	0.25
4	m	0	0	0.0705907	0.0705907
5	m1	0.0405	0.0003976	0.0405	0.0003976
6	n	1.	1.	1.	1.
7	s	1.	1.	1.	1.
8	t	0	0	8.	8.
9	Y1	0	0	3.325638	3.02E-10
10	z	0	–0.851258	2.625338	–0.4325338
11	z1	0	0	0.9105228	0.4355978

Differential equations

1 d(z1)/d(t) = (exp(t/(1 + 0.067*t)))*(exp(A*(z^s)))*((exp(0.1*m)))*m1*z^n

Explicit equations

1 m = (0.0405*t – 0.01126*t^2 + 0.001462*t^3 – 0.00006868*t^4)
2 m1 = (0.0405 – 0.02252*t^1 + 0.004386*t^2 – 0.0002747*t^3)
3 E = (1–(tanh(3*(t – 4))))*0.5
4 E1 = 0.625 – 0.375*tanh(5*(t – 4))
5 z = t*E–z1*((exp(–0.1*m)))
6 n = 1.0
7 s = 1.0
8 A = 0.4
9 Y1 = t*E

Model: z = a* x*(1–(tanh(3*(x – 4))))*0.5

Variable	Initial guess	Value
a	1.	0.808

Model: z = σ = 0.404*θ*(1–(tanh(3*(θ – 4)))) (3.55)

Example 3.5

Calculated values of DEQ variables [A = 0.6];

	Variable	Initial value	Minimal value	Maximal value	Final value
1	A	0.6	0.6	0.6	0.6
2	E	1.	3.775E-11	1.	3.775E-11
3	E1	1.	0.25	1.	0.25
4	m	0	0	0.0705907	0.0705907
5	m1	0.0405	0.0003976	0.0405	0.0003976
6	n	1.	1.	1.	1.
7	s	1.	1.	1.	1.
8	t	0	0	8.	8.
9	Y1	0	0	3.333753	3.02E-10
10	z	0	–1.123229	2.393607	–0.6745744
11	z1	0	0	1.186097	0.6793531

Differential equations

1 d(z1)/d(t) = (exp(t/(1 + 0.067*t)))*(exp(A*(z^s)))*((exp(0.1*m)))*m1*z^n

Explicit equations

1 m = (0.0405*t – 0.01126*t^2 + 0.001462*t^3 – 0.00006868*t^4)
2 m1 = (0.0405 – 0.02252*t^1 + 0.004386*t^2 – 0.0002747*t^3)
3 E = (1–(tanh(3*(t – 4))))*0.5
4 z = t*E–z1*((exp(–0.1*m)))
5 n = 1.0
6 s = 1.0
7 A = 0.6
8 Y1 = t*E

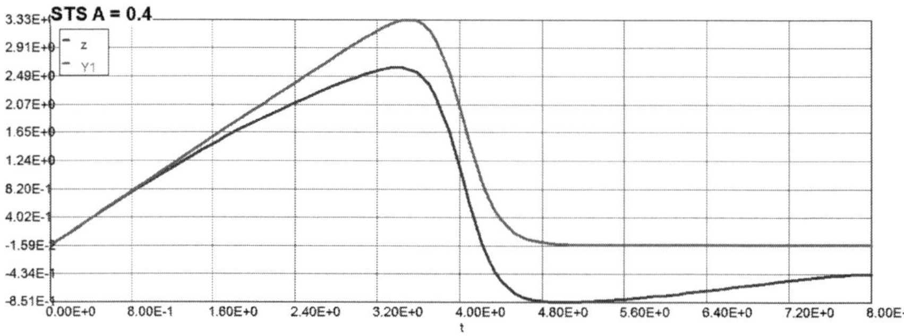

Figure 3.13: Stress-temperature-strain diagram A = 0.4.

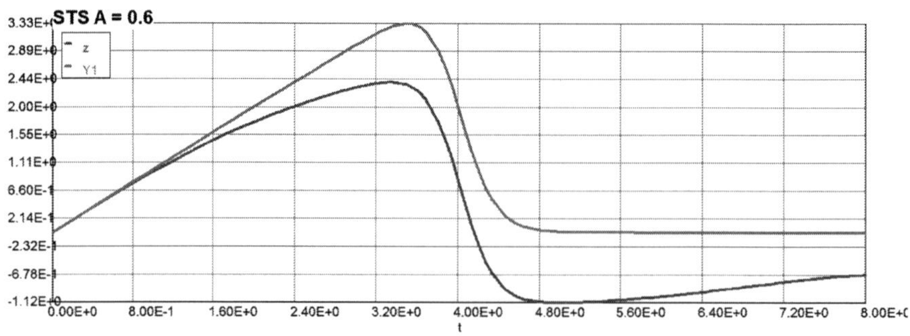

Figure 3.14: Stress-temperature-strain diagram A = 0.6.

Model: z = a* x*(1–(tanh(3*(x – 4))))*0.5

Variable	Initial guess	Value
a	1.	0.746

Model: z = σ = 0.373*θ*(1–(tanh(3*(θ – 4)))) (3.56)

Example 3.6

Calculated values of DEQ variables [A = 0.8]; E = (a–b*(tanh(c*(θ – θ_g))))

	Variable	Initial value	Minimal value	Maximal value	Final value
1	A	0.8	0.8	0.8	0.8
2	E	1.	3.775E-11	1.	3.775E-11
3	E1	1.	0.25	1.	0.25
4	m	0	0	0.0705907	0.0705907
5	m1	0.0405	0.0003976	0.0405	0.0003976
6	n	1.	1.	1.	1.
7	s	1.	1.	1.	1.
8	t	0	0	8.	8.
9	Y1	0	0	3.331388	3.02E-10
10	z	0	–1.388104	2.159872	–0.9935905
11	z1	0	0	1.447426	1.000629

Differential equations

1 d(z1)/d(t) = (exp(t/(1+0.067*t)))*(exp(A*(z^s)))*((exp(0.1*m)))*m1*z^n

Explicit equations

1 m = (0.0405*t – 0.01126*t^2 + 0.001462*t^3 – 0.00006868*t^4)
2 m1 = (0.0405 – 0.02252*t^1 + 0.004386*t^2 – 0.0002747*t^3)
3 E = (1–(tanh(3*(t – 4))))*0.5
4 E1 = 0.625 – 0.375*tanh(5*(t – 4))
5 z = t*E–z1*((exp(–0.1*m)))
6 n = 1.0
7 s = 1.0
8 A = 0.8
9 Y1 = t*E

Model: z = a* x*(1–(tanh(3*(x – 4))))*0.5

Variable	Initial guess	Value
a	1.	0.678

Model: z = σ = 0.339*θ*(1 (tanh(3*(θ 4)))) (3.57)

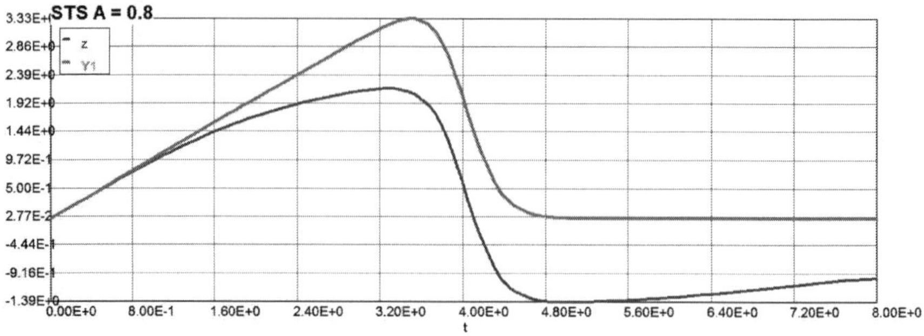

Figure 3.15: Stress-temperature-strain diagram A = 0.8.

Example 3.7

Calculated values of DEQ variables [A = 1.0] E = (a–b*(tanh(c*(θ – θ_g))))

	Variable	Initial value	Minimal value	Maximal value	Final value
1	A	1.	1.	1.	1.
2	E	1.	3.775E-11	1.	3.775E-11
3	E1	1.	0.25	1.	0.25
4	m	0	0	0.0705907	0.0705907
5	m1	0.0405	0.0003976	0.0405	0.0003976
6	n	1.	1.	1.	1.
7	s	1.	1.	1.	1.
8	t	0	0	8.	8.
9	Y1	0	0	3.328159	3.02E-10
10	z	0	–1.622472	1.948363	–1.334546
11	z1	0	0	1.673797	1.344

Differential equations

1 d(z1)/d(t) = (exp(t/(1 + 0.067*t)))*(exp(A*(z^s)))*((exp(0.1*m)))*m1*z^n

Explicit equations

1 m = (0.0405*t – 0.01126*t^2 + 0.001462*t^3 – 0.00006868*t^4)

2 m1 = (0.0405 – 0.02252*t^1 + 0.004386*t^2 – 0.0002747*t^3)

3 E = (1–(tanh(3*(t – 4))))*0.5

4 E1 = 0.625 – 0.375*tanh(5* (t – 4))

5 z = t*E–z1*((exp(–0.1*m)))

6 n = 1.0
7 s = 1.0
8 A = 1.0
9 Y1 = t*E

Model: z1 = a* x*(1–(tanh(3*(x – 4))))*0.5

Variable	Initial guess	Value
a	1.	0.616

Model: $z = \sigma = 0.308*\theta*(1-(\tanh(3*(\theta - 4))))$ (3.58)

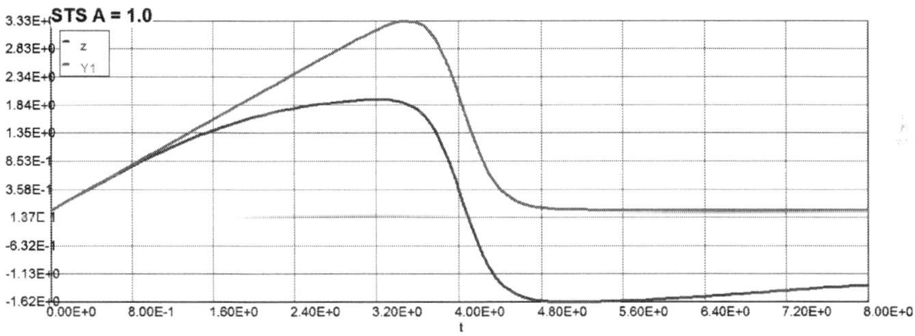

Figure 3.16: Stress-temperature-strain diagram A = 1.0.

Table 3.1: Maximum creep stresses vs. parameter A.

A	0	0.1	0.2	0.4	0.6	0.8	1.0
σ_{max}	3.328	2.914	2.824	2.625	2.394	2.160	1.948

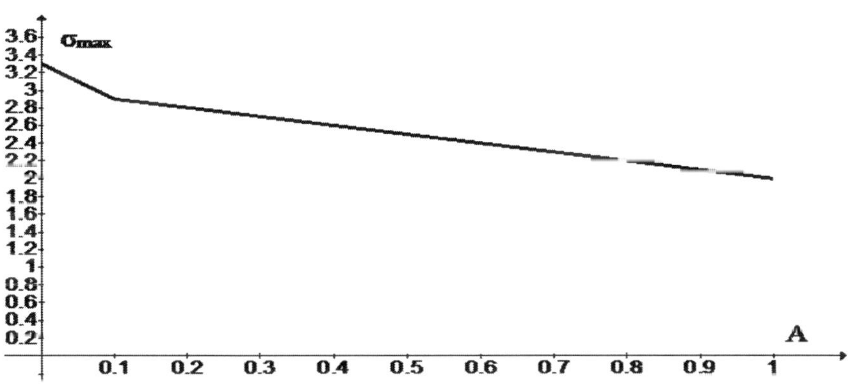

Figure 3.17: Maximum creep stresses vs. parameter A.

$T = \beta T_*\theta + T_* = 0.067(400)\ \theta + 400 = 26.8\ \theta + 400;\ \theta_{max} = 3.5$
$\varepsilon = \alpha_0\ [T - T_*] = \alpha_0[\beta T_*\theta] = (10^{-4})(0.067)(400)\ \theta = 26.8\ \theta\ (10^{-4})$
$\sigma = 26.8\ \theta\ (10^{-4})E_0 = 26.8\ (\theta)(10^{-4})1\ [GPa] = 9.38\ [MPa] = 1.362$ ksi

Example 3.8

Calculate allowable creep stress and maximum temperature if A = 0.4

Solution

σ_{all} - ? → A = 0.4 $\sigma_{all} = 2.625(26.8)3.5(10^{-1}) = 24.62$ [MPa] = 3.57 ksi

$T_{max} = 0.067(400)\ 3.5 + 400 = 493.8°\ K \approx 221°C$

Composite materials affected by high temperature loading conditions usually exhibit quite high level of nonlinear behavior (n >> 1). The dimensionless type of Equation (3.52) allows analyzing the effect of this parameter on the final result: stress-temperature-strain (STS) diagram. Below are the solution of Equation (3.52) with different values of n = 2; 3;...;10.

Example 3.9

Data: A = 0.4; β = 0.067; n = 2; E = a(1–b*(tanh(c*(θ – θ_g))))

Calculated values of DEQ variables

	Variable	Initial value	Minimal value	Maximal value	Final value
1	A	0.4	0.4	0.4	0.4
2	E	1.	3.775E-11	1.	3.775E-11
3	m	0	0	0.0705907	0.0705907
4	m1	0.0405	0.0003976	0.0405	0.0003976
5	n	2.	2.	2.	2.
6	s	1.	1.	1.	1.
7	t	0	0	8.	8.
8	Y1	0	0	3.333999	3.02E-10
9	z	0	–3.143764	2.230913	–3.143764
10	z1	0	0	3.166035	3.166035

Differential equations

1 d(z1)/d(t) = (exp(t/(1 + 0.067*t)))*(exp(A*(z^s)))*((exp(0.1*m)))*m1*z^n

Explicit equations

1 m = (0.0405*t – 0.01126*t^2 + 0.001462*t^3 – 0.00006868*t^4)
2 m1 = (0.0405 – 0.02252*t^1 + 0.004386*t^2 – 0.0002747*t^3)
3 E = (1–(tanh(3*(t – 4))))*0.5

4 z = t*E–z1*((exp(–0.1*m)))
5 n = 2.0
6 s = 1.0
7 A = 0.4
8 Y1 = t*E

Model: z = a* x*(1–(tanh(3*(x – 4))))*0.5

Variable	Initial guess	Value
a	1.	0.706

Model: $z = \sigma = 0.353*\theta*(1–(\tanh(3*(\theta – 4))))$ (3.59)

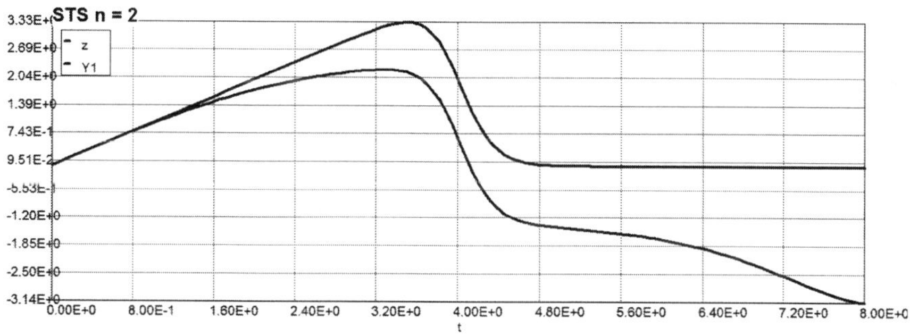

Figure 3.18: Stress-temperature-strain diagram n = 2.

Example 3.10

Data: A = 0.4; $\beta = 0.067$; n = 3; $E = a(1–b*(\tanh(c*(\theta – \theta_g))))$

Calculated values of DEQ variables

	Variable	Initial value	Minimal value	Maximal value	Final value
1	A	0.4	0.4	0.4	0.4
2	E	1.	3.775E-11	1.	3.775E-11
3	m	0	0	0.0705907	0.0705907
4	m1	0.0405	0.0003976	0.0405	0.0003976
5	n	3.	3.	3.	3.
6	s	1.	1.	1.	1.
7	t	0	0	8.	8.
8	Y1	0	0	3.332574	3.02E-10
9	z	0	–1.563249	1.90245	–0.7963757
10	z1	0	0	1.702771	0.8020173

Differential equations

1 d(z1)/d(t) = (exp(t/(1 + 0.067*t)))*(exp(A*(z^s)))*((exp(0.1*m)))*m1*z^n

Explicit equations

1 m = (0.0405*t – 0.01126*t^2 + 0.001462*t^3 – 0.00006868*t^4)
2 m1 = (0.0405 – 0.02252*t^1 + 0.004386*t^2 – 0.0002747*t^3)
3 E = (1–(tanh(3*(t – 4))))*0.5
4 z = t*E–z1*((exp(–0.1*m)))
5 n = 3.0
6 s = 1.0
7 A = 0.4
8 Y1 = t*E

Model: z = a* x*(1–(tanh(3*(x – 4))))*0.5

Variable	Initial guess	Value
a	1.	0.614

Model: z = σ = 0.307*θ*(1–(tanh(3*(θ – 4)))) (3.60)

Example 3.11

Data: A = 0.4; β = 0.067; n = 5; E = a(1–b*(tanh(c*(θ – θ_g))))

Calculated values of DEQ variables

	Variable	Initial value	Minimal value	Maximal value	Final value
1	A	0.4	0.4	0.4	0.4
2	E	1.	3.775E-11	1.	3.775E-11
3	m	0	0	0.0705907	0.0705907
4	m1	0.0405	0.0003976	0.0405	0.0003976
5	n	5.	5.	5.	5.
6	s	1.	1.	1.	1.
7	t	0	0	8.	8.
8	Y1	0	0	3.333534	3.02E-10
9	z	0	–1.705718	1.53886	–0.7898746
10	z1	0	0	2.018042	0.7954701

Differential equations

1 d(z1)/d(t) = (exp(t/(1 + 0.067*t)))*(exp(A*(z^s)))*((exp(0.1*m)))*m1*z^n

Explicit equations

1 m = (0.0405*t – 0.01126*t^2 + 0.001462*t^3 – 0.00006868*t^4)
2 m1 = (0.0405 – 0.02252*t^1 + 0.004386*t^2 – 0.0002747*t^3)
3 E = (1–(tanh(3*(t – 4))))*0.5
4 z = t*E–z1*((exp(–0.1*m)))
5 n = 5.0
6 s = 1.0
7 A = 0.4
8 Y1 = t*E

Model: z = a* x*(1–(tanh(3*(x – 4))))*0.5

Variable	Initial guess	Value
a	1.	0.50

Model: $z = \sigma = 0.25*\theta*(1-(\tanh(3*(\theta - 4))))$ (3.61)

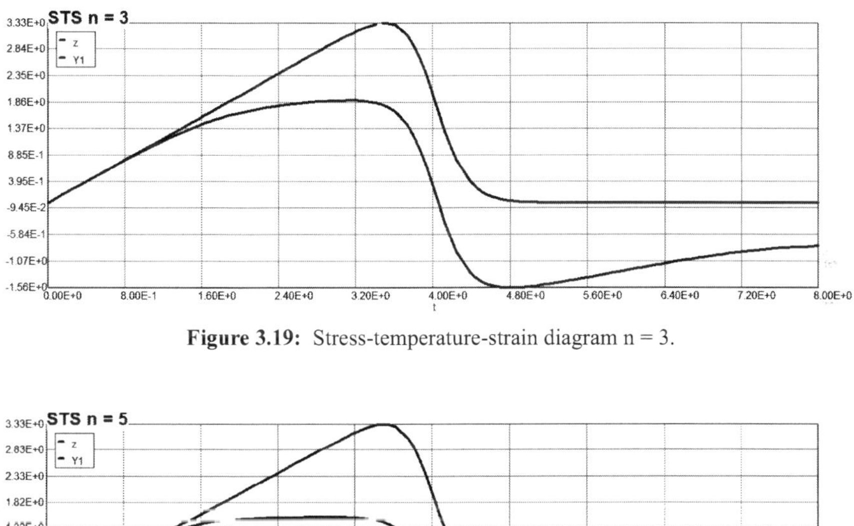

Figure 3.19: Stress-temperature-strain diagram n = 3.

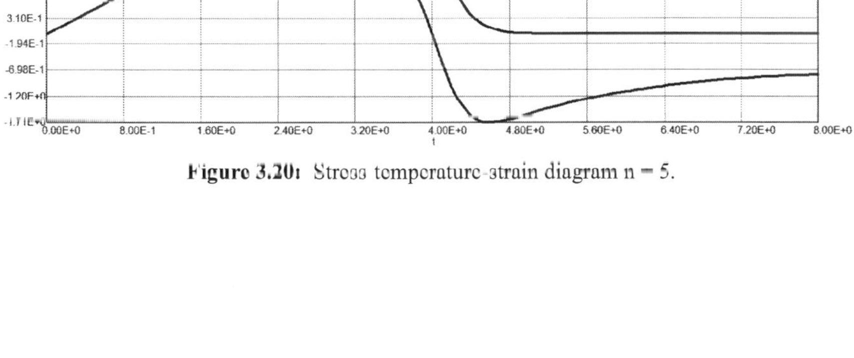

Figure 3.20: Stress-temperature-strain diagram n = 5.

Example 3.12

Data: A = 0.4; β = 0.067; n = 6; θ_{max} = 4; E = a(1–b*(tanh(c*(θ – θ_g))))

Calculated values of DEQ variables

	Variable	Initial value	Minimal value	Maximal value	Final value
1	A	0.4	0.4	0.4	0.4
2	E	1.	0.5	1.	0.5
3	m	0	0	0.0578259	0.0578259
4	m1	0.0405	0.0030152	0.0405	0.0030152
5	n	6.	6.	6.	6.
6	s	1.	1.	1.	1.
7	t	0	0	4.	4.
8	Y1	0	0	3.333525	2.
9	z	0	–0.0799443	1.442635	–0.0799443
10	z1	0	0	2.092007	2.092007

Differential equations

1 d(z1)/d(t) = (exp(t/(1 + 0.067*t)))*(exp(A*(z^s)))*((exp(0.1*m)))*m1*z^n

Explicit equations

1 m = (0.0405*t – 0.01126*t^2 + 0.001462*t^3 – 0.00006868*t^4)
2 m1 = (0.0405 – 0.02252*t^1 + 0.004386*t^2 – 0.0002747*t^3)
3 E = (1–(tanh(3*(t – 4))))*0.5
4 z = t*E–z1*((exp(–0.1*m)))
5 n = 6.0
6 s = 1.0
7 A = 0.4
8 Y1 = t*E

Model: z = a1*t + a2*t^2 + a3*t^3 + a4*t^4

Variable	Value
a1	1.580277
a2	–0.8760136
a3	0.3151484
a4	–0.0475337

z = σ = 1.58*θ – 0.876*θ^2 + 0.315*θ^3 – 0.0475*θ^4 (3.62)

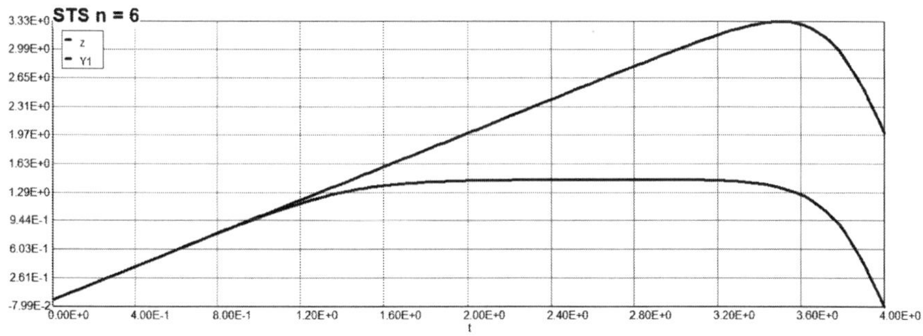

Figure 3.21: Stress-temperature-strain diagram n = 6.

Example 3.13

Data: A = 0.4; β = 0.067; n = 8; θ_{max} = 4; E = a(1–b*(tanh(c*($\theta - \theta_g$))))

Calculated values of DEQ variables

	Variable	Initial value	Minimal value	Maximal value	Final value
1	A	0.4	0.4	0.4	0.4
2	E	1.	0.5	1.	0.5
3	m	0	0	0.0578259	0.0578259
4	m1	0.0405	0.0030152	0.0405	0.0030152
5	n	8.	8.	8.	8.
6	s	1.	1.	1.	1.
7	t	0	0	4.	4.
8	Y1	0	0	3.333967	2.
9	z	0	–0.1651243	1.324539	–0.1651243
10	z1	0	0	2.177681	2.177681

Differential equations

1 d(z1)/d(t) = (exp(t/(1 + 0.067*t)))*(exp(A*(z^s)))*((exp(0.1*m)))*m1*z^n

Explicit equations

1 m = (0.0405*t – 0.01126*t^2 + 0.001462*t^3 – 0.00006868*t^4)

2 m1 = (0.0405 – 0.02252*t^1 + 0.004386*t^2 – 0.0002747*t^3)

3 E = (1–(tanh(3*(t – 4))))*0.5

4 z = t*E–z1*((exp(–0.1*m)))

5 n = 8.0

6 s = 1.0

7 A = 0.4

8 Y1 = t*E

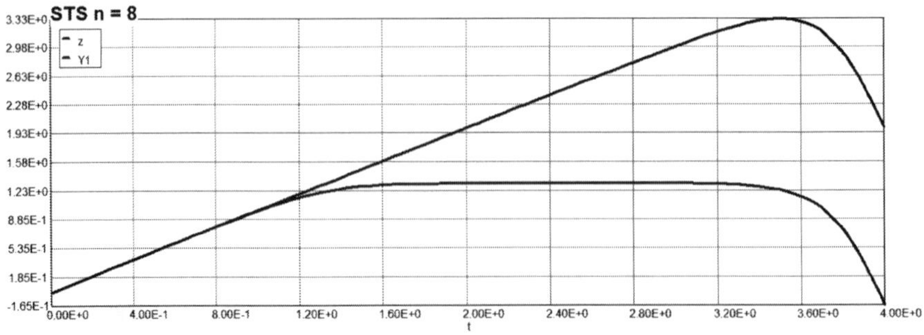

Figure 3.22: Stress-temperature-strain diagram n = 8.

Model: z = a1*x + a2*x^2 + a3*x^3 + a4*x^4

Variable	Value
a1	0.834
a2	–0.254
a3	0.0449
a4	–0.00322

$$z = \sigma = 0.834*\theta - 0.254*\theta\hat{}2 + 0.0449*\theta\hat{}3 - 0.00322*\theta\hat{}4 \quad (3.63)$$

Example 3.14

Data: A = 0.4; β = 0.067; n = 10; θ_{max} = 4; E = a(1–b*(tanh(c*($\theta - \theta_g$))))

Calculated values of DEQ variables

	Variable	Initial value	Minimal value	Maximal value	Final value
1	A	0.4	0.4	0.4	0.4
2	E	1.	0.5	1.	0.5
3	m	0	0	0.0578259	0.0578259
4	m1	0.0405	0.0030152	0.0405	0.0030152
5	n	10.	10.	10.	10.
6	s	1.	1.	1.	1.
7	t	0	0	4.	4.
8	Y1	0	0	3.333919	2.
9	z	0	–0.2117418	1.255689	–0.2117418
10	z1	0	0	2.224568	2.224568

Differential equations

1 d(z1)/d(t) = (exp(t/(1 + 0.067*t)))*(exp(A*(z^s)))*((exp(0.1*m)))*m1*z^n

Explicit equations

1 m = (0.0405*t – 0.01126*t^2 + 0.001462*t^3 – 0.00006868*t^4)
2 m1 = (0.0405 – 0.02252*t^1 + 0.004386*t^2 – 0.0002747*t^3)
3 E = (1–(tanh(3*(t – 4))))*0.5
4 z = t*E–z1*((exp(–0.1*m)))
5 n = 10.0
6 s = 1.0
7 A = 0.4
8 Y1 = t*E

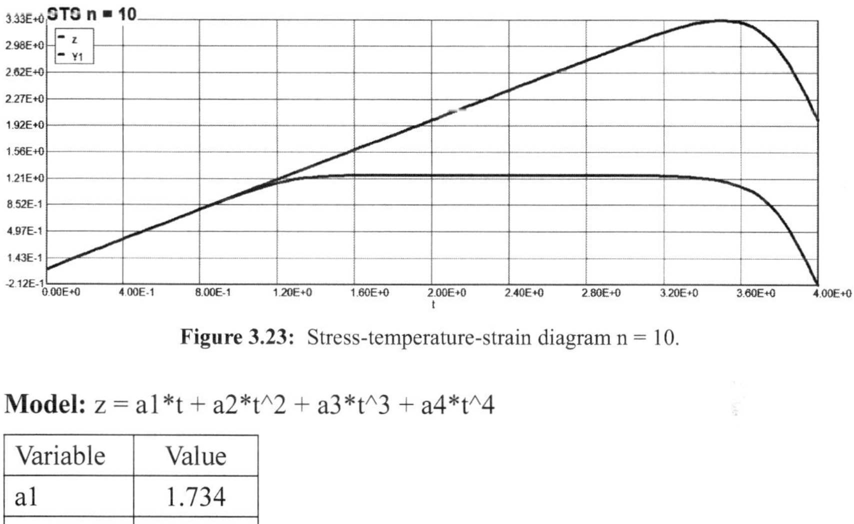

Figure 3.23: Stress-temperature-strain diagram n = 10.

Model: z = a1*t + a2*t^2 + a3*t^3 + a4*t^4

Variable	Value
a1	1.734
a2	–1.130
a3	0.402
a4	–0.0562

$$z = \sigma = 1.734*\theta - 1.13*\theta^2 + 0.0402*\theta^3 - 0.0562*\theta^4 \quad (3.64)$$

Table 3.2: Maximum creep stresses vs. parameter n.

n	1	2	3	5	6	8	10
σ_{max}	2.625	2.231	1.902	1.539	1.443	1.325	1.256

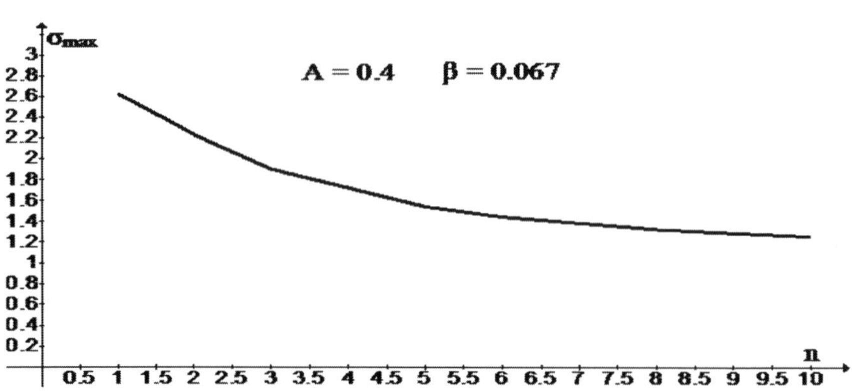

Figure 3.24: Maximum creep stresses vs. parameter n.

Model: $\sigma_{max} = 93*(1-(\tanh(0.05*(n + 41.7))))$ (3.65)

Model: $\sigma_{max} = a(1-(\tanh(b*(n + c))))$

Variable	Initial guess	Value
a	45.	93.0
b	0.05	0.05
c	–34.	–41.7

Example 3.15

Calculate allowable creep stress and maximum temperature if A = 0.4 and n = 8

Solution

σ_{all} - ? → A = 0.4 $\sigma_{all} = 1.325(26.8)3.5(10^{-1}) = 12.43$ [MPa] = 1.802 ksi

$$T_{max} = 0.067(400)\ 3.5 + 400 = 493.8^{o}\ K \approx 221^{o}C$$

Different values of dimensionless parameters 'c' and θ_g (dimensionless transitional temperature) obviously also affect the STS diagram (see examples below). Overall, the proper choice of these parameters that are usually obtained from simple creep testing data could offer a good balance of strength and stiffness of composites.

Example 3.16

Data: A = 0.4; β = 0.067; n = 2; $\theta_g = 4$; c = 3; $E = (a-b*(\tanh(c*(t - \theta_g))))$

Calculated values of DEQ variables

	Variable	Initial value	Minimal value	Maximal value	Final value
1	A	0.4	0.4	0.4	0.4
2	E	1.	0.25	1.	0.25
3	m	0	0	0.0705949	0.0676285
4	m1	0.0405	–0.0071703	0.0405	–0.0071703
5	n	2.	2.	2.	2.
6	s	1.	1.	1.	1.
7	t	0	0	9.	9.
8	Y1	0	0	3.382295	2.25
9	z	0	–0.9345511	1.733073	0.0246378
10	z1	0	0	2.242709	2.240463

Differential equations

1 d(z1)/d(t) = (exp(t/(1 + 0.067*t)))*(exp(A*t*(z^s)))*((exp(0.1*m)))*m1*z^n

Explicit equations

1 m = (0.0405*t – 0.01126*t^2 + 0.001462*t^3 – 0.00006868*t^4)

2 m1 = (0.0405 – 0.02252*t^1 + 0.004386*t^2 – 0.0002747*t^3)

3 E = (0.625 – 0.375*(tanh(3*(t – 4))))

4 z = t*E–z1*((exp(–0.1*m)))

5 n = 2.0

6 s = 1.0

7 A = 0.4

8 Y1 = t*E

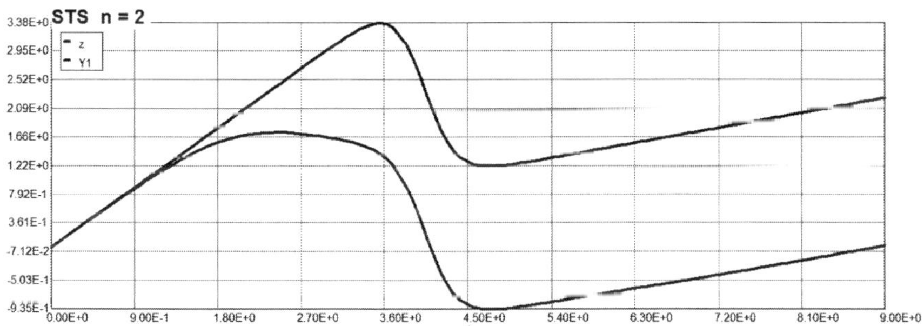

Figure 3.25: Creep stresses diagram $\theta_g = 2$.

Model: z = a1*t + a2*t^2 + a3*t^3 + a4*t^4

Variable	Value
a1	2.367
a2	–1.023
a3	0.135
a4	–0.0055

Model: z = σ = 2.367*θ – 1.023*θ^2 + 0.135*θ^3 – 0.0055θ^4 (3.66)

Example 3.17

Data: A = 0.4; β = 0.067; T_* = 400° K; T_{max} = 600° K; c = 3; E = (a–b*(tanh (c*(θ – θ_g))))

θ_g = 2 instead of 4

Calculated values of DEQ variables

	Variable	Initial value	Minimal value	Maximal value	Final value
1	A	0.4	0.4	0.4	0.4
2	E	0.9999954	0.25	0.9999954	0.25
3	m	0	0	0.0706026	0.0702896
4	m1	0.0405	–0.0020076	0.0405	–0.0020076
5	n	2.	2.	2.	2.
6	s	1.	1.	1.	1.
7	t	0	0	8.4	8.4
8	Y1	0	0	2.1	2.1
9	z	0	0	1.373791	0.7886489
10	z1	0	0	1.54178	1.320601

Differential equations

1 d(z1)/d(t) = (exp(t/(1 + 0.067*t)))*(exp(A*t*(z^s)))*((exp(0.1*m)))*m1*z^n

Explicit equations

1 m = (0.0405*t – 0.01126*t^2 + 0.001462*t^3 – 0.00006868*t^4)
2 m1 = (0.0405 – 0.02252*t^1 + 0.004386*t^2 – 0.0002747*t^3)
3 E = (0.625 – 0.375*(tanh(3*(t – 2))))
4 z = t*E–z1*((exp(–0.1*m)))

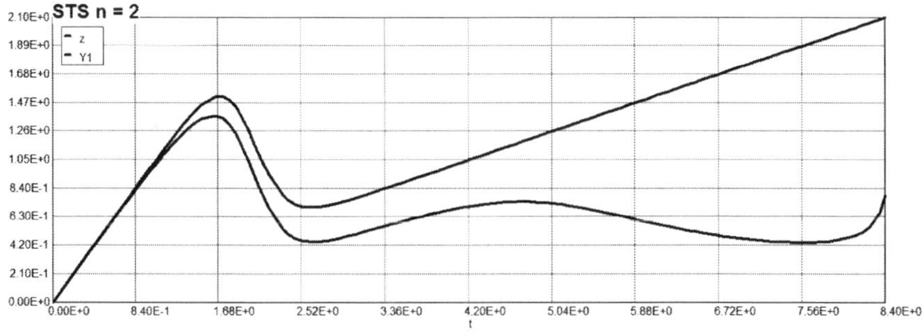

Figure 3.26: Creep stresses diagram $\theta_g = 4$ (Y1-STS without creep; z-STS with creep).

5 n = 2.0
6 s = 1.0
7 A = 0.4
8 Y1 = t*E

Model: z = a1*t + a2*t^2 + a3*t^3 + a4*t^4

Variable	Value
a1	1.290291
a2	–0.5657943
a3	0.089362
a4	–0.0047262

$$z = \sigma = 1.29\,\theta - 0.566\,\theta^2 + 0.0894\,\theta^3 - 0.0047\,\theta^4 \quad (3.67)$$

Examples (3.1–3.17) show that the material responses are qualitatively and quantitatively different depending on the values of dimensionless parameters A; c; n, and transitional temperature θ_g. As noted above, the number of dimensionless parameters in the integral creep equation obtained on the basis of appropriate experimental studies should be minimal. However, the qualitative nature of the experimentally obtained creep curve can turn out to be much more complicated (for example, the CTC diagram can have sections characterizing the plasticity of the composite). Therefore, a somewhat more general form of the integral creep equation (with a somewhat larger number of dimensionless parameters) is given below, which makes it possible to describe the nonlinear behavior of the composite practically from the very beginning of the creep process.

3.9 General form of Equation (3.37)

Consider now the more general form of Equation (3.37):

$$E(\theta)[\theta] = \sigma(\theta) + \int_0^\theta \tau^p e^{\frac{\tau}{1+\beta\tau} + A\tau^k\sigma^s} K(\theta,\tau)\sigma^n(\tau)\,m\,\mathbf{1}d\tau$$

$$K(\theta,\tau) = \varphi(\theta)f(\tau) = m\mathbf{1}(\tau)\sum_{i=1}^{N} \exp(-\alpha_i\, m(\theta))\exp(\alpha_i\, m(\tau)) \tag{3.68}$$

$$E = E_0(a - b*(\tanh(c*(\theta - \theta_g))))$$

$$p \geq \mathbf{0}; k \geq \mathbf{0}; s \geq \mathbf{0}; n \geq \mathbf{0}; A \geq \mathbf{0}$$

The STS relationship is mostly affected by changing parameter 'A' in Equation (3.68). A is a constant for mechanics analysis in a modified form of creep constitutive equation to account for varying stresses. For small values of 'A' ($0 < A < 1$—see above) the linear elastic part of the curve (see Fig. 3.5 through 3.10)—the viscous part of creep process has practically negligible effect— the allowable stresses increase with the temperature increase. However when the temperatures are in transitional zone (lathery phase) the stresses decrease with the rate proportional to parameters 'c' and 'n' until new equilibrium occurs (the viscous part of deformation is governing); and after that ($\theta >> \theta_g$) the stresses increase again due to the recrystallization process (with almost steady rate) until the material entries the tertiary phase. For larger "A" ($A > 1$) the stages in general remain the same, but the STS diagram reflects substantial changes in shape (almost elastic behavior at the beginning following a 'plastic' deformation part afterwards; changing to substantial drop in stresses at the lathery phase; and finally a new increasing part of STS diagram up until the beginning of the tertiary stage that can be considered as a failure point (a strength limit or lifetime of solids from the engineering creep theory point of view). Failure is most often treated as a separate issue from the determination of properties of materials. In fact, most failure laws are derived empirically from observations related to a catastrophic event such as yielding or rupture. As a result, a great deal of testing and data analysis is necessary to establish an appropriate failure law for a material. On the other hand, constitutive laws are derived by more rational means of relating deformations to the forces which produce them. Failure, however it is defined, should be a part of a complete constitutive description of a material. In other words, the key to dealing effectively with failure lies in treating its behavior as a termination of the nonlinear viscoelastic process. The examples below illustrate this phenomenon.

Example 3.18

Data: A = 1.0; β = 0.067; α_1 = 0.1; T_* = 400° K; $T_{max} \cong$ 600° K; c = 3; a = 0.625; b = 0.375; n = 2; s = 1; k = p = 0; θ_{max} = 8; E = E_0(a–b*(tanh(c*(θ – θ_g)))); θ_g = 2; E_0 = 1 [GPa]

Calculated values of DEQ variables

	Variable	Initial value	Minimal value	Maximal value	Final value
1	A	1.	1.	1.	1.
2	E	0.9999954	0.25	0.9999954	0.25
3	k	0	0	0	0
4	m	0	0	0.0705907	0.0705907
5	m1	0.0405	0.0003976	0.0405	0.0003976
6	n	2.	2.	2.	2.
7	p	0	0	0	0
8	s	1.	1.	1.	1.
9	t	0	0	8.	8.
10	Y1	0	0	2.	2.
11	z	0	0	1.299265	0.6588526
12	z1	0	0	1.350648	1.350648

Differential equations

1 d(z1)/d(t) = (exp(t/(1 + 0.067*t)))*(t^p)*(exp(A*(t^k)*(z^s)))*((exp(0.1*m)))*m1*z^n

Explicit equations

1 m = (0.0405*t – 0.01126*t^2 + 0.001462*t^3 – 0.00006868*t^4)

2 m1 = (0.0405 – 0.02252*t^1 + 0.004386*t^2 – 0.0002747*t^3)

3 E = (0.625 – 0.375*(tanh(3*(t – 2))))

4 z = t*E–z1*((exp(–0.1*m)))

5 n = 2.0

6 s = 1.0

7 A = 1

8 k = 0

9 Y1 = t*E

10 p = 0

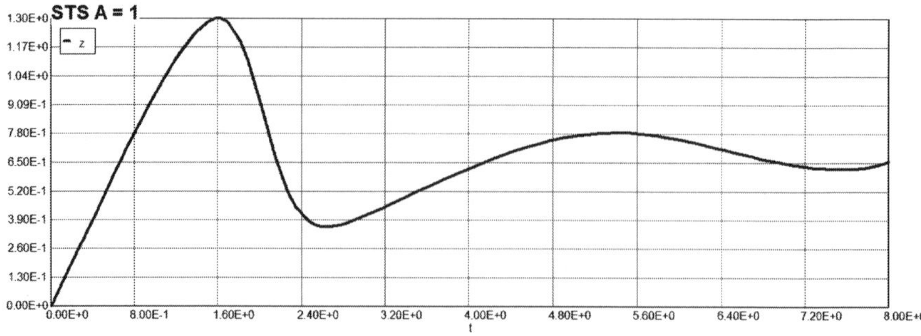

Figure 3.27: STS diagram A = 1.

Model: z = a1*t + a2*t^2 + a3*t^3 + a4*t^4

Variable	Value
a1	1.424
a2	–0.722
a3	0.131
a4	–0.0078

Model: $z = \sigma = 1.424*\theta - 0.722*\theta\text{^}2 + 0.131*\theta\text{^}3 - 0.0078*\theta\text{^}4$ (3.69)

Example 3.19

Data: $A = 3.0$; $\beta = 0.067$; $\alpha_1 = 0.1$; $T_* = 400^o$ K; $T_{max} \cong 600^o$ K; $c = 3$; $a = 0.625$; $b = 0.375$; $n = 2$; $s = 1$; $k = s = 0$; $\theta_{max} = 8$; $E = E_0(a\text{–}b*(\tanh(c*(\theta - \theta_g))))$; $\theta_g = 2$; $E_0 = 1$ [GPa]

Calculated values of DEQ variables

	Variable	Initial value	Minimal value	Maximal value	Final value
1	A	3.	3.	3.	3.
2	E	0.9999954	0.25	0.9999954	0.25
3	k	0	0	0	0
4	m	0	0	0.0705907	0.0705907
5	m1	0.0405	0.0003976	0.0405	0.0003976
6	n	2.	2.	2.	2.
7	p	0	0	0	0
8	s	1.	1.	1.	1.
9	t	0	0	8.	8.
10	Y1	0	0	2.	2.
11	z	0	–0.0592095	0.9335587	0.4671968
12	z1	0	0	1.543662	1.543662

Differential equations

1 d(z1)/d(t) = (exp(t/(1 + 0.067*t)))*(t^p)*(exp(A*(t^k)*(z^s)))*((exp(0.1*m)))*m1*z^n

Explicit equations

1 m = (0.0405*t – 0.01126*t^2 + 0.001462*t^3 – 0.00006868*t^4)

2 m1 = (0.0405 – 0.02252*t^1 + 0.004386*t^2 – 0.0002747*t^3)

3 E = (0.625 – 0.375*(tanh(3*(t – 2))))

4 z = t*E–z1*((exp(–0.1*m)))

5 n = 2.0

6 s = 1.0

7 A = 3

8 k = 0

9 Y1 = t*E

10 p = 0

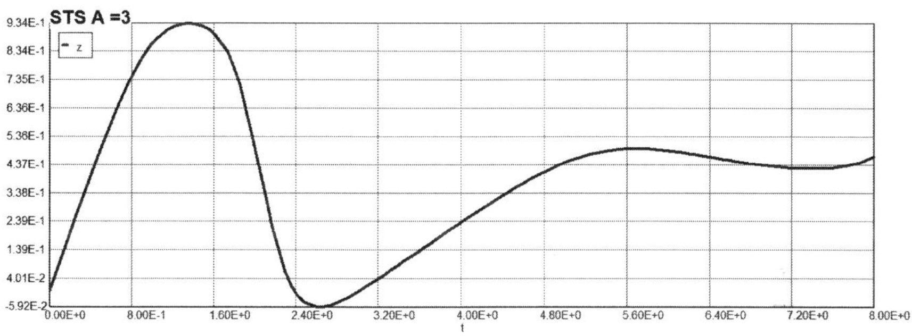

Figure 3.28: STS diagram A = 3.

Model: z = a1*t + a2*t^2 + a3*t^3 + a4*t^4

Variable	Value
a1	1.197
a2	–0.709
a3	0.138
a4	–0.0085

Model: z = σ = 1.197*θ – 0.709*θ^2 + 0.138*θ^3 – 0.0085*θ^4 (3.70)

Example 3.20

Data: $A = 5.0$; $\beta = 0.067$; $\alpha_1 = 0.1$; $T_* = 400^o$ K; $T_{max} \cong 600^o$ K; $c = 3$; $a = 0.625$; $b = 0.375$; $n = 2$; $s = 1$; $k = s = 0$; $\theta_{max} = 8$; $E = E_0(a–b*(tanh(c*(\theta - \theta_g))))$; $\theta_g = 2$; $E_0 = 1$ [GPa]

Calculated values of DEQ variables

	Variable	Initial value	Minimal value	Maximal value	Final value
1	A	5.	5.	5.	5.
2	E	0.9999954	0.25	0.9999954	0.25
3	k	0	0	0	0
4	m	0	0	0.0705907	0.0705907
5	m1	0.0405	0.0003976	0.0405	0.0003976
6	n	2.	2.	2.	2.
7	p	0	0	0	0
8	s	1.	1.	1.	1.
9	t	0	0	8.	8.
10	Y1	0	0	2.	2.
11	z	0	–0.2671324	0.7048392	0.3806049
12	z1	0	0	1.630867	1.630867

Differential equations

1 d(z1)/d(t) = (exp(t/(1 + 0.067*t)))*(t^p)*(exp(A*(t^k)*(z^s)))*((exp (0.1*m)))*m1*z^n

Explicit equations

1 m = (0.0405*t – 0.01126*t^2 + 0.001462*t^3 – 0.00006868*t^4)

2 m1 = (0.0405 – 0.02252*t^1 + 0.004386*t^2 – 0.0002747*t^3)

3 E = (0.625 – 0.375*(tanh(3*(t – 2))))

4 z = t*E–z1*((exp(–0.1*m)))

5 n = 2.0

6 s = 1.0

7 A = 5

8 k = 0

9 Y1 = t*E

10 p = 0

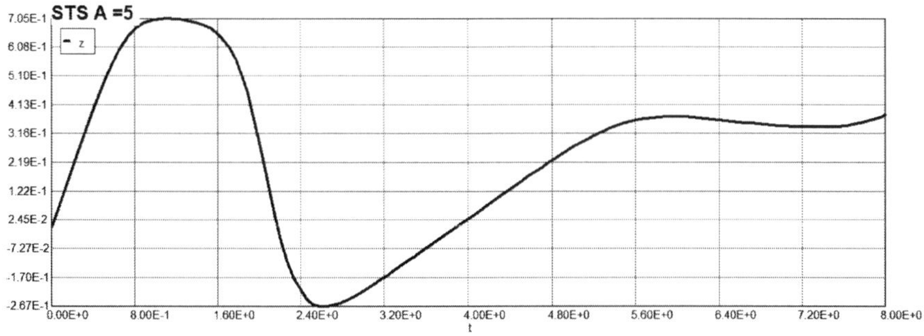

Figure 3.29: STS diagram A = 5.

Model: z = a1*t + a2*t^2 + a3*t^3 + a4*t^4

Variable	Value
a1	1.0
a2	–0.656
a3	0.134
a4	–0.0084

Model: $z = \sigma = 1.0*\theta - 0.656*\theta^\wedge 2 + 0.134*\theta^\wedge 3 - 0.0084*\theta^\wedge 4$ (3.71)

Example 3.21

Data: $A = 7.0$; $\beta = 0.067$; $\alpha_1 = 0.1$; $T_* = 400^o$ K; $T_{max} \cong 600^o$ K; $c = 3$; $a = 0.625$; $b = 0.375$; $n = 2$; $s = 1$; $k = s = 0$; $\theta_{max} = 8$; $E = E_0(a-b*(tanh(c*(\theta - \theta_g))))$; $\theta_g = 2$; $E_0 = 1$ [GPa]

Calculated values of DEQ variables

	Variable	Initial value	Minimal value	Maximal value	Final value
1	A	7.	7.	7.	7.
2	E	0.9999954	0.25	0.9999954	0.25
3	k	0	0	0	0
4	m	0	0	0.0705907	0.0705907
5	m1	0.0405	0.0003976	0.0405	0.0003976
6	n	2.	2.	2.	2.
7	p	0	0	0	0
8	s	1.	1.	1.	1.
9	t	0	0	8.	8.
10	Y1	0	0	2.	2.
11	z	0	–0.3780442	0.5722127	0.3272713
12	z1	0	0	1.684578	1.684578

Differential equations

1 d(z1)/d(t) = (exp(t/(1 + 0.067*t)))*(t^p)*(exp(A*(t^k)*(z^s)))*((exp (0.1*m)))*m1*z^n

Explicit equations

1 m = (0.0405*t – 0.01126*t^2 + 0.001462*t^3 – 0.00006868*t^4)

2 m1 = (0.0405 – 0.02252*t^1 + 0.004386*t^2 – 0.0002747*t^3)

3 E = (0.625 – 0.375*(tanh(3*(t – 2))))

4 z = t*E–z1*((exp(–0.1*m)))

5 n = 2.0

6 s = 1.0

7 A = 7

8 k = 0

9 Y1 = t*E

10 p = 0

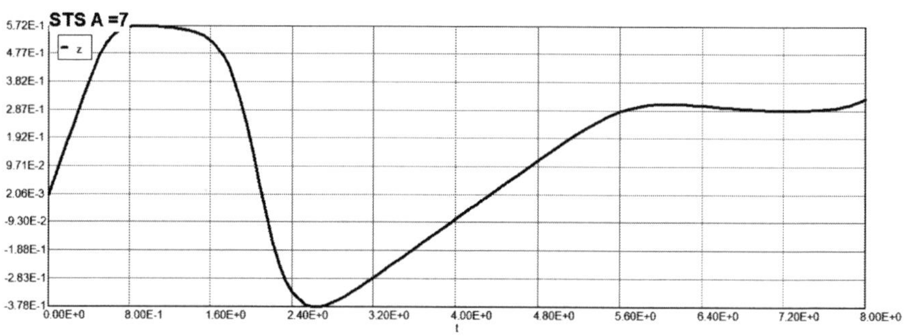

Figure 3.30: STS diagram A = 7.

Model: z = a1*t + a2*t^2 + a3*t^3 + a4*t^4

Variable	Value
a1	0.871
a2	–0.611
a3	0.128
a4	–0.0082

Model: z = σ = 0.871*θ – 0.611*θ^2 + 0.128*θ^3 – 0.0082*θ^4 (3.72)

Example 3.22

Data: A = 10.0; β = 0.067; α_1 = 0.1; T_* = 400° K; T_{max} ≅ 600° K; c = 3; a = 0.625; b = 0.375; n = 2; s = 1; k = s = 0; θ_{max} = 8; E = E_0(a–b*(tanh(c*(θ – θ_g)))); θ_g = 2; E_0 = 1 [GPa]

Calculated values of DEQ variables

	Variable	Initial value	Minimal value	Maximal value	Final value
1	A	10.	10.	10.	10.
2	E	0.9999954	0.25	0.9999954	0.25
3	k	0	0	0	0
4	m	0	0	0.0705907	0.0705907
5	m1	0.0405	0.0003976	0.0405	0.0003976
6	n	2.	2.	2.	2.
7	p	0	0	0	0
8	s	1.	1.	1.	1.
9	t	0	0	8.	8.
10	Y1	0	0	2.	2.
11	z	0	–0.4744615	0.4528926	0.2749389
12	z1	0	0	1.737282	1.737282

Differential equations

1 d(z1)/d(t) = (exp(t/(1+0.067*t)))*(t^p)*(exp(A*(t^k)*(z^s)))*((exp (0.1*m)))*m1*z^n

Explicit equations

1 m = (0.0405*t – 0.01126*t^2 + 0.001462*t^3 – 0.00006868*t^4)

2 m1 = (0.0405 – 0.02252*t^1 + 0.004386*t^2 – 0.0002747*t^3)

3 E = (0.625 – 0.375*(tanh(3*(t – 2))))

4 z = t*E–z1*((exp(–0.1*m)))

5 n = 2.0

6 s = 1.0

7 A = 10

8 k – 0

9 Y1 = t*E

10 p = 0

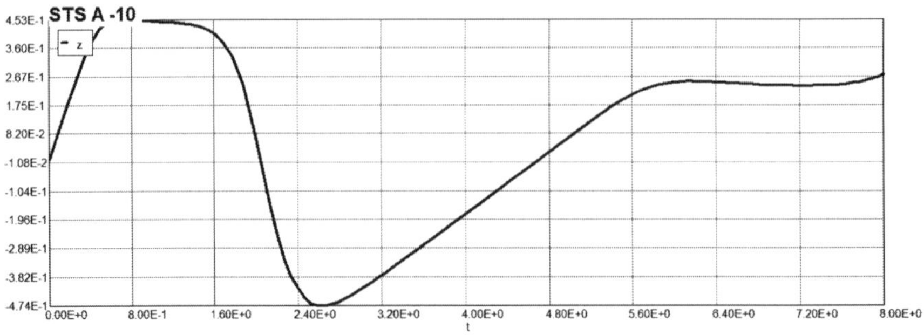

Figure 3.31: STS diagram A = 10.

Model: z1 = a1*t + a2*t^2 + a3*t^3 + a4*t^4

Variable	Value
a1	0.866
a2	–0.208
a3	0.0177
a4	–0.00020

Model: $z = \sigma = 0.866*\theta - 0.208*\theta\text{^}2 + 0.0177*\theta\text{^}3 - 0.0002*\theta\text{^}4$ (3.73)

3.10 Phenomenological creep models of composites with dispersed filler

The new phenomenological model for composites is formulated through the following assumptions:

- The composite is an isotropic, nonlinear viscoelastic body whose rheological properties are described using Gibbs theory, i.e., Equation (3.37).
- A Prony series given by Equation (3.37) is used as the relaxation function.

Gibbs free energy formulation of the Schapery equations

The formulation of Gibbs free energy is particularly useful, since it allows the treatment of temperature and stress tensor components as independent variables. The Gibbs free energy G is defined as

$$G = U - TS - \frac{1}{\rho} Q_m q_m \tag{3.74}$$

where U is the specific internal energy, T the absolute temperature, S the specific entropy, ρ the density, q_m (m = 1,2,...,k) a set of k state variables consisting of observed variables (strains), and Q_m (m = 1,2,...,k) is a set of k generalized forces defined by the virtual work condition $S_w = Q_i\, q_i$.

Equation (3.1) shows that free energy can be accumulated by increasing internal energy; by decreasing entropy or through a potential energy loss of the external loading. For detailed information on this subject see [20].

A strong dependence on temperature (time) of the properties of soft composites (PMC) exists compared with those of other materials such as metals and ceramics (MMC and CMC). This strong dependence is due to the viscoelastic nature of PMC. Generally, PMC behave in a more elastic fashion in response to a rapidly applied temperature rise (force) and in a more viscous fashion in response to a slowly applied temperature increase.

In the previously considered examples, it was assumed that the free Gibbs energy is part of the total activation H, and therefore was included in the index of the exponential function (3.37). However, this approach can be considered as rather formal and less justified from the physical point of view, since, according to the Gibbs theory, internal stresses (as well as strain, temperature and external loads) do the work that is *additive* to the potential energy of the thermal mechanical system. Therefore, instead of one integral, there must be at least two integrals in the creep Equation (3.1). This formulation (and solution) of the problem is presented below in the following examples. The creep deformation process is clearly dependent on the final outcome of a conjunction between deviatoric free energy and dissipated energy. Thus the effect of strain history on the creep phenomenon follows from this model in a natural way. This formulation of a single integral creep constitutive equation (and solution) of the problem is presented below in the following examples.

$$E(\theta)[\theta] = \sigma(\theta) + \int_0^\theta \Phi_1 \sigma^n(\tau)\, m\, \mathbf{1} d\tau + \int_0^\theta \Phi_2 \sigma^n(\tau)\, m\, \mathbf{1} d\tau$$

$$K_1(\theta,\tau) = \varphi_1(\theta) f_1(\tau) = m\mathbf{1}(\tau) \sum_{i=1}^{N} \exp(-\alpha_i\, m(\theta)) \exp(\alpha_i\, m(\tau))$$

$$K_2(\theta,\tau) = \varphi_2(\theta) f_2(\tau) = m\mathbf{1}(\tau) \sum_{i=1}^{N} \exp(-\beta_i\, m(\theta)) \exp(\beta_i\, m(\tau)) \qquad (3.75)$$

$$f(\sigma) = A_2 \sigma^s; \quad s = \mathbf{1}, \mathbf{2}, \mathbf{3} ..., M \text{ or: } f(\sigma) = \ln A_2 (\sigma^s)$$

$$\Phi_1 = e^{\frac{\tau}{1+\beta\tau}} K_1(\theta,\tau); \quad \Phi_2 = A_2 e^{f(\sigma)} K_2(\theta,\tau) \text{ - Case \# 1}$$

$$\text{or: } \Phi_2 = A_2 \sigma^s K_2(\theta,\tau) \text{ - Case \# 2}$$

This formulation is particularly useful, since it allows the treatment of temperature and stress tensor components as independent variables. Thus the entropy of a composite element changes with time for two reasons. First, because entropy flows into this element due to temperature increase, second because there is an entropy source due to irreversible phenomena (distortional energy) inside the volume element. This entropy source is always a non-negative quantity. Equation (3.75) states that the entropy is an extensive property, i.e., the entropy of the system is the sum of the entropies of the subsystems: $dS_{I+II} = dS_I + dS_{II}$. This form of the constitutive creep equation will allow to describe micro-structured rearrangements related to straightening and relative sliding of long chain molecules. The variables θ and σ are thus eventually observable and certainly equally controllable. With the use of mathematical expressions derived from consideration of the models it is possible to derive constants that can be used as a basis for comparison or prediction. *This formulation of creep constitutive law, Equation (3.75), can now be developed to give the basis for measurement of the mechanical properties of a material at a variety of temperatures, which yields the relative contributions of the internal and entropic components of the total creep deformation.* The integral creep equation incorporates essentially two different phenomenological models, depending on the degree of stress influence on the process of creep strains. In the first case, this dependence is strong (exponential), and in the second case —weak (power law). Of course, the choice of the model should be based on the intrinsic factors and experimental data of the composite. For example, for the so-called soft composites, the second model is likely to be suitable, and for metallic composites—the first one. Below are examples using both creep constitutive models, as well as qualitative and quantitative comparison of the results (for example, the maximum allowable creep stress).

Example 3.23 Case #1

Data: $A_2 = 1.0$; $\beta = 0.067$; $\alpha_1 = 0.1$; $T_* = 400^o$ K; $T_{max} \cong 600^o$ K; c = 3; a = 0.625; b = 0.375; n = 1;

Calculated values of DEQ variables

	Variable	Initial value	Minimal value	Maximal value	Final value
1	A2	1.	1.	1.	1.
2	E	1.	0.25	1.	0.25
3	k	0	0	0	0
4	m	0	0	0.0705907	0.0705907
5	m1	0.0405	0.0003976	0.0405	0.0003976
6	n	1.	1.	1.	1.
7	p	0	0	0	0
8	s	1.	1.	1.	1.
9	t	0	0	8.	8.
10	Y1	0	0	2.436604	2.
11	z	0	–0.0410232	1.662955	0.3305589
12	z1	0	0	0.4905164	0.4905164
13	z2	0	0	1.190751	1.190751

Differential equations

1 d(z1)/d(t) = (exp(t/(1 + 0.067*t)))*((exp(0.1*m)))*m1*z^n

2 d(z2)/d(t) = (exp(t/(1 + 0.067*t)))*(exp(A2*(z^s)))*((exp(0.1*m)))*m1*z^n

Explicit equations

1 m = (0.0405*t – 0.01126*t^2 + 0.001462*t^3 – 0.00006868*t^4)

2 m1 = (0.0405 – 0.02252*t^1 + 0.004386*t^2 – 0.0002747*t^3)

3 E = (0.625 – 0.375*(tanh(3*(t – 3))))

4 z = t*E–(z1 + z2)*((exp(–0.1*m)))

5 n = 1.0

6 s = 1.0

7 Y1 = t*E

8 A2 = 1

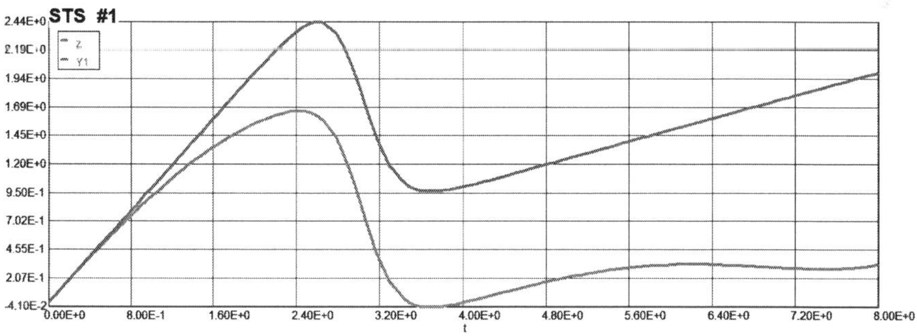

Figure 3.32: STS diagram case #1.

Model: z = a1*t + a2*t^2 + a3*t^3 + a4*t^4 + a5*t^5

Variable	Value
a1	1.677085
a2	–0.5382875
a3	–0.0217411
a4	0.0202491
a5	–0.0015492

z = σ = 1.677*θ – 0.538*θ^2 – 0.0217*θ^3 + 0.0202*θ^4 – 0.00155*θ^5 (3.76)

Example 3.23 Case #2

Data: A_2 = 1.0; β = 0.067; α_1 = 0.1; T_* = 400° K; T_{max} ≅ 600° K; c = 3; a = 0.625; b = 0.375; n = 1;

Calculated values of DEQ variables

	Variable	Initial value	Minimal value	Maximal value	Final value
1	A2	1.	1.	1.	1.
2	E	1.	0.25	1.	0.25
3	m	0	0	0.0705907	0.0705907
4	m1	0.0405	0.0003976	0.0405	0.0003976
5	n	1.	1.	1.	1.
6	p	0	0	0	0
7	s	1.	1.	1.	1.
8	t	0	0	8.	8.
9	Y1	0	0	2.43641	2.
10	z	0	0	1.99545	0.5349166
11	z1	0	0	0.8021905	0.8021905
12	z3	0	0	0.6732717	0.6732717

Differential equations

1 d(z1)/d(t) = (exp(t/(1 + 0.067*t)))*((exp(0.1*m)))*m1*z^n

2 d(z3)/d(t) = (exp(t/(1 + 0.067*t)))*(A2*(z^s))*((exp(0.1*m)))*m1*z^n

Explicit equations

1 m = (0.0405*t – 0.01126*t^2 + 0.001462*t^3 – 0.00006868*t^4)

2 m1 = (0.0405 – 0.02252*t^1 + 0.004386*t^2 – 0.0002747*t^3)

3 E = (0.625 – 0.375*(tanh(3*(t – 3))))

4 z = t*E–(z1 + z3)*((exp(–0.1*m)))

5 n = 1.0

6 s = 1.0

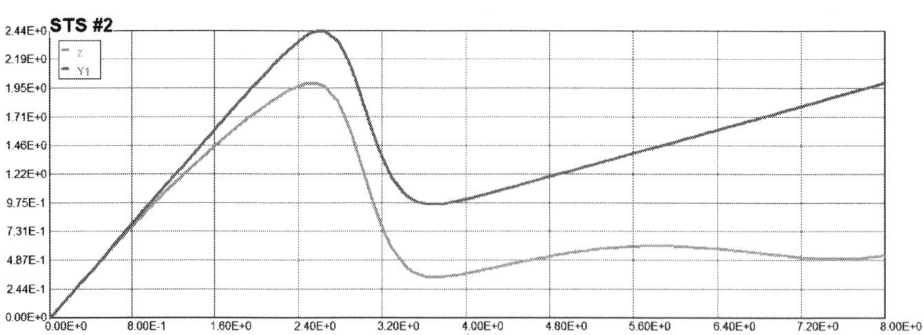

Figure 3.33: STS diagram case #2.

7 Y1 = t*E
8 p = 0
9 A2 = 1

Model: z = a1*t + a2*t^2 + a3*t^3 + a4*t^4 + a5*t^5

Variable	Value
a1	1.493827
a2	–0.243225
a3	–0.1165928
a4	0.031979
a5	–0.0020592

$$z = \sigma = 1.494*\theta - 0.243*\theta^{\wedge}2 - 0.116*\theta^{\wedge}3 + 0.032*\theta^{\wedge}4 - 0.00206*\theta^{\wedge}5 \quad (3.77)$$

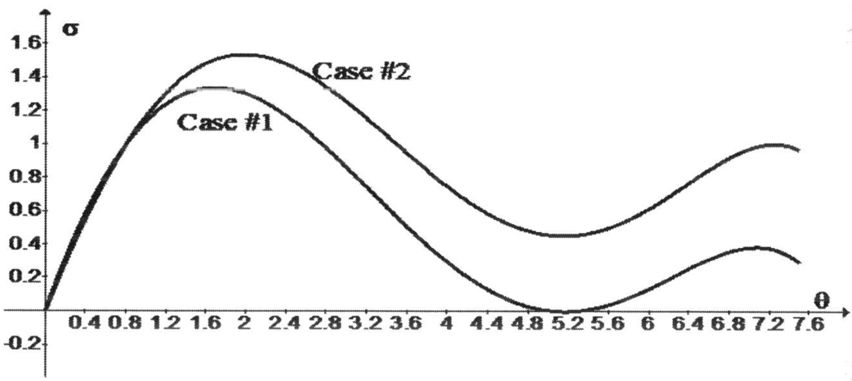

Figure 3.34: STS diagram cases #1 and #2.

As can be seen from Fig. 3.34, the power law leads to a larger maximum value of the allowable creep stress. The power law, as is well known, has wide application in studies related to the nonlinear creep of composites. Therefore, in the future, only the power law will be used in the integral expression for creep. The technique for analyzing the effect of various dimensionless parameters characterizing the mechanical and physical properties of the composite on the creep process as a whole (such as the glass transition temperature, the nonlinearity index of the power law, and so on) is analogous to the above technique. Therefore, in order to save space, only the final results are given below, and the intermediate calculations are omitted. It should be noted that the final results can be used at the preliminary design stage of the composite structural system, witting out all obviously not suitable design options. They can also be used to solve the so-called inverse problems: finding the optimal mechanical and physical parameters of the composite (taking into account the technology of its production) according to a predetermined stress-temperature-strain (STS) diagram. The computer code is as follows:

```
m = (0.0405*t – 0.01126*t^2 + 0.001462*t^3 – 0.00006868*t^4)
m1 = (0.0405 – 0.02252*t^1 + 0.004386*t^2 – 0.0002747*t^3)
t(0) = 0
t(f) = 8.0
z = t*E–(z1 + z2)*((exp(–0.1*m)))
d(z1)/d(t) = (exp(t/(1 + 0.067*t)))*((exp(0.1*m)))*m1*z^n
E = (0.625 – 0.375*(tanh(3*(t – 3))))
n = 1.0; s = 1.0
z1(0) = 0
Y1 = t*E
d(z2)/d(t) = (exp(t/(1 + 0.067*t)))*(A2*(z^s))*((exp(0.1*m)))*m1*z^n
z2(0) = 0
A2 = 1
```

Table 3.3: Maximum creep stress as function of parameter A.

A	0	1	2	3	4	5	6	7	8
σ_{max}	2.245	1.995	1.808	1.665	1.548	1.453	1.372	1.303	1.243

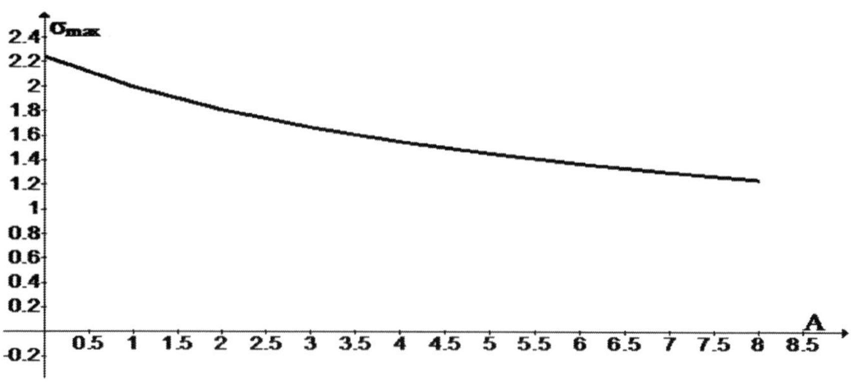

Figure 3.35: Maximum creep stress as function of parameter A.

$$\sigma_{max} = 2.2\text{–}0.2*A + 0.0095*A\hat{}2 \tag{3.78}$$

Table 3.4: Maximum creep stress as function of power law index.

n	1	2	4	6	8	10	12	14	15
σ_{max}	1.995	1.824	1.530	1.365	1.274	1.218	1.182	1.155	1.145

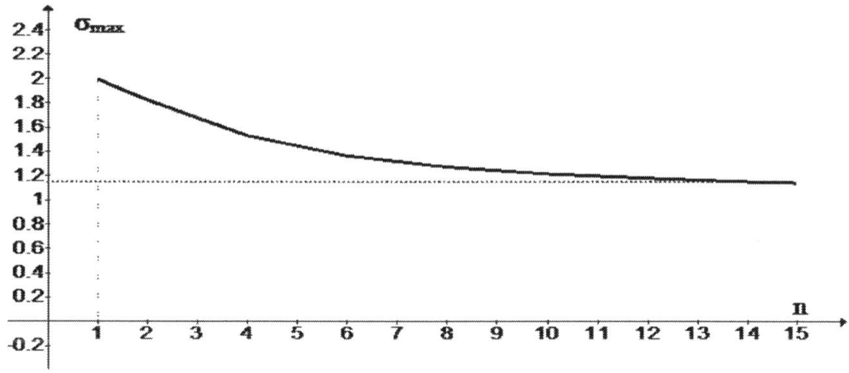

Figure 3.36: Maximum creep stress as function of power law index.

$$\sigma_{max} = 0.85[1.4 + \exp(-0.33*(n-1))] \tag{3.79}$$

References

[1] Boyle, J.T. and Spence, J. 1983. Stress Analysis for Creep. Butterworths, London.

[2] Batchelor, G.K. 2000. An Introduction to Fluid Dynamic. Cambridge Mathematical Library Series. Cambridge University Press, Cambridge, UK.

[3] Panton, R.L. 2013. Incompressible Flow (Fourth ed.). John Wiley & Sons, Hoboken, 114 p.

[4] Dodge, D.W. 1959. Fluid systems. Industrial and Engineering Chemistry 51(7): 839–840.

[5] Wilkinson, W.L. 1960. Non-newtonian fluids. Hydromechanics, Mixing and Heat Transfer, Pergamon Press, Oxford, UK.

[6] Kaye, A. 1962. Non-Newtonian Flow in Incompressible Fluids, Note No. 134. The College of Aeronautics, Cranfield, UK.

[7] Casson, N. 1959. A flow equation for pigment-oil suspensions of the printing ink./N. Casson. pp. 84–104. *In*: C.C. Mill (ed.). Rheology of Disperse Systems. Pergamon Press, London, UK.

[8] Revere, A. and Guibad, G. 2005. Viscosity evolution of anaerobic granular sludge. Biochemical Engineering Journal 315–322.

[9] Chou, T.W. 1992. Microstructural Design of Fiber Composites. Cambridge University Press, Cambridge, UK.

[10] Rosensaft, M. and Marom, G. 1985. Evaluation of bending test methods for composite materials. J. Compos. Technol. Res. 7(1): 12–6.

[11] ASTM. 1994. Plastics (II), 08.02, ASTM, Philadelphia, PA, USA.

[12] Yi, Xiao-Su, Shanyi Du and Litong Zhang (eds.). 2017. Composite materials engineering. Volume 1: Fundamentals of Composite Materials, Springer, N.Y.

[13] Hart-Smith, L.J. 1973. Non-Classical Adhesive-Bonded Joints in Practical Aerospace Construction. NASA CR-I 12238.

[14] Pipkin, A.C. and Rogers, T.G. 1968. A nonlinear integral representation for viscoelastic behavior. J. of the Mechanics and Physics of Solids 16: 59.
[15] Rabotnov, Yu.N. 1948. Some problems of creep theory, Vestnik Moskovsky. Mathematic and Mekhanic 10: 81–91 (in Russian).
[16] Hiel, C., Cardon, A.H. and Brinson, H.F. 1984. The Nonlinear Viscoelastic Response of Resin Matrix Composite Laminates. National Aeronautics and Space Administration (NASA), Scientific and Technical Information Branch, NASA Contractor Report 3772.
[17] Schapery, R.A. 1965. A theory of nonlinear thermo elasticity based on irreversible thermodynamics. Purdue Univ. Report No. AAES 65–7.
[18] Schapery, R.A. 1966. An engineering theory of nonlinear viscoelasticity with applications. Int. J. Solids Structure 2: 407–41.
[19] Schapery, R.A. 1969. On the characterization of nonlinear viscoelastic materials. Polym. Eng. Sci. 9: 295–300.
[20] Razdolsky, L. 2012. Structural Fire Loads: Theory and Principles. McGraw—Hill Co. N.Y.
[21] Shay, Jr., R.M. 1990. Doctoral Thesis, Purdue University.
[22] Shay, Jr., R.M. and Caruthers, J.M. 1984. A free volume model for yield stress. Proceedings of IXth Int. Congr. Rheol. I 549–553.
[23] Shay, Jr., R.M. and Caruthers, J.M. 1986. A new nonlinear viscoelastic constitutive equation for predicting yield in amorphous solid polymers. J. Rheol. 30: 781–827.
[24] Shay, Jr., R.M. and Caruthers, J.M. 1987. Proceedings of 20th Mid-western Mechanics Conference. Purdue University 14b: 493.
[25] Knauss, W.G. and Emri, I.J. 1983. Nonlinear viscoelasticity based on free volume consideration. Computer and Structures 13: 123–128.
[26] Wineman, A.S. and Waldron, Jr., W.K. 1993. Interaction of nonhomogeneous shear, nonlinear viscoelasticity, and yield of a solid polymer. Engineering and Science 33(1): 1217–1228.
[27] Razdolsky, L. 2017. Probability Based High Temperature Engineering. Springer Nature Publishing Co., AG Switzerland.
[28] Razdolsky, L. 2017. Phenomenological high temperature creep models of composites and nanomaterials. Proceedings of the AIAA SPACE 2017, Orlando, FL., USA.
[29] Wentzel, L.O. 1986. Applied Problems in Probability Theory. Translated from Russian by Irene Aleksandrova, Mir Publishers, Moscow.
[30] Razdolsky, L. 2014. Probability-Based Structural Fire Load. Cambridge University Press, UK.
[31] Rzhanitsin, A.R. 1986. Multilayered Beams and Plates Stroiizdat. Moscow (in Russian).
[32] Voigt, W. 1889. Ueber die Beziehung zwischen den beiden Elasticitätsconstanten isotroper Körper. Annalen der Physik 274: 573–587.
[33] Reuss, A. 1929. Berechnung der Fließgrenze von Mischkristallen auf Grund der Plastizitätsbedingung für Einkristalle. Zeitschrift für Angewandte Mathematik und Mechanik 9: 49–58.
[34] Coleman, B.D. 1956. Time dependence of mechanical breakdown phenomena. In J. Appl. Phys. 27: 862.
[35] Razdolsky, L. 2015. High temperature creep and structural fire resistance. Proceedings of SEI/ASCE 2015 Conference, Philadelphia.
[36] Kirby, S.II. and Kronenberg, A.K. 1987. Rheology of the Lithosphere' Selected Topics. Reviews of Geophysics. U.S. National Report to International Union of Geodesy and Geophysics 25(6): 1219–1244.
[37] Fung, Y.C. 1981. Biomechanics: Mechanical Properties of Living Tissues. Springer, NY.

4

Creep Models of Nanocomposites
Deterministic Approach

4.1 Introduction

Most experts in the field of scientific and technical policy, strategic planning and investment are confident that in the next decade, nano-revolution is expected in all areas of science, production, defense, and the social sphere. Large-scale and systemic intrusion of nanostructured materials, products and methods for their production will literally enter all spheres of life. That is why nanomaterials and nanotechnologies are among the priority areas for the development of modern materials science [1]. The features of nanoscale objects are related to the fact that a decrease in the particle size of a solid substance below a certain threshold leads to a significant change in their properties. The threshold particle size at which a discontinuous change in properties occurs—the size effect—for most materials known to date varies from 1 to 100 nm [2]. The historical formation and development of independent fundamental directions of nano-science and the prospects for their application in various fields of nanotechnology were considered in [3, 4]. Some of the results achieved in recent years in the above areas are reflected in reviews and monographs [5, 6]. The properties, including the surface characteristics of the nanoparticles, responsible for their interaction with the environment, depend crucially on the production method. On the other hand, it is known that it is the interphase interactions in the nanocomposites that are the most significantly influences on their properties [2].

Nanotechnologies aimed at obtaining nanostructured materials can be conditionally divided into two groups: "bottom-up" and "top down". This classification is carried out taking into account the key stage (or process) of nanotechnology, on which a nanostructure is formed. In the first group, methods are used in which nanoparticles are formed from atoms and molecules, i.e., the initial particles are enlarged to nanometer sizes; in the second, methods in which the nanometer sizes of particles are achieved by grinding large particles, powders or grains. The methods of the first group of nanotechnologies are mainly based on the chemical approach, and the second on the physical one. Often, using these two fundamentally different groups of technologies, nanomaterials of the same chemical composition, but with different properties are acquired.

One of the simplest ways of obtaining nanoparticles is to condense the vapor substances in an inert atmosphere. This method can produce nanoparticles of both simple and complex substances. If nanoparticles of metal compounds, for example, oxides, nitrides, carbides, etc., are needed, then an appropriate reaction gas must be added to the atmosphere—oxygen, nitrogen, carbon dioxide, methane, etc. In order to obtain a vapor of matter, the easiest way is to use the evaporation process. Atoms of matter that have passed into vapor, because of collisions with atoms of an inert gas, rapidly lose kinetic energy and form nanoparticles. In case of metal compounds, the metal also reacts with the reaction gas. To form particles of the right size, it is necessary to select the pressure of the inert gas. The shape of the nanoparticles obtained by gas-phase synthesis depends on their size: nanoparticles < 20 nm in size have a spherical shape, which is caused by a change in the relative contribution of the surface energy to the total energy of the nanoparticles as its size decreases. Larger particles have a facet shape. Another interesting fact for small size of isolated nanocrystals and associated with it energy, is the absence of dislocations in them.

Plasma-chemical synthesis is the most common method of obtaining highly disperses powders of borides, carbides, nitrides and oxides. In this method, a low-temperature nitrogen, ammonia, hydrogen, hydrocarbon or argon plasma is used, which are created by means of arc, glow, high- or super-high-frequency discharges. The characteristics of the powders obtained depend on the raw materials used, the synthesis technology and the type of reactor. Particles of such powders are most often single crystals with sizes from 10 to 100–200 nm or more. At a high plasma temperature, all the starting materials pass into the gaseous ionized state [7]. The presence of ions leads to high rates of interaction and a short (10^{-3}–10^{-6} s) reaction time. Nanoparticles synthesized by the plasma chemical method have a large excess energy, so their chemical and phase composition may not correspond to the equilibrium phase diagram.

The method of chemical precipitation from colloidal solutions is widely used for the synthesis of highly disperse powders. To obtain nanoparticles from colloidal solutions, the chemical reaction between the components of the solution is interrupted at a certain point in time, after which the system is transferred from the liquid (colloid) to the solid (dispersed) state. In the preparation of nanocrystalline metal powders and their compounds by paralysis (thermal decomposition), the starting materials are usually complex elements and metal-organic compounds, polymers, hydroxides, carbonyls, formats, nitrates, oxalates, amides, imides. These substances contain all or almost all of the chemical elements that must be present in the product. When heating to a certain temperature, the starting materials decompose to form the product synthesized and release gaseous compounds. Mechanic-synthesis is one of the most productive "dry" chemical technologies that do not require (or minimize) the use of solvents for chemical reactions. When mechanical action on solid mixtures occurs, the material is milled, there is acceleration of mass transfer and mixing of the components of the mixtures into atomic level resulting in the activation of their chemical interaction.

Mechanisms of mechanic-chemical reactions are multistage; their most important stages are as follows: initial deformation of the crystal structure of the reagents, formation, accumulation and interaction of point and linear defects, grinding substances to individual aggregates, the formation of metastable states in the contact zone of different phases, chemical homogenization and relaxation of products reaction to the equilibrium state [8]. There is another type of mechanical action, at which conditions are simultaneously created, both for crushing the starting materials, and for the synthesis of the final product, the impact of the shock wave, detonation of explosives—the energy of the explosion—is widely used for the implementation of phase transitions and the synthesis of new compounds. Detonation synthesis of nano-powders proceeds under dynamic conditions, in which kinetic processes play an important role. Detonation synthesis of diamond by shock-wave action on rhombohedra graphite (the pressure was developed up to 30 GPa) was described in [9].

Recently, thanks to their unique properties, metal nanoparticles have been increasingly used. A consequence of this was the active development of methods for obtaining them, based on the effects of a different nature: physical, chemical, biochemical, etc. [10]. Numerous techniques for the production of metallic nanoparticles in condensed media based on physical methods such as metal evaporation, laser ablation, photolysis, radiolysis, etc. are used. The assignment of physical techniques in a number of cases is conditional because due to the physical influence on the system it is possible to change the chemical reaction processes leading to the formation nanocrystals. Using physical methods, chemical reactions that lead to the formation of metallic nanoparticles can be initiated in condensed media. However, the nature of the main process is chemical, based on the reaction of metal reduction with the aid of an oxidizer. Nanocomposites are objects where nanoparticles are packed together into a macroscopic sample in which interparticle and interphase interactions become strong and mask the properties of isolated particles. For each type of interaction, it is important to know how the properties of the material change due to its size. It should be noted that all successes in the development of nanochemistry do not yet allow answering in general terms the question of the relationship between the size of the material particles and their properties.

The authors [11, 12] have described the characteristics of nanoparticles and the effect of nanoparticles sizes on the mechanical properties of composite nanomaterials (strength, elongation, Young's modulus). Applied interest in nanomaterials is due to the possibility of significant modification or even fundamental change in the properties of known materials, as well as the new opportunities that nanotechnology opens up in the creation of materials and products from nano-sized material structural elements. Management of the fundamental properties of solids (semiconductors, metals, polymers, etc.), based on synthesizing in their volume nano-sized inclusions, crystallites, defect structures or the formation of nanoscale films and structures on the surface, is currently one of the main problems of leading scientific centers of the world working in the field of nanotechnology. The basis for the development of a number of modern promising areas (micro system equipment and nano electronics, optics, power engineering, biotechnology, etc.) is already using the functional materials and structures created by microelectronics, nanotechnologies and a number of other methods that provide processing at the micro and nanoscale levels. At the same time, the assortment of used materials has sharply expanded and requirements for their parameters have increased, in particular, the provision of a set of functional characteristics of the materials used is becoming a prerequisite—crystalloid-chemical and thermo-chemical compatibility; mechanical, thermal and electrical resistance; low mechanical and thermo-chemical fatigue and electrical degradation. The solution of these problems requires fundamentally new approaches both in the field of materials science and in the field of synthesis and formation technology. The transition to the submicron and nanometer range of element sizes requires consideration of scaling factors that reflect the influence of geometric dimensions on the properties of the material. By reducing the size of the particles and composing the nanocomposite material results in the fact that the surface properties of the material begin to be affected by its surface characteristics, and from certain dimensions (10–100 nm) surface properties begin to dominate over the bulk ones and mainly determine the properties of the system as a whole. It becomes possible to vary the energy characteristics of the system being created in a wide range by structured and dimensional parameters (thickness of films, size of nanotubes, the structure of interfaces, etc.), and therefore, to obtain the necessary combination of various properties (mechanical, optical and many others). There is a real opportunity to change the properties of the material, giving it characteristics unattainable for bulk materials.

4.1.1 Small scaled materials

Small scaled materials with particle diameters in the nanometer range have been in use for a long time. A generalization of the reasons for the change in the physical, electronic or chemical properties of a material in the nanometer range cannot be made, as diverse phenomena play important roles. Nevertheless, one common denominator can be determined when looking at the peculiarity of nanoscale material. *Traditionally, in a bulk crystal, the properties of the material are independent of the size and are only dependent on the chemical composition, as most atoms of a material are located in the bulk of a particle and hence exceed the influence of surface atoms on the material's properties*. As the size of the crystal decreases to the nanometer regime, the small size of the particle begins to modify the properties of the crystal. The surface area increases, which results in a larger contribution of the surface energy to the overall energy of the whole system and, thus, results in a reduction of the impact of the bulk atoms on the properties of a material. The significant shift towards surface atoms of nanomaterials is illustrated in Fig. 4.1 where the percentage of surface atoms in a particle is plotted against the particle size.

The percentage of surface atoms in a particle is strongly dependent on the particle size. Particles below 10–20 nm start having a relevant part of their atoms on the surface. This results in surface atoms contributing substantially to the properties of the bulk material. There is a large energy associated with this surface. In most of their potential applications, the surface of nanoparticles will undoubtedly play the crucial role in the determination of the nano particulate properties, including the stability and reactivity which will be discussed in the following sections of this chapter.

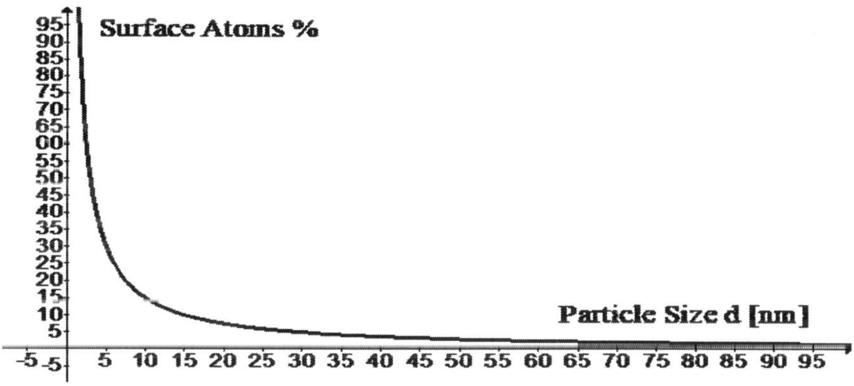

Figure 4.1: Surface areas vs. particle size.

4.1.2 Approaches to larger surface area

The relation of single crystal and polycrystalline strength properties was a major topic of research in the 20th century; and now in the 21st century constitutes a major effort directed toward producing nanocrystalline materials and understanding their mechanical properties. There have been many different attempts to create surface area by decreasing the particle size of materials. An engineer would naturally prefer to start with the bulk material and make it smaller, thus breaking up larger particles by the use of physical processes like crushing, milling or grinding. A chemist on the other hand is more familiar with building small scaled materials starting from the atomic scale by traditional chemistry methods and stopping the reactions at the right time to inhibit further particle growth. With either of the methods the goal of larger surface area can be reached whilst keeping the chemical composition of the material constant. As mentioned above, there are two approaches towards the synthesis of nano-sized materials:

1) Top-down: size reduction from bulk materials
2) Bottom-up: material synthesis from atomic level

Top-down approaches such as milling have been known and applied within living memory. Modern milling techniques include the ball milling that can yield nano particulate materials. Unfortunately, there are a few drawbacks, e.g., the low energy efficiency; the susceptibility to impurities from abrasion and the broader particle size distributions. Bottom-up techniques include plasma, laser, liquid phase or flame spray synthesis. These "self-assembly" preparation methods generally result in well controlled nanoparticles which are built on smaller building blocks allowing for the synthesis of more complex materials or the fabrication of nanoparticles with a very narrow size distribution.

Deep knowledge of the physical and mechanical properties of nanomaterials is crucial in the design and reliability of structural components. However, such property experimental measurements generally are time consuming and require special expertise. Accordingly, the development cycle for new materials and processes has not kept pace with the structural component's development cycle. Computational tools provide a stratagem for shortening the material and process development time. Such a computational tool allows material scientists to simulate physical properties of complex microstructures from a simple testing data of that microstructure. Examples of applications of such computer models that elucidate influences of micro structural features are presented below.

The main objectives of this chapter are as follows:

1. Elucidate the role of heterogeneous microstructures on bulk physical and mechanical properties and damage evolution processes.

2. Correlate physical properties with the materials development cycle.
3. Improve bulk materials processing and enable more reliable structural design by developing computational tools for simulating properties of nanomaterials and corresponding creep constitutive phenomenological law.

4.1.3 Size effect and the nanomaterials properties

Traditionally, Material Science studies the behavior of materials at different scales in order to observe and quantify the chemo-physical processes at the underlying micro-mechanical scales, molecular and atomistic levels. Multi-scale material modeling is scaling up these processes onto the macroscopic level. The first step is to decompose the entire range of scales into sub-ranges, from the view point of characterizing and designing engineering materials:

- Meter level: Practical civil, mechanical and aerospace structural analysis and design of structural systems.
- Millimeter Level: This constitutes the macro-scale level at which materials treated as a homogeneous continua after homogenizing all constituents into so-called "effective" properties for laboratory testing purpose.
- Micrometer Level: Micro-structural features such as micro-defects, the grain size of polycrystalline materials are observed at this scale. Materials at this level may be treated as heterogeneous composites, e.g., metal matrix in which particle inclusions are bonded to the matrix by cohesive/frictional interface layers.
- Nanometer Level: Molecular and atomistic processes that includes the molecular chaining bonds of polymers and behavior of single crystals. Many diffusion mechanisms and aggressive chemicals are considered to be active at this level, which includes, e.g., the transport process of chemical compounds and heat.

4.2 Physical aspects of nanocomposite structures

It is known that the surface always has a greater energy than the volume, and at the same time it retains its characteristics. It is difficult to accurately determine the boundary between the surface and volume. It is believed that the surface properties of a solid begin to appear at the interface between two media and end at a depth of about 100 nm. The width of the interface for various compact materials starts from 0.4–1.5 nm. Film materials in the range 0.5–100 nm are also characterized by a strong dependence of properties on thickness [13]. Therefore, the surface can be regarded as a physical object whose properties are determined by a combination of physical phenomena of a three-dimensional volume, a two-dimensional surface, and in some cases a one-dimensional component. Such an approach allows to consider and reveal a whole class of structures having a fractional dimension in the approximate range from 0.5 to 3.3, whose properties substantially exceed the properties of bulk materials. In connection with this, the question naturally arises of the properties dependence of a solid body (hardness, mechanical characteristics, etc.) and total surface energy. The specific surface energy determined by the Gibbs-Curie principle and Wulff's law [14–16] is calculated as follows:

$$\sigma_{hkl} = \delta_{hkl} - T\left(\frac{d\sigma_{hkl}}{dT}\right)_p \tag{4.1}$$

where δ_{hkl} is the specific total surface energy; T—temperature. B.F. Ormond [17] has shown that it depends on the atomization energy Ω (the atomization energy means the energy expended on the decomposition of the structure into its constituent atoms) that is computed as follows:

$$\sigma_{hkl} = M_{\delta_{hkl}} \frac{\Omega}{a^2} \tag{4.2}$$

where $M_{\delta hkl}$ is the structural constant of the surface energy, depending on the structural type and the face symbol (hkl); a—period of identity. Equations (4.1) and (4.2) allow concluding that the atomization energy, which determines the mechanical properties of the material, also determines the specific surface energy, which is confirmed by the experimental results. It can be seen that by changing the value σ_{hkl}, one can change the enthalpy H several times. This conclusion coincides with the traditional approaches to the problem of the strength of a solid, in accordance with which the limiting stored energy of the crystal lattice under mechanical and other effects (for example, thermal) is determined by the enthalpy of the metal at the melting point. The question of the dependence of hardness and mechanical strength of solids on energy characteristics has long attracted the attention of many scientists in view of its theoretical and practical importance. So, the connection between the energy and mechanical properties of solids is proven. Then, using the generalized equation of the first and second laws of thermodynamics for the surface of a solid, it can be written

$$dU_F = TdS + \sigma dF + \sum_i \mu_i dn_i \tag{4.3}$$

where U is the internal energy; S—entropy; σ—specific surface energy (surface tension); F—area; μ_i—chemical potential; n_i—mole fraction of the substance, we can conclude that the greater the internal energy of a solid, the higher its mechanical characteristics (yield stress, hardness, wear resistance, etc.). In a three-dimensional structure in an ideal crystal lattice, entropy and internal energy can only be increased by increasing the bond energy and the number of these bonds (i.e., by changing the composition and structure). In this case, as the hardness of the material increases, its fragility increases. In a two-dimensional system, in accordance with Equation (4.3), the internal energy can also increase due to the increase in entropy. This fact is confirmed by experimental observations of phase transitions on the surface of a solid and measurements of thermal conductivity in the bulk and on the surface. Considering that:

$$S = \int_0^T \frac{C}{T} dt \tag{4.4}$$

where C is the heat capacity; T—temperature, it is clear that the heat capacity, and hence the entropy of the surface is much higher than the entropy of the volume.

Differentiating Equation (4.4) and substituting in the Equation (4.3), one can obtain:

$$dU_F = CdT + \sigma dF + \sum_i \mu_i dn_i \tag{4.5}$$

Experimental data confirm the fact that in the transition from a massive material to nanocomposites and with a decrease in the particle size, the heat capacity increases on average of 1.2–2 times, and consequently, the internal energy and entropy of the system increases. Thus, it can be said that the plasticity at the expense of the growth of the entropy factor can increase at the surface with increasing internal energy and increasing hardness. This provision has also received experimental confirmation.

The change in the ratio of volume to surface occurs due to changes in the *shape and structure* of the nanoparticles entering the nanocomposites, and due to a change in the structure and properties of the interface. Experimentally, this statement is confirmed by the dependence of the properties of film materials on the thickness. With a decrease in thickness, the maximum failure stress of nanocomposites and the longevity increase, i.e., their mechanical durability also increases. Returning to the generalized equation of the first and second laws of thermodynamics (4.3), one should pay attention to the third term on the right hand side of the equation, which unequivocally indicates an increase in the internal energy of the system when a chemical potential occurs. The driving force of a chemical reaction is related to the concentrations of reactants and products is: $a[A] + b[B] \rightarrow c[C] + d[D]$. The ratio of products to reactants is expressed by the equation: $[C]^c [D]^d/[A]^a [B]^b = Q$. At equilibrium this ratio is equal to the equilibrium constant. For a closed system at constant stress and constant temperature, the criterion for equilibrium is that the total free energy of the system (G_{tot}) is at minimum value. The Total Free Energy is the sum of the free energies of each component.

$$G_{tot} = n_A G_A + n_B G_B + n_C G_C + n_D G_D$$

where G_i = Free energy/mole of (i); n_i = number of moles of (i).

For a reaction proceeding in incremental amounts toward equilibrium, the change in G_{tot} is proportional to dG, where $\Delta G = (\Sigma \upsilon_i G_i)$ products–reactants, where: υ_i = the stochiometric coefficient; G_i = the free energy per mole. Applying to the base equation: $\Delta G = \Delta H - T\Delta S$ (at constant T and P): (a) $\Delta G < 0$, and G_{tot} decreases as the reaction proceeds, if the reaction is spontaneous; (b) $\Delta G = 0$, therefore the reaction is at equilibrium, and G_{tot} is at a minimum; (c) $\Delta G > 0$, the reaction is not spontaneous as written, but proceeds in the opposite direction. Values of ΔG for a reaction can predict whether or not reactions are possible.

4.3 Chemical aspects of nanocomposite structures and reaction kinetics

Chemical reaction kinetics deals with the rates of chemical processes. Any chemical process may be broken down into a sequence of one or more single-step processes known either as elementary processes, elementary reactions, or elementary steps. Elementary reactions usually involve either a single reactive collision between two molecules (which we refer to as a bimolecular step), or dissociation of a single reactant molecule, which we refer to as a unimolecular step.

An important point to recognize is that many reactions that are written as a single reaction equation in actual fact consist of a series of elementary steps. This becomes extremely important in the theory of complex chemical reaction rates. As a general rule, elementary processes involve a transition between two atomic or molecular states separated by a potential barrier. The potential barrier constitutes the activation energy of the process, and determines the rate at which it occurs. When the barrier is low, the thermal energy of the reactants will generally be high enough to surmount the barrier and move over to products, and the reaction will be fast. However, when the barrier is high, only a few reactants will have sufficient energy, and the reaction will be much slower. The presence of a potential barrier to reaction is also the source of the temperature dependence of reaction rates.

A study into the kinetics of a chemical reaction is usually carried out with one or both of two main goals in mind:

1. Analysis of the sequence of elementary steps giving rise to the overall reaction, i.e., the reaction mechanism.
2. Determination of the absolute rate of the reaction and/or its individual elementary steps.

4.3.1 *Rate of reaction*

The rate law is an expression relating the rate of a reaction to the concentrations of the chemical species present, which may include reactants, products, and catalysts. Many reactions follow a simple rate law, which takes the form:

$$v = k[A]^a[B]^b[C]^c \tag{4.6}$$

i.e., the rate is proportional to the concentrations of the reactants each raised to some power. The constant of proportionality, k, is called the rate constant. The power a particular concentration is raised to is the order of the reaction with respect to that reactant. Note that the orders do not have to be integers. The sum of the powers is called the overall order. Even reactions that involve multiple elementary steps often obey rate laws of this kind, though in these cases the orders will not necessarily reflect the stochiometry of the reaction equation. Other reactions follow complex rate laws. These often have a much more complicated dependence on the chemical species present, and may also contain more than one rate constant. Complex rate laws always imply a multi-step reaction mechanism.

4.3.2 *Integrated rate laws*

In many simple cases, the rate law may be integrated analytically. Otherwise, numerical (computer-based) techniques may be used. Three of the simplest rate laws are given below in both their differential and integrated form:

$A \rightarrow P$ zeroth $d[A]/dt = -k$

$A \rightarrow P$ first $d[A]/dt = -k[A]$

$A + A \rightarrow P$ second $d[A]/dt = -k[A]^2$

A final important point about rate laws is that overall rate laws for a reaction may contain reactant, product and catalyst concentrations, but must not contain concentrations of reactive intermediates (these will of course appear in rate laws for individual elementary steps).

If the heat generated in a reaction due to the reaction exothermicity cannot be dissipated sufficiently rapidly, the temperature of the reaction mixture increases. This increases the rate constant, and therefore the reaction rate, producing more heat and accelerating the reaction rate still further, and so on until an explosion results. Such explosions are known as thermal explosions, and in principle may occur whenever the rate of heat production by a reaction mixture exceeds the rate of heat loss to the surroundings (often the walls of the reaction vessel). The second category of explosions arise from chain branching within a chain reaction, and are known as chain branching explosions (or somtimes, somewhat misleadingly, isothermal explosions). In this case, one or more steps in the reaction mechanism produce two or more chain carriers from one chain carrier, increasing the number of chain carriers, and therefore the overall reaction rate. In practice, both mechanisms often occur simultaneously, since any acceleration in the rate of an exothermic reaction will eventually lead to an increase in temperature. When the Arrhenius equation [Rate (k) ≈ $\exp(-E_a/RT)$] is applied to the overall kinetics of a multi-step reaction, E_a simply becomes an experimental parameter describing the temperature dependence of the overall reaction rate. E_a may vary with temperature, and may take positive or negative values. In this context, the activation energy is defined as:

$$E_a = RT^2 \frac{d(\ln k)}{dT} \tag{4.7}$$

There are a few observations that follow from Equation (4.7).

1. The higher the activation energy, the stronger the temperature dependence of the rate constant.
2. A reaction with no temperature dependence has an activation energy of zero (this is common in ion-molecule reactions and radical-radical recombination).
3. Negative activation energy implies that the rate decreases as the temperature increases, and always indicates a complex reaction mechanism. Simple collision theory provides a good first attempt at rationalizing the Arrhenius temperature dependence seen for many reaction rate constants. However, at a quantitative level the predictions of the theory are far from accurate. If it is considered that a chemical energy arises at the interface between the two media, then the internal energy of the nanocomposites will naturally grow. Structural changes on the surface entail a change in surface energy, so the development and modification of the surface, which entails an increase in surface energy, leads to an increase in the mechanical characteristics of the material. Dependence of the physical and mechanical properties of a solid on the value of internal energy allows us to draw the following conclusions:
   - the change in the surface structure over a wide range allows one to change the energy of the surface, and hence the mechanical and physical properties of nanocomposites. The range of properties can be significantly higher than for bulk materials;
   - changing the thickness of the film also makes it possible to significantly change the surface energy, and hence the properties of the film.

Thus, there is a real technological opportunity to change the properties of the material, giving it characteristics unattainable for bulk materials. In practice, the increase in internal energy is solved by increasing deformation of the bond and the creation of point defects by doping using implantation or diffusion methods or, for example, coating a material that, a priori has a higher internal energy, and hence mechanical properties. Based on the foregoing, it can be concluded that from a certain geometric size (about 100 nm or less), all materials have significantly higher energy than in the bulk material. Therefore, nanoscale materials and structures are self-organizing stable systems that provide cooperative behavior of electronic, phonon and other subsystems and possess mechanical and dynamic characteristics that are unattainable for massive structures. Formation of a heterogeneous metallic phase in a chemical recovery in solution occurs in two main stages: nucleation (nucleation) and growth of the cluster formed. The nucleation of a new phase proceeds during an oxidation-reduction reaction, the driving force of which is the positive value of the difference between the two potentials—a reducing agent and an oxidizer. The initial stage (induction period) is characterized by a low rate. The dependence of the conversion of metal ions on time has the form of an S-shaped curve on which the site of induction, acceleration and deceleration of the reaction is distinguished (see Fig. 4.2) [18].

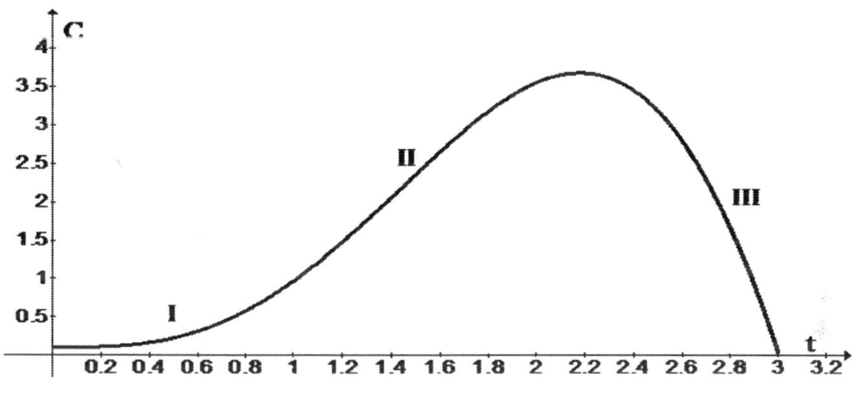

Figure 4.2: Induction, acceleration and deceleration of the reaction.

4.3.3 Dependence of the conversion of metal ions on time

The first section (I) is the induction section. It is characterized by low speed reaction. In this case, small stable particles are formed that have catalytic activity. Next is the acceleration section (II). In the third section (III) with the consumption of reagents and the stabilization of particles, the reaction rate decreases. The activation energy for the formation of stable nuclei of the new phase is sufficiently large, since the induction period proceeds under homogeneous conditions. The duration of the induction period depends on the nature of the reagents, their concentration and ratio, temperature, the presence of various catalytic impurities, etc. For many systems, the conversion of metal ions for the induction period is usually not more than 5%. Strong reducing agents (alkali metal tetrahydroborates, hypophosphite, hydrazine, hydroxylamine, etc.) reduce the induction period, while weaker ones (formaldehyde, glucose, alcohols, etc.) increase it. Stable particles have a size of at least 1 nm and are formed by successively increasing small particles with reduced atoms metal [18]. In the transition from a homogeneous regime to a heterogeneous one (formation of nuclei new phase), at which the activation energy of the reaction is much lower, acceleration of oxidation-reduction reaction is observed. Homogeneous nucleation becomes energetically unstable, and the reduction of metal ions proceeds mainly on the already formed clusters (the growth stage). The clusters serve as carriers of electrons from the molecules (ions) of the reducing agent to the metal ions that are reconstructed on their surface, thereby catalyzing the reduction process.

The growth of new phase nuclei can also precede as a result of their aggregation (coagulation) processes. Stabilizing ability of polymers, which determines the speed interaction of nuclei with a growing particle, is of great importance at the stage of growth. Therefore, a higher rate of binding of the polymer to the growing particle as compared to the growth rate of the particles themselves is one of the conditions for obtaining a finely dispersed nanoparticles size distribution in the polymer matrix. In this case, a protective adsorption polymer layer is formed, which prevents particle coagulation [19]. Thus, during the induction stage, the aggregation processes formed at the stage of induction of nuclei, as well as the rate of interaction of the polymer and metallic components determine the dimensional characteristics of crystallization process. An increase in the duration of the induction stage promotes a greater appearance of new phase nuclei in the reaction medium. An increase in the concentration of nuclei promotes the growth processes due to their aggregation, the suppression of which is also a necessary condition for obtaining narrow-dispersed nanoparticles. General approaches to the synthesis and investigation of nanocomposites, including nanoparticles of metals and their oxides dispersed in polymeric material are described in a number of works [20–23].

4.3.4 Physicochemical aspects of formation of structured nanocomposites

Chemical thermodynamics is a branch of physical chemistry, in which thermodynamic methods (general thermodynamics) are used for analysis of chemical and physicochemical phenomena: chemical reactions, phase transitions and processes in solutions. Chemical thermodynamics uses for calculations parameters that are known from experience—data about the initial and final state of the system and the conditions under which chemical process is evolving (temperature, pressure, etc.). Consequently, chemical energy is an integral part of the general Gibbs theory (and hence the integral creep equation), for example, in the case of the creation of nanocomposites materials (the 'nucleation and growth of clusters') by the synthesis of chemical elements or by an autocatalytic chemical reaction. Many of nano-chemical processes can be described (at least in engineering applications) as a first order chemical reaction, except for autocatalytic reactions. Autocatalysis is the process of catalytic acceleration of a chemical reaction by one of its products. The kinetic curve of the product of the autocatalytic reaction has a characteristic S-shape. The rate of these equations for autocatalytic reactions are fundamentally nonlinear. There are many methods for *Measurement of Reaction Rates*: one might monitor the concentrations; the total volume or pressure if these are related in a simple way to the concentrations. Whatever the method, the result is usually something like that illustrated in Fig. 4.3.

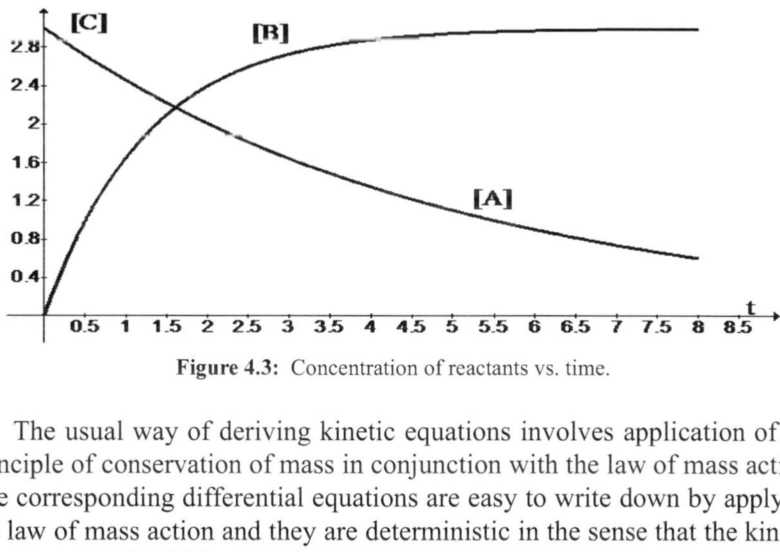

Figure 4.3: Concentration of reactants vs. time.

The usual way of deriving kinetic equations involves application of the principle of conservation of mass in conjunction with the law of mass action. The corresponding differential equations are easy to write down by applying the law of mass action and they are deterministic in the sense that the kinetic parameters and initial state of the reaction system completely determine the future states. The simple fundamental equation that describes an autocatalytic process is as following:

$$\frac{d[C]}{d\tau} = \gamma\delta[C]^p(1-[C])^q Ze^{-\frac{E_a}{RT}} \tag{4.8}$$

[C] is the product concentration fraction at any specific time, t. The derivative, d[C]/dt, is the rate of the reaction. E_a is the "Arrhenius activation energy," and Z is the "*Arrhenius pre-exponential*", it applies only to a single specific reactant being studied. The value of the "rate constant," k = d[C]/dt is different at each specific temperature: It is a constant only at one temperature, and it applies only to one specific reaction. All of these values have fundamental meaning in the chemical reaction. The exponent indexes p and q allow the prediction of the position of the maximum rate in an autocatalytic process, i.e., the amount reacted at the maximum rate—at given temperature. An example of autocatalytic rate curve is shown in the Fig. 4.4. Notice that the rate increases with time in the autocatalytic curve, at constant temperature, until it reaches a maximum reaction rate. Then the rate decreases. However, the initial rate at any temperature is much lower than the maximum rate.

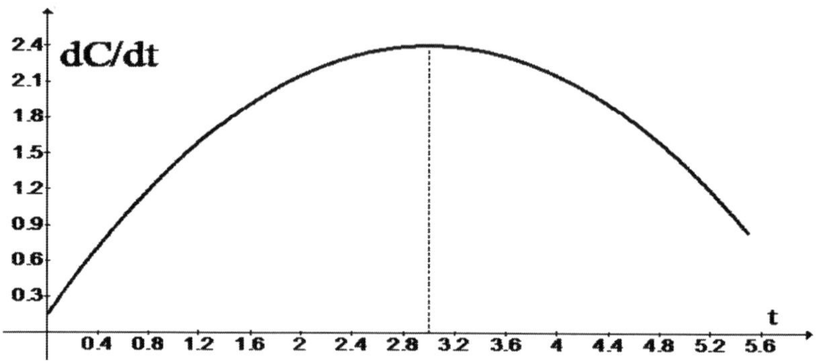

Figure 4.4: Maximum reaction rate.

The rate of these equations for autocatalytic reactions should be presented by using the conservation of mass differential equation similar to Equation (4.8). In order to simplify the heat and mass conservation equations the following assumptions are made in this book:

- The heat transfer due to conduction can be neglected due to nanometer scale of elements.
- The increase in energy flux (in addition to heat combustion release) is due to natural convection.
- The loss of energy is due to irradiation only (the conductive heat loss to the atmosphere are neglected due to much weaker dependence on temperature).

Let's consider now the spatial averaging of temperature and unsteady first-order chemical reaction rate. The Equation (4.8) are written now in dimensionless form and simplified further due to assumptions made above [24]:

$$\begin{aligned} \frac{\partial \theta}{\partial \tau} &= \delta(1-C)^k \exp\left(\frac{\theta}{1+\beta\theta}\right) - \alpha\theta^4 \\ \frac{\partial C}{\partial \tau} &= \gamma\delta(1-C)^k \exp\left(\frac{\theta}{1+\beta\theta}\right) \end{aligned} \tag{4.9}$$

For an autocatalytic process the simplified Equation (4.8) has the form as follows

$$\frac{\partial\theta}{\partial\tau} = \delta C^{p}(1-C)^{q}\exp\left(\frac{\theta}{1+\beta\theta}\right) - \alpha\theta^{4}$$
$$\frac{dC}{d\tau} = \gamma\delta C^{p}(1-C)^{q} e^{\frac{\theta}{1+\beta\theta}} \tag{4.10}$$

The solutions of Equations (4.9) and (4.10) are presenting temperature-time functions m; m1; m2 and m21 that are used in constitutive creep equations for nanomaterials. The more accurate analysis of temperature-time functions obviously can be obtained if the second Equation in (4.9) and (4.10) will be substituted by the corresponding solution of kinetic equation (for a given chemical reaction of a particular nanomaterials. For the detailed analysis of chemical energy and application of it to the nanocomposites behavior under high temperature loading conditions see Chapter 5.

4.4 Mechanical properties of nanocrystalline metals and alloys

In this section, we review the principal mechanical properties of nanocrystalline metals. At the outset, it should be emphasized that porosity is of utmost importance and can mask and/or distort properties. The early "bottom-up" synthesis methods often resulted in porosity and incomplete bonding among the grains. This decrease in Young's modulus with porosity is well known and is indeed expressed in many mechanics simulations. One of the equations is Wachtman and Mackenzie [25, 26]:

$$E = E_0(1 - f_1{*}pr - f_2{*}pr^2) \tag{4.11}$$

where pr is the porosity and f1 and f2 are equal to 1.9 and 0.9, respectively. For relatively low porosity, p2 can be neglected and we have, approximately $E = E_0(1 - f_1{*}pr)$. The yield stress and tensile ductility are simultaneously affected. The decrease in strength is obvious. The existing pores provide initiation sites for failure.

4.4.1 *Yield strength*

Grain size is known to have a significant effect on the mechanical behavior of materials, in particular, on the yield stress. The dependence of yield stress on grain size in metals is well established in the conventional polycrystalline range (micrometer and larger sized grains). Yield stress, σ_y, for materials with grain size d, is found to follow the Hall–Petch relation [27, 28]:

$$\sigma_y = \sigma_0 + kd^{-1/2} \tag{4.12}$$

where σ_0 is the friction stress and k is a constant. Equation (4.12) is approximate, and a more general formulation is to use a power law expression with exponent n, where $0.3 < n < 0.7$. The mechanical properties of metals with nano-range grain sizes have been estimated from uniaxial tension/compression tests and micro- or nano-indentation. Often micro-size tensile samples are used to avoid the influence of imperfections, e.g., voids that might adversely influence the mechanical response of the material. Nanomaterials with particle size of 1–100 nm are of current interest because they show noble physical and chemical properties that may differ from those of the corresponding bulk counterparts. Nano-solids can be the spherical nano-solids, nano-wires or nano-films. Many physical properties such as hardness, melting temperature, etc., may be dependent upon the particle size. Conventional methods of strengthening metallic materials involve enhancing resistance to dislocation motion by refining length scales of intrinsic microstructures, e.g., grain or particle size and spacing between solute atoms, second phases or particles, etc. In fibrous or particle-reinforced polymer nanocomposites (PNCs), dispersion of the nanoparticles and adhesion at the particle–matrix interface play crucial roles in determining the mechanical properties of the nanocomposites.

4.5 Creep of small coarse grained and nanocrystalline materials

4.5.1 Creep mechanisms in small coarse grained materials

Creep in coarse grained materials has been studied for many years and phenomenological models exist that explain mechanisms involved therein. Creep analysis in nanocrystalline materials are much more complex due to the following: (1) there are the limitation of synthesizing bulk nanomaterials free of defects (porosity and impurities) that could provide reliable creep deformation data; (2) significant increase in the volume fraction of grain boundaries that renders complication in the creep micromechanics and leads to challenges in developing a corresponding phenomenological model that could explain the creep deformation process; (3) grain growth process is a combination of heat flux delivered to the system and the heat flux due to autocatalytic chemical reaction, which in itself presenting a very complex micro-scaled process with many stages and uncertain parameters. Phenomenological models presented in this book reflect the relative simplicity of the collective creep behavior of nanocomposites, automatically including the minimum amount of dimensionless parameters to avoid over-fitting but lacking all intermediate chemical kinematics and mechanics explanations. Thus, phenomenological models exploit correlations among observed data to make predictions about similar experimental data.

4.5.2 Phenomenological creep models of nanocrystalline materials

The integral representation of creep process is a very appealing theoretical concept, since it is not limited to a particular material or class of materials. This chapter reviews the thermodynamic development with a special view to developing a better understanding of the nature of the four nonlinear parameters which appear, as well as their physical origin. A mathematical expression for the phenomenological creep models of nanocrystalline materials that include the entropy of the whole system (sum of entropies of the subsystems) will have the form [29]:

$$E(t)[\varepsilon(t)] - \sigma(t) + \int_{-\infty}^{t} K_1[t-\tau,\sigma(\tau)]d\tau + \int_{-\infty}^{t} K_2[t-\tau,\sigma(\tau)]d\tau + \int_{-\infty}^{t} K_3[t-\tau,\sigma(\tau)]d\tau \quad (4.13)$$

First integral on the right hand side of the Equation (4.13) represents the entropy flows into the volume element; second represents the entropy source due to creep dependence on stress and temperature that are lumped together, and finally the third represents the entropy source due to micro-structural rearrangements related to creep dependence on thermally activated structure evolution (dislocation density; chemical kinetics and autocatalytic processes) of nanomaterials only. Based on the simplifications introduced in the author's previous work [30], and a system of dimensionless parameters and variables (stress and temperature) integral Equation (4.13) with variant kernels K_1; K_2 and K_3 is solvable for creep stress now. This together with the two dimensionless temperatures-time equations (one for the entropy flows into the volume element, and the other one for the entropy flow due to chemical energy thermally activated by nanocrystalline structure evolution) obtained from the corresponding conservation of energy and mass equations are as follows [30]:

$$\begin{aligned}
&\tau = m = (0.0405*\theta - 0.01126*\theta\hat{}2 + 0.001462\theta*\hat{}3 - 0.00006868*\theta\hat{}4)\\
&m1 = (0.0405 - 0.02252*\theta\hat{}1 + 0.004386*\theta\hat{}2 - 0.0002747*\theta\hat{}3)\\
&\tau = m2 = 1.266\theta + 2.12\theta^2 - 3.077\theta^3 + 1.085\theta^4\\
&m21 = 1.266 + 4.24\,\theta - 9.231\,\theta^2 + 4.34\,\theta^3
\end{aligned} \quad (4.14)$$

The former Equation (4.13) can be explicitly rewritten with the reduced times and their first derivatives (m; m1 and m2; m21) as:

$$E(\theta)[\theta] = \sigma(\theta) + \int_0^\theta e^{\frac{\tau}{1+\beta\tau}} K_1(\theta,\tau)\sigma^n(\tau)\,m\mathbf{1}d\tau + \int_0^\theta e^{\frac{\tau}{1+\beta\tau}} f_1[\sigma(\tau)]K_2(\theta,\tau)\sigma^n(\tau)\,m\mathbf{1}d\tau +$$

$$+\int_0^\theta \frac{E_0}{E_2} e^{\frac{\tau}{1+\beta\tau}} f_2[d^{-1}(\tau)]K_3(\theta,\tau)\sigma(\tau)\,m\mathbf{21}d\tau$$

$$K_1(\theta,\tau) = m\mathbf{1}(\tau)\sum_{i=1}^{N} \exp(-\alpha_i\, m(\theta))\exp(\alpha_i\, m(\tau))$$

$$K_2(\theta,\tau) = m\mathbf{1}(\tau)\sum_{i=1}^{N} e^{\frac{\tau}{1+\beta\tau}} \exp(-\beta_{i1}\, m(\theta))\exp(\beta_{i1}\, m(\tau))$$

$$K_3(\theta,\tau) = m\mathbf{21}(\tau)\sum_{i=1}^{N} e^{\frac{\tau}{1+\beta\tau}} \exp(-\beta_{i2}\, m\mathbf{2}(\theta))\exp(\beta_{i2}\, m\mathbf{2}(\tau))$$

$$f_1(\sigma) = A_2\sigma^s;\quad s = \mathbf{1,2,3}...,M;\quad f_2(d) = A_3[d^q(1+d^q)];\ 0 < q < 1 \tag{4.15}$$

Consider now f1 and f2 as increasing and decreasing functions respectfully. Equation (4.15) in this case is as follows:

$$E(\theta)[\theta] = \sigma(\theta) + \int_0^\theta e^{\frac{\tau}{1+\beta\tau}} K_1(\theta,\tau)\sigma^n(\tau)\,m\mathbf{1}d\tau + \int_0^\theta A_2 f_2[\sigma(\tau)]K_2(\theta,\tau)\sigma^n(\tau)\,m\mathbf{1}d\tau +$$

$$+\int_0^\theta \frac{E_0}{E_1} A_3 f_3[d(\tau)]K_3(\theta,\tau)\sigma(\tau)\,m\mathbf{21}d\tau$$

$$K_1(\theta,\tau) = \varphi_1(\theta)f_1(\tau) = m\mathbf{1}(\tau)\sum_{i=1}^{N} \exp(-\alpha_i\, m(\theta))\exp(\alpha_i\, m(\tau))$$

$$K_2(\theta,\tau) = e^{\frac{\tau}{1+\beta\tau}} m\mathbf{1}(\tau)\sum_{i=1}^{N} \exp(-\beta_{i2}\, m(\theta))\exp(\beta_{i2}\, m(\tau))$$

$$K_3(\theta,\tau) = e^{\frac{\tau}{1+\beta\tau}} m\mathbf{21}(\tau)\sum_{i=1}^{N} \exp(-\beta_{i3}\, m\mathbf{2}(\theta))\exp(\beta_{i3}\, m\mathbf{2}(\tau))$$

$$f_2(\sigma) = \sigma^s;\quad s = \mathbf{1,2,3}...,M;\quad f_3(\theta) = A_3\ [d^{-0.5}(1+d^{-0.5})];$$
$$d = \varphi(\theta);\quad k = E_0/E_1;\quad A_2 = \text{const.};\quad A_3(\theta) = A_3 f(\theta) \tag{4.16}$$

4.5.3 Nucleation and growth process of nanoparticles

Recently, much attention has been paid to the study of regularities and mechanisms of diffusion-controlled grain growth during the recrystallization of nanostructured materials. The process of growth of nanoparticles controlled by bulk diffusion of carbon or the chemical reaction of their formation on a semi-coherent matrix (particle) interface can determine the kinetics and temperature dependence of the growth rate of grains up to the temperature of the onset of secondary recrystallization, when the boundaries of individual grains playing the role of clusters with this kind of recrystallization, break away from particles. Investigation of the patterns of degradation of nanocomposites at elevated temperatures will allow predicting the development of structured degradation processes at variable high temperatures acting for long and short times. $A_3 = f(\theta)$ reflects the laws of phase and structured transformations of nanomaterials elucidated from microscopic and micro structured analysis. Due to the fact that the temperature effect on the process of degradation of the nanocomposites structure depends strictly on the chemical composition (nature) of this particular material, the function $f(\theta)$ can have a different form (continuous or piecewise continuous function), and, therefore, must be determined (at least from qualitative side) on the basis of microscopic experimental analysis and data. As noted above, the main attention from the practical point of view should be given to the process of transition from one phase state to another as the temperature varies with time. A simple example of such process is the process of changing the size of a piece of ice placed in a vessel with water when the vessel is heated by a variable with time temperature. Another simple example is a system of dilute molecules diffusing in a homogeneous environment. In this system, the molecules tend to move from areas with high concentration to low concentration, until eventually the concentration is the same everywhere. The microscopic explanation for this is based on kinetic theory and the random motion of molecules. However, it is simpler to describe the process in terms of chemical potentials: for a given temperature and time, a particle has a higher chemical potential if a concentration is higher, and a lower chemical potential if concentration is lower. Movement of particles from higher chemical potential to lower chemical potential is accompanied by a release of free energy.

4.5.4 *Modeling of nucleation and growth process of nanoparticles*

Formation of metal-containing nanocomposites using high-molecular compounds increased the interest of researchers, as it is associated with broad practical possibilities of their application. A fundamental understanding of the formation and stabilization of metallic nanoparticles by polymer compounds is fundamental to the creation of metal-polymer nanocomposites systems with desired flexibly configurable properties. As stabilizing matrices in such compounds, polymers are capable of interacting with the emerging nanoparticles metals. Of particular interest are high molecular compounds, containing in their composition heterogeneous in terms of functionality groups, capable of stabilizing metal nanoparticles, preventing their aggregation. In this connection, homo-polymers and copolymers of 1-vinyl-1,2,4-triazole are promising which, due to the presence of the nitrogen-containing heterocyclic moiety and the functional groups of the co-monomers, are able to effectively interact with metallic nanoparticles at the earliest stages of their formation, regulating their growth, aggregation, dimensionality and uniformity of distribution in the polymer matrix. The most common kinetic law of the formation of nanoscale particles (NP) is the combination of a high rate of nucleation of a metal-containing phase with a low rate of its growth. Two main approaches to obtaining an NP-condensation (physical and chemical methods) and dispersing have emerged. The first is connected with the "assembly" of the NP from individual metal atoms during the phase transformation, the second—with the grinding of coarsely dispersed particles to the size of the NP. In chemical methods, chemical transformations are the main "supplier" of the material being formed, but the formation of a new phase is necessarily associated with a phase transition (physical process). Significantly less frequently, NPs are obtained by mechanically dispersing massive particles, but acoustic dispersion of solids is intensively developed. These methods make it possible to obtain metal-containing particles of different levels of dispersion with a variety of physicochemical properties. The identification and control of numerous growths parameters is crucial for the development and optimization of the synthesis process, study of growth mechanisms and fabrication. Therefore, the detailed study of the effect of experimental conditions such as time, temperature, catalyst, carbon gas precursor, among others, is of great relevance and now it is the subject of many experimental researches. The development of technologies involving nanomaterials can result in nanocomposites structural elements characterized by high performance, low mass and low consumption of energy. Thermal chemical vapor deposition (CVD) has been shown to be an efficient and versatile technique for the chemical synthesis processes. However, many challenges (in particular, the lack of precise control of the nanoparticles growth during the CVD process) still have to be addressed.

The growth processes of nanoparticles in thermal synthesis usually are investigated numerically. Considering the free energy gradients for particle formation in a binary system, the model simulates the collective and simultaneous combined process of binary nucleation and binary co-condensation of high temperature vapors, and coagulation among nanoparticles. The gradients of the free energy of particle formation, W, composed of the chemical potentials and the surface energy are evaluated for nanoparticles with each size and each composition. Modeling of nucleation and growth processes are performed by the molecular dynamics method. Time of nucleation and speed of clusters growth is obtained for different densities and temperatures. The bases of classical nucleation theory usually are tested and recommendations for multi scaling modeling of nucleation process are performed.

The following are the features of the behavior of the system when the density and temperature change should be considered.

1. The dependence of the rate of nucleation and growth on the density is a power law, which allows to state that the growth of clusters can be described through Brownian collisions, the number of which per unit time is proportional to the square of the density. In this case, the coefficients of adherence of atoms, fusion and destruction of clusters, of course, cannot be considered constant, but a function of time and temperature.
2. A linear, with good accuracy, form of the dependence of the nucleation time and growth of clusters on temperature suggests that the coefficients governing the adhesion of atoms, clusters, and the decay of clusters in the process of Brownian collisions also have linear temperature dependence. This allows constructing a model for gas-dynamic theory, which quantitatively describes the nucleation and growth of clusters. The results obtained in molecular dynamics (MD) showed that the nucleation theory in the Frenkel [31]; Zel'dovich [32] and Brennen [33] interpretation correctly determines the mechanisms of nucleation and growth of clusters, but the quantitative parameters for describing these processes cannot be obtained analytically due to the large number of factors influencing them. To use the results of molecular dynamics simulation in modeling by the methods of continuous medium dynamics, it is necessary to construct a quantitatively correct macroscopic model. In this model, it is reasonable to assume that:
   1. all processes (growth and decay) occur in the collision of monomers and/or clusters. Clusters can be considered spherical, and the scattering cross section is determined by the classical ratio:

      $$\sigma = \pi\left(8\frac{3\nu_{at}n}{4\pi}\right)^{2/3} \tag{4.17}$$

      where n is the number of atoms in the cluster, ν_{at} is the volume of one atom. The inaccuracy of this representation is corrected by the sticking parameters, which depend on the size of the cluster and the conditions in the system.
   2. Clusters "all with all" are taken into account for clusters and monomers. In this case, two clusters always stick together (for a small concentration of clusters this process is unlikely, so this assumption does not affect the nucleation process). In addition, the process of merging clusters is energetically beneficial, which leads to almost 100% of the probability of sticking.
   3. The interaction of clusters with monomers and small clusters is determined by the growth coefficient $\beta_n = b(n)/a(n)$ is obtained from

molecular dynamics modeling. Here b (n) is the probability of growth (sticking coefficient), a(n)—probability of cluster decay. The growth coefficient determines the stability or instability of the cluster. The growth coefficient determines the stability ($\beta_n > 1$) or instability ($\beta_n < 1$).

4.6 Basic creep equations and nanomaterials parameters relations

Consider the creep deformation of a nanocomposite material under conditions of high temperature load. The elastic components of the deformation are omitted, since it is assumed that the main deformations are due to high thermal external load. It is also assumed that the heat impact induces the short-term creep under conditions of viscous or viscoelastic nonlinear flow (see Chapter 3). The single integral type nonlinear creep equation (taking into account the grain size effects, nucleation and growth process of nanoparticles and its dependence on temperature) describing the behavior of the nano material obeying the generalized Gibbs energy theory, explained above, can be written in the form similar to the integral creep Equation (4.16) for bulk homogeneous isotropic material as follows:

$$E(\theta)[\theta] = \sigma(\theta) + \int_0^\theta e^{\frac{\tau}{1+\beta\tau}} K_1(\theta,\tau)\sigma^n(\tau)\,m\mathbf{1}d\tau + \int_0^\theta e^{f_2[\sigma(\tau)]} K_2(\theta,\tau)\sigma^n(\tau)\,m\mathbf{1}d\tau +$$

$$+\int_0^\theta \frac{E_0}{E_2} e^{f_3[d(\tau)]} K_3(\theta,\tau)\sigma(\tau)\,m\mathbf{21}d\tau$$

$$K_1(\theta,\tau) = \varphi_1(\theta)f_1(\tau) = m\mathbf{1}(\tau)\sum_{i=1}^{N}\exp(-\alpha_i\, m(\theta))\exp(\alpha_i\, m(\tau))$$

$$K_2(\theta,\tau) = e^{\frac{\tau}{1+\beta\tau}} m\mathbf{1}(\tau)\sum_{i=1}^{N}\exp(-\beta_{i2}\, m(\theta))\exp(\beta_{i2}\, m(\tau)) \qquad (4.18)$$

$$K_3(\theta,\tau) = e^{\frac{\tau}{1+\beta\tau}} m\mathbf{21}(\tau)\sum_{i=1}^{N}\exp(-\beta_{i3}\, m\mathbf{2}(\theta))\exp(\beta_{i3}\, m\mathbf{2}(\tau))$$

$$f_2(\sigma) = A_2\sigma^s;\quad s = \mathbf{1,2,3}\ldots, M;\quad f_3(\theta) = A_3\ [d^{-0.5}(1+d^{-0.5})];\ d = k\theta$$

The computer code presented below allows to analyze and compare the combination of different factors affecting creep deformation process of nanocomposites (such as: thermal stresses in a given structural engineering system; different length and time scaling factors; chemical composition of nanocomposites and their transformation with time and temperature changes, etc.); as well as a separate influences of each component on creep. This allows us not only to see the reaction of a nanocomposites on high temperature load, but to optimize the mechanical properties of such materials based on given working loading conditions. Since the integral creep Equation (4.18) is written in dimensionless form, the numerical analyses below are presented for various parameters A2; A3; θ^*; θ_g; θ; p as well as functions f_2 and f_3 describing the nucleation and growth process of nanocomposites. In this book the various types of functions $f_3(\theta)$ are considered (continuous, monotonic, piecewise-continuous, and so on) in order to study their qualitative influence on the process of *engineering* creep of the nanocomposites, more precisely, on the nature of the change in the stress-temperature-strain (STS) diagram. The solutions of Equation (4.18) for some particular expressions of integrants are presented below via Examples. The actual forms of integrants in Equation (4.18) obviously should be based on simple uniaxial creep tests data and corresponding temperature-time functions of external heat transfer to the system as well as internal heat of chemical reaction itself (temperature-time relationship m2 and m21) is defined in this case by the solution of corresponding chemical kinetic equation—see Chapter 5.

4.6.1 *Effect of function f3 (nucleation and growth process of nanoparticles) type on creep process*

An analysis of the effect of the type of function f3 on the character of the creep process described by the single integral Equation (4.18) begins with the simplest case when A2 = A3 = 0. Obviously, in this case this the equation becomes a well-known creep equation for uniaxial tension of an isotropic material with one difference: the instant modulus of elasticity of the composite is determined by the "rule of mixture" (matrix and uniformly dispersed nanomaterials).

Example 4.1:

$A_2 = 0$; $A_3 = 0$; $\mu_f = 0.1$; $E_0/E_2 = k = 0.1$

E_0 and E_2 are modulus of elasticity of matrix and nanomaterials respectfully

Calculated values of DEQ variables

	Variable	Initial value	Minimal value	Maximal value	Final value
1	A2	0	0	0	0
2	A3	0	0	0	0
3	E	1.9	0.475	1.9	0.475
4	E1	1.	0.25	1.	0.25
5	f3	0	0	1.	1.
6	f31	5.0E-06	0	1.	0
7	f32	0.999999	0.449329	0.999999	0.449329

8	f33	6.162E-11	6.162E-11	0.9998458	1.015E-05
9	f34	1.	0	1.	0
10	μ_f	0.1	0.1	0.1	0.1
11	k	0.1	0.1	0.1	0.1
12	m	4.05E-07	4.05E-07	0.0705907	0.0705907
13	m1	0.0404998	0.0003976	0.0404998	0.0003976
14	m2	1.266E-05	1.266E-05	3014.544	3014.544
15	m21	1.266042	0.5327369	1666.482	1666.482
16	n	1.	1.	1.	1.
17	p	–0.5	–0.5	–0.5	–0.5
18	s	1.	1.	1.	1.
19	t	1.0E-05	1.0E-05	8.	8.
20	Y1	1.9E-05	1.9E-05	5.591693	3.8
21	z	1.9E-05	1.9E-05	4.660971	1.409941
22	z1	0	0	2.391747	2.391747
23	z11	0	0	1.156661	1.156661
24	z111	1.0E-05	1.0E-05	2.392408	0.844155
25	z2	0	0	0	0
26	z22	0	0	0	0
27	z3	0	0	0	0

Differential equations

1 d(z1)/d(t) = (exp(t/(1 + 0.067*t)))*((exp(0.01*m)))*m1*z^n

2 d(z2)/d(t) = (exp(t/(1 + 0.067*t)))*(A2*(z^s))*((exp(0.01*m)))*m1*z^n

3 d(z3)/d(t) = k*(exp(t/(1 + 0.067*t)))*A3*(f3*(t^p/(1 + 0.1*t^p)))*((exp(0.01*m2)))*m21*z^n

4 d(z11)/d(t) = (exp(t/(1 + 0.067*t)))*((exp(0.01*m)))*m1*z111^n

5 d(z22)/d(t) = (exp(t/(1 + 0.067*t)))*(A2*(z111^s))*((exp(0.01*m)))*m1 *z111^n

Explicit equations

1 m = (0.0405*t – 0.01126*t^2 + 0.001462*t^3 – 0.00006868*t^4)

2 m1 = (0.0405 – 0.02252*t^1 + 0.004386*t^2 – 0.0002747*t^3)

3 m2 = 1.266*t + 2.12*t^2 – 3.077*t^3 + 1.085*t^4

4 μ_f = 0.1

```
k = 0.1
n = 1.0
s = 1.0
E = (0.625 – 0.375*(tanh(5*(t – 3))))*(1 – fi)+(1/k)* (0.625 –
  0.375*(tanh(5*(t – 5))))*(fi)
A2 = 0
A3 = 0
p = –0.5
z = t*E–(z1+z2)*((exp(–0.01*m)))–z3*(exp(–0.01*m2))
m21 = 1.266 + 4.24 *t – 9.231 *t^2 + 4.34 *t^3
f3 = if (t < 4) then (0) else (1)
f31 = (1/16)*t*(8 – t)
f32 = (exp(–0.1*t))
f33 = (sin(3.14*t/4))^2
f34 = if (t < 4) then (1) else (0)
E1 = (0.625 – 0.375*(tanh(5*(t – 3))))*1.0
z111 = t*E1–(z11+z22)*((exp(–0.01*m)))–z3*(exp(–0.01*m2))*0
Y1 = t*E
```

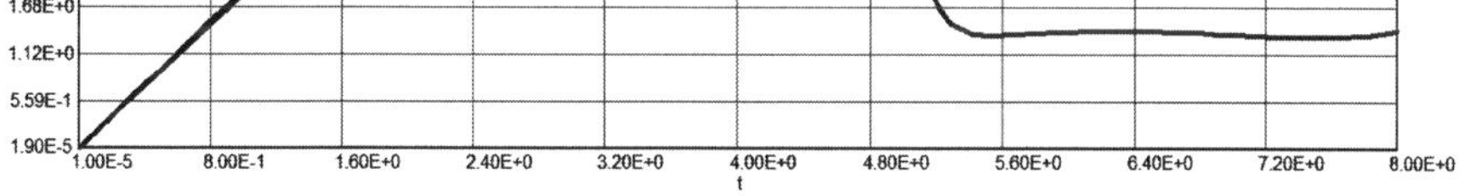

Figure 4.5: STS diagram: A2 = A3 = 0.

Upper curve–STS relationship in Fig. 4.5 of a composite is due to instantaneous thermal stress (without creep effect). Lower curve–STS relationship in Fig. 4.5 of nanocomposites is due to instantaneous thermal stress; internal stress effect on creep deformation process (with creep effect)

Approximation by polynomial (STS without creep)–upper curve

Model: Y1 = a1*t + a2*t^2 + a3*t^3 + a4*t^4 + a5*t^5

Variable	Value
a1	1.172343
a2	1.110332
a3	–0.4789155
a4	0.0601987
a5	–0.0023704

$$\sigma = 1.172\theta + 1.11\,\theta^2 - 0.479\,\theta^3 + 0.06\,\theta^4 - 0.00237\,\theta^5 \tag{4.19}$$

Approximation by polynomial (STS with creep)–lower curve

Model: z = a1*t + a2*t^2 + a3*t^3 + a4*t^4 + a5*t^5

Variable	Value
a1	1.22796
a2	0.9839048
a3	–0.4510116
a4	0.0566827
a5	–0.0022081

$$\sigma = 1.228\theta + 0.984\theta^2 - 0.451\theta^3 + 0.0567\theta^4 - 0.0022\theta^5 \tag{4.20}$$

As can be seen from the above diagrams (see Figs. 4.1–4.5), the allowable stresses that are taking into account the creep of the composite, are less than the corresponding allowable stresses without considering the creep process of the composite, that is, in this case the stresses are simply equal to instantaneous stresses that vary with time, due to the change in the modulus of elasticity with temperature (time).

Of course, taking into account the influence of stresses caused by a high temperature load leads to a further decrease in the allowable stress, and the use of nanoparticles as a filler of the composite should lead to an increase in the strength of the nanocomposites. Such problems of creep deformations are considered below.

Depending on the specific chemical composition and the phase state of the nanomaterials, the second part of the function f3 can be monotonically increasing ($p > 0$) or monotonically decreasing ($p < 0$) (see Fig. 4.6). For simplicity's sake, consider the case ($p < 0$), but different forms of the first part of the function f3. This part of the function f3 is responsible for the description of the growth process of nanomaterials grains under the influence of a high temperature load and their dependence on time (and hence on temperature), as well as mechanical bonding to the matrix material. Such dependence can be constant (the nanomaterials and matrix are rigidly connected during the entire duration of the action of the temperature load $[0 < \theta < \theta_{max}]$ without the formation of micro cracks); it can be variable (and the instantaneous modulus of elasticity of the nanocomposites is calculated by the "rule of the mixture"). For the convenience of numerical analysis, all 5 (five) of these kind of functions f3 are given in the general form of a computer code and can be easily interchanged in the analysis of a particular creep process for a particular type of nanomaterials.

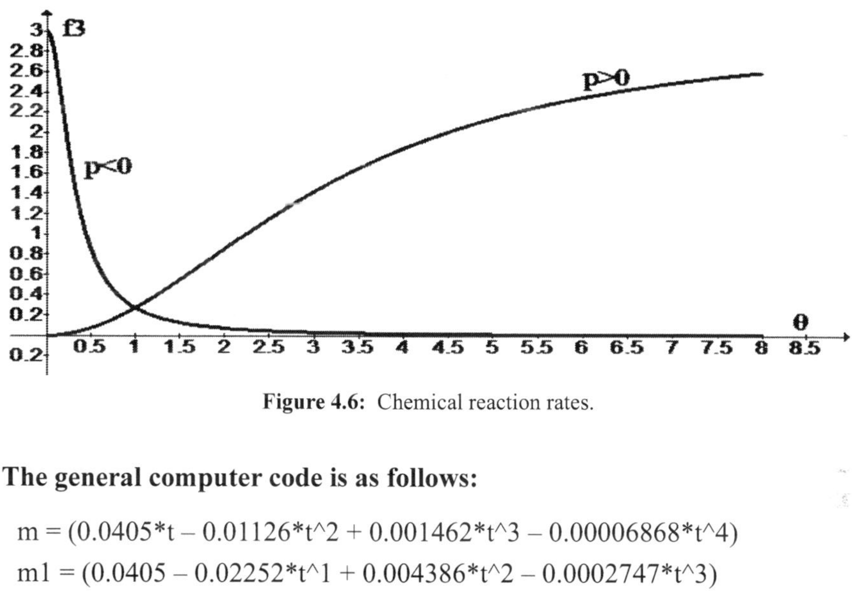

Figure 4.6: Chemical reaction rates.

The general computer code is as follows:

```
m = (0.0405*t – 0.01126*t^2 + 0.001462*t^3 – 0.00006868*t^4)
m1 = (0.0405 – 0.02252*t^1 + 0.004386*t^2 – 0.0002747*t^3)
t(0) = 0.00001; t(f) = 8.0
z = t*E–(z1+z2)*((exp(–0.01*m)))–z3*(exp(–0.01*m2))
d(z1)/d(t) = (exp(t/(1 + 0.067*t)))*((exp(0.01*m)))*m1*z^n
E = (0.625 – 0.375*(tanh(5*(t – 3))))*(1 – fi)+(1/k)*
  (0.625 – 0.375*(tanh(5*(t – 5))))*(fi)
k = 0.1; n = 1.0; s = 1.0; p = –2; fi = 0.1; Y1 = t*E
z1(0) = 0
d(z2)/d(t) = (exp(t/(1 + 0.067*t)))*(A2*(z^s))*((exp(0.01*m)))*m1*z^n
z2(0) = 0
A2 = 1; A3 = 1
```

d(z3)/d(t) = k*(exp(t/(1 + 0.067*t)))*A3*(f35*(t^p/(1 + 0.1*t^p)))*((exp(0.01*m2)))*m21*z^n

z3(0) = 0

m2 = 1.266*t + 2.12*t^2 – 3.077*t^3 + 1.085*t^4

m21 = 1.266 + 4.24 *t – 9.231 *t^2 + 4.34 *t^3

f3 = if (t < 0.9) then (0) else (1)

f31 = (1/16)*t*(8 – t)

f32 = (exp(–0.1*t))

f33 = (sin(3.14*t/4))^2

f34 = if (t < 7.9) then (1) else (0)

f35 = (1/8)*t^1

z111 = t*E1–(z11 + z22)*((exp(–0.01*m)))

E1 = (0.625 – 0.375*(tanh(5*(t – 3))))*1.0

d(z11)/d(t) = (exp(t/(1 + 0.067*t)))*((exp(0.01*m)))*m1*z111^n

d(z22)/d(t) = (exp(t/(1 + 0.067*t)))*(1*(z111^s))*((exp(0.01*m)))*m1*z111^n

z22(0) = 0; z11(0) = 0

Consider now the first example of function f3 = f3(**θ**) = if (t < 0.1) then (0) else (1).

Example 4.2:

A_2 = 1; A_3 = 1; p = –1; f3(**θ**) = if (t < 0.1) then (0) else (1) φ_f = 0.1; E_0/E_2 = 0.1 (ratio of initial value of matrix modulus of elasticity and initial value of nanomaterials modulus of elasticity)

Calculated values of DEQ variables

	Variable	Initial value	Minimal value	Maximal value	Final value
1	A2	1.	1.	1.	1.
2	A3	1.	1.	1.	1.
3	E	1.9	0.475	1.9	0.475
4	E1	1.	0.25	1.	0.25
5	f3	0	0	1.	1.
6	f31	5.0E-06	0	0.9999999	0
7	f32	0.999999	0.449329	0.999999	0.449329
8	f33	6.162E-11	6.162E-11	0.9999963	1.015E-05
9	f34	0	0	1.	1.
10	f35	1.25E-06	1.25E-06	1.	1.

11	fi	0.1	0.1	0.1	0.1
12	k	0.1	0.1	0.1	0.1
13	m	4.05E-07	4.05E-07	0.0705907	0.0705907
14	m1	0.0404998	0.0003976	0.0404998	0.0003976
15	m2	1.266E-05	1.266E-05	3014.544	3014.544
16	m21	1.266042	0.545494	1666.482	1666.482
17	n	1.	1.	1.	1.
18	p	–1.	–1.	–1.	–1.
19	s	1.	1.	1.	1.
20	t	1.0E-05	1.0E-05	8.	8.
21	Y1	1.9E-05	1.9E-05	5.592214	3.8
22	z	1.9E-05	–0.0611572	2.213671	0.0143329
23	z1	0	0	0.2350457	0.2350457
24	z11	0	0	0.79102	0.79102
25	z111	1.0E-05	1.0E-05	2.109287	0.5333333
26	z2	0	0	0.3451801	0.3451801
27	z22	0	0	0.6766824	0.6766824
28	z3	0	0	3.962E+13	3.962E+13

Differential equations

1 d(z1)/d(t) = (exp(t/(1 + 0.067*t)))*((exp(0.01*m)))*m1*z^n

2 d(z2)/d(t) = (exp(t/(1 + 0.067*t)))*(A2*(z^s))*((exp(0.01*m)))*m1*z^n

3 d(z3)/d(t) = k*(exp(t/(1 + 0.067*t)))*A3*(f3*(t^p/(1 + 0.1*t^p)))*((exp (0.01*m2)))*m21*z^n

4 d(z11)/d(t) = (exp(t/(1 + 0.067*t)))*((exp(0.01*m)))*m1*z111^n

5 d(z22)/d(t) = (exp(t/(1 + 0.067*t)))*(1*(z111^s))*((exp(0.01*m)))*m1* z111^n

Explicit equations

1 m = (0.0405*t – 0.01126*t^2 + 0.001462*t^3 – 0.00006868*t^4)

2 m1 = (0.0405 – 0.02252*t^1 + 0.004386*t^2 – 0.0002747*t^3)

3 m2 = 1.266*t + 2.12*t^2 – 3.077*t^3 + 1.085*t^4

4 fi = 0.1

5 k = 0.1

6 n = 1.0

7 s = 1.0

8 E = (0.625 – 0.375*(tanh(5*(t – 3))))*(1 – fi)+(1/k)* (0.625 – 0.375*(tanh(5*(t – 5))))*(fi)

9 A2 = 1

10 A3 = 1

11 p = –1

12 z = t*E–(z1 + z2)*((exp(–0.01*m)))–z3*(exp(–0.01*m2))

13 m21 = 1.266 + 4.24 *t – 9.231 *t^2 + 4.34 *t^3

14 f3 = if (t < 0.1) then (0) else (1)

15 f31 = (1/16)*t*(8 – t)

16 f32 = (exp(–0.1*t))

17 f33 = (sin(3.14*t/4))^2

18 f34 = if (t < 8) then (0) else (1)

19 f35 = (1/8)*t^1

20 E1 = (0.625 – 0.375*(tanh(5*(t – 3))))*1.0

21 z111 = t*E1– (z11 + z22)*((exp(–0.01*m)))

22 Y1 = t*E

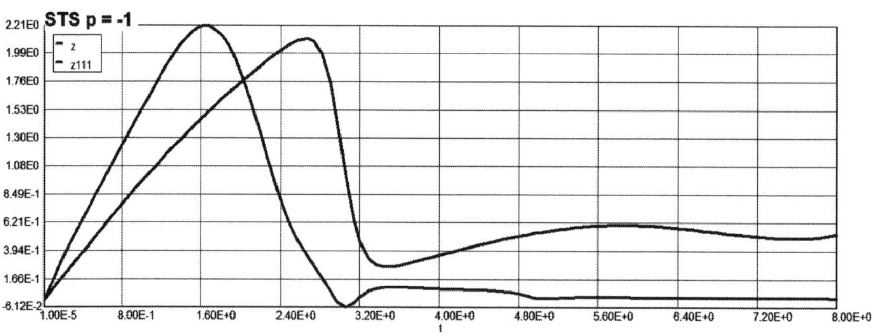

Figure 4.7: STS with nanomaterials creep-z; STS without nanomaterials creep-z111.

Approximation by polynomial (STS with nanomaterials creep)–upper curve

Model: z = a1*t + a2*t^2 + a3*t^3 + a4*t^4 + a5*t^5

Variable	Value
a1	3.759902
a2	–2.592692
a3	0.6510857
a4	–0.0707857
a5	0.0028206

$$\sigma = 3.760\theta - 2.593\theta^2 + 0.651\theta^3 - 0.708\theta^4 + 0.002820\theta^5 \quad (4.21)$$

Approximation by polynomial (STS without nanomaterials creep)–lower curve

Model: z111 = a1*t + a2*t^2 + a3*t^3 + a4*t^4 + a5*t^5

Variable	Value
a1	1.516851
a2	–0.2439001
a3	–0.1240914
a4	0.0339512
a5	–0.0021948

$$\sigma = 1.517\theta - 0.244\theta^2 - 0.124\theta^3 + 0.034\theta^4 - 0.0022\theta^5 \quad (4.22)$$

As can be seen from Fig. 4.9 (z-STS, the diagram of the nanocomposites, and z111-STS diagram of the matrix material without nanoparticles), the maximum value of the nanocomposites creep stress is greater than that of the matrix material, as expected. Moreover, this difference (as follows from the examples below) increases significantly with increasing exponent "p", which in turn characterizes the allowable creep stresses dependence vs. growth rate of nuclei and grain sizes. It should also be noted that the STS diagram of the nanocomposites is shifted to the left in comparison with the "standard homogeneous" diagram of the matrix material. This fact by itself may indicate a less plasticity (greater brittleness) of the nanomaterials at failure.

Example 4.3:

Data: see previous Example 4.2, but p = –2

Calculated values of DEQ variables

	Variable	Initial value	Minimal value	Maximal value	Final value
1	A2	1.	1.	1.	1.
2	A3	1.	1.	1.	1.
3	E	1.9	0.475	1.9	0.475
4	E1	1.	0.25	1.	0.25
5	f3	0	0	1.	1.
6	f31	5.0E-06	0	0.999986	0
7	f32	0.999999	0.449329	0.999999	0.449329
8	f33	6.162E-11	6.162E-11	0.9999999	1.015E-05
9	f34	0	0	1.	1.

10	f35	1.25E-06	1.25E-06	1.	1.
11	fi	0.1	0.1	0.1	0.1
12	k	0.1	0.1	0.1	0.1
13	m	4.05E-07	4.05E-07	0.0705907	0.0705907
14	m1	0.0404998	0.0003976	0.0404998	0.0003976
15	m2	1.266E-05	1.266E-05	3014.544	3014.544
16	m21	1.266042	0.5477732	1666.482	1666.482
17	n	1.	1.	1.	1.
18	p	–2.	–2.	–2.	–2.
19	s	1.	1.	1.	1.
20	t	1.0E-05	1.0E-05	8.	8.
21	Y1	1.9E-05	1.9E-05	5.59327	3.8
22	z	1.9E-05	–0.0151178	2.264155	0.1020796
23	z1	0	0	0.3735657	0.3735657
24	z11	0	0	0.79102	0.79102
25	z111	1.0E-05	1.0E-05	2.110689	0.5333333
26	z2	0	0	0.4421103	0.4421103
27	z22	0	0	0.6766824	0.6766824
28	z3	0	0	3.563E+13	3.563E+13

Differential equations

1 d(z1)/d(t) = (exp(t/(1 + 0.067*t)))*((exp(0.01*m)))*m1*z^n

2 d(z2)/d(t) = (exp(t/(1 + 0.067*t)))*(A2*(z^s))*((exp(0.01*m)))*m1*z^n

3 d(z3)/d(t) = k*(exp(t/(1 + 0.067*t)))*A3*(f3*(t^p/(1 + 0.1*t^p)))*((exp (0.01*m2)))*m21*z^n

4 d(z11)/d(t) = (exp(t/(1 + 0.067*t)))*((exp(0.01*m)))*m1*z111^n

5 d(z22)/d(t) = (exp(t/(1 + 0.067*t)))*(1*(z111^s))*((exp(0.01*m)))*m1* z111^n

Explicit equations

1 m = (0.0405*t – 0.01126*t^2 + 0.001462*t^3 – 0.00006868*t^4)

2 m1 = (0.0405 – 0.02252*t^1 + 0.004386*t^2 – 0.0002747*t^3)

3 m2 = 1.266*t + 2.12*t^2 – 3.077*t^3 + 1.085*t^4

4 fi = 0.1

5 k = 0.1

6 n = 1.0

```
s = 1.0
E = (0.625 – 0.375*(tanh(5*(t – 3))))*(1 – fi)+(1/k)* (0.625 –
  0.375*(tanh(5*(t – 5))))*(fi)
A2 = 1
A3 = 1
p = –2
z = t*E–(z1 + z2)*((exp(–0.01*m)))–z3*(exp(–0.01*m2))
m21 = 1.266 + 4.24 *t – 9.231 *t^2 + 4.34 *t^3
f3 = if (t < 0.1) then (0) else (1)
f31 = (1/16)*t*(8 – t)
f32 = (exp(–0.1*t))
f33 = (sin(3.14*t/4))^2
f34 = if (t < 8) then (0) else (1)
f35 = (1/8)*t^1
E1 = (0.625 – 0.375*(tanh(5*(t – 3))))*1.0
z111 = t*E1 – (z11 + z22)*((exp(–0.01*m)))
Y1 = t*E
```

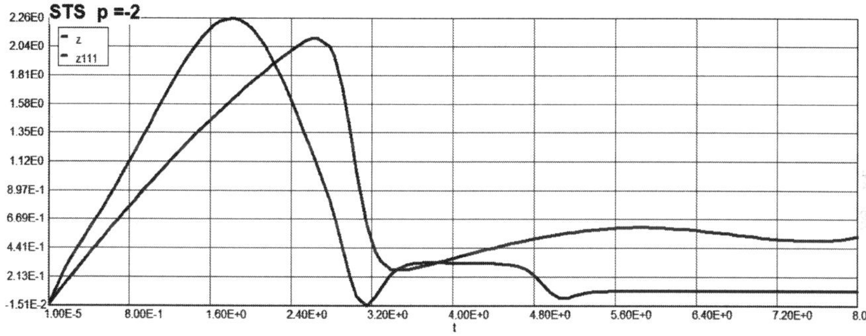

Figure 4.8: Stress-temperature-strain diagram A3 = $f_3(t)$; p = –2.

Model: z = a1*t + a2*t^2 + a3*t^3 + a4*t^4 + a5*t^5

Variable	Value
a1	3.176296
a2	–1.807464
a3	0.366913
a4	–0.0308301
a5	0.0008742

Approximation by polynomial (STS with nanomaterials creep)–upper curve

$$\sigma = 3.176\theta - 1.807\theta^2 + 0.367\theta^3 - 0.0308\theta^4 + 0.00087\theta^5 \quad (4.23)$$

Approximation by polynomial (STS without nanomaterials creep)–lower curve

Model: z111 = a1*t + a2*t^2 + a3*t^3 + a4*t^4 + a5*t^5

Variable	Value
a1	1.522525
a2	–0.2520544
a3	–0.1209322
a4	0.0334823
a5	–0.0021709

$$\sigma = 1.522\theta - 0.252\theta^2 - 0.121\theta^3 + 0.034\theta^4 - 0.0022\theta^5 \quad (4.24)$$

Example 4.4:

Data: see previous Example 4.3, but p = –4

Calculated values of DEQ variables

	Variable	Initial value	Minimal value	Maximal value	Final value
1	A2	1.	1.	1.	1.
2	A3	1.	1.	1.	1.
3	E	1.9	0.475	1.9	0.475
4	E1	1.	0.25	1.	0.25
5	f3	0	0	1.	1.
6	f31	5.0E-06	0	0.999999	0
7	f32	0.999999	0.449329	0.999999	0.449329
8	f33	6.162E-11	6.162E-11	0.9999592	1.015E-05
9	f34	0	0	1.	1.
10	f35	1.25E-06	1.25E-06	1.	1.
11	fi	0.1	0.1	0.1	0.1
12	k	0.1	0.1	0.1	0.1
13	m	4.05E-07	4.05E-07	0.0705907	0.0705907
14	m1	0.0404998	0.0003976	0.0404998	0.0003976
15	m2	1.266E-05	1.266E-05	3014.544	3014.544

16	m21	1.266042	0.5324871	1666.482	1666.482
17	n	1.	1.	1.	1.
18	p	–4.	–4.	–4.	–4.
19	s	1.	1.	1.	1.
20	t	1.0E-05	1.0E-05	8.	8.
21	Y1	1.9E-05	1.9E-05	5.59378	3.8
22	z	1.9E-05	1.9E-05	2.731871	0.756509
23	z1	0	0	1.139514	1.139514
24	z11	0	0	0.79102	0.79102
25	z111	1.0E-05	1.0E-05	2.110478	0.5333333
26	z2	0	0	1.573532	1.573532
27	z22	0	0	0.6766824	0.6766824
28	z3	0	0	4.108E+12	4.108E+12

Differential equations

1 d(z1)/d(t) = (exp(t/(1 + 0.067*t)))*((exp(0.01*m)))*m1*z^n

2 d(z2)/d(t) = (exp(t/(1 + 0.067*t)))*(A2*(z^s))*((exp(0.01*m)))*m1*z^n

3 d(z3)/d(t) = k*(exp(t/(1 + 0.067*t)))*A3*(f3*(t^p/(1 + 0.1*t^p)))*((exp (0.01*m2)))*m21*z^n

4 d(z11)/d(t) = (exp(t/(1 + 0.067*t)))*((exp(0.01*m)))*m1*z111^n

5 d(z22)/d(t) = (exp(t/(1 + 0.067*t)))*(1*(z111^s))*((exp(0.01*m)))*m1* z111^n

Explicit equations

1 m = (0.0405*t – 0.01126*t^2 + 0.001462*t^3 – 0.00006868*t^4)

2 m1 = (0.0405 – 0.02252*t^1 + 0.004386*t^2 – 0.0002747*t^3)

3 m2 = 1.266*t + 2.12*t^2 – 3.077*t^3 + 1.085*t^4

4 fi = 0.1

5 k = 0.1

6 n = 1.0

7 s = 1.0

8 E = (0.625 – 0.375*(tanh(5*(t – 3))))*(1 – fi)+(1/k)* (0.625 – 0.375*(tanh(5*(t – 5))))*(fi)

9 A2 = 1

10 A3 = 1

```
p = –4
z = t*E–(z1 + z2)*((exp(–0.01*m)))–z3*(exp(–0.01*m2))
m21 = 1.266 + 4.24 *t – 9.231 *t^2 + 4.34 *t^3
f3 = if (t < 0.1) then (0) else (1)
f31 = (1/16)*t*(8 – t)
f32 = (exp(–0.1*t))
f33 = (sin(3.14*t/4))^2
f34 = if (t < 8) then (0) else (1)
f35 = (1/8)*t^1
E1 = (0.625 – 0.375*(tanh(5*(t – 3))))*1.0
z111 = t*E1–(z11+z22)*((exp(–0.01*m)))
Y1 = t*E
```

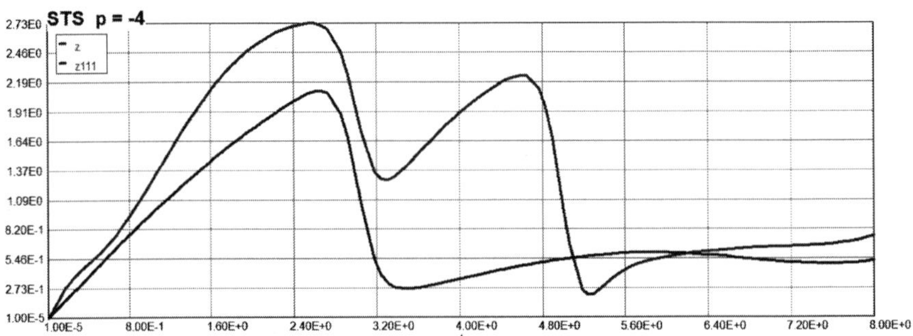

Figure 4.9: Stress-temperature-strain diagram A3 = $f_3(t)$; p = –4.

Model: z = a1*t + a2*t^2 + a3*t^3 + a4*t^4 + a5*t^5

Variable	Value
a1	1.41647
a2	0.1714371
a3	–0.2266863
a4	0.0384787
a5	–0.0019254

Approximation by polynomial (STS with nanomaterials creep)–upper curve

$$\sigma = 1.416\theta + 0.171\theta^2 - 0.227\theta^3 + 0.0385\theta^4 - 0.00192\theta^5 \qquad (4.25)$$

Approximation by polynomial (STS without nanomaterials creep)–lower curve

Model: z111 = a1*t + a2*t^2 + a3*t^3 + a4*t^4 + a5*t^5

Variable	Value
a1	1.527485
a2	–0.2605253
a3	–0.1171783
a4	0.0328656
a5	–0.0021369

$$\sigma = 1.527\theta - 0.261\theta^2 - 0.117\theta^3 + 0.0328\theta^4 - 0.00214\theta^5 \tag{4.26}$$

Example 4.5:

Data: see previous Example 4.5, but p = –6

Calculated values of DEQ variables

	Variable	Initial value	Minimal value	Maximal value	Final value
1	A2	1.	1.	1.	1.
2	A3	1.	1.	1.	1.
3	E	1.9	0.475	1.9	0.475
4	E1	1.	0.25	1.	0.25
5	f3	0	0	1.	1.
6	f31	5.0E-06	0	0.9999463	0
7	f32	0.999999	0.449329	0.999999	0.449329
8	f33	6.162E-11	6.162E-11	0.9997609	1.015E-05
9	f34	0	0	1.	1.
10	f35	1.25E-06	1.25E-06	1.	1.
11	fi	0.1	0.1	0.1	0.1
12	k	0.1	0.1	0.1	0.1
13	m	4.05E-07	4.05E-07	0.0705907	0.0705907
14	m1	0.0404998	0.0003976	0.0404998	0.0003976
15	m2	1.266E-05	1.266E-05	3014.544	3014.544
16	m21	1.266042	0.5344467	1666.482	1666.482

17	n	1.	1.	1.	1.
18	p	–6.	–6.	–6.	–6.
19	s	1.	1.	1.	1.
20	t	1.0E-05	1.0E-05	8.	8.
21	Y1	1.9E-05	1.9E-05	5.596162	3.8
22	z	1.9E-05	–0.1229624	3.263817	0.6610635
23	z1	0	0	1.047441	1.047441
24	z11	0	0	0.79102	0.79102
25	z111	1.0E-05	1.0E-05	2.108342	0.5333333
26	z2	0	0	2.089159	2.089159
27	z22	0	0	0.6766824	0.6766824
28	z3	0	0	5.624E+10	5.624E+10

Differential equations

1 d(z1)/d(t) = (exp(t/(1 + 0.067*t)))*((exp(0.01*m)))*m1*z^n

2 d(z2)/d(t) = (exp(t/(1 + 0.067*t)))*(A2*(z^s))*((exp(0.01*m)))*m1*z^n

3 d(z3)/d(t) = k*(exp(t/(1 + 0.067*t)))*A3*(f3*(t^p/(1 + 0.1*t^p)))*((exp(0.01*m2)))*m21*z^n

4 d(z11)/d(t) = (exp(t/(1 + 0.067*t)))*((exp(0.01*m)))*m1*z111^n

5 d(z22)/d(t) = (exp(t/(1 + 0.067*t)))*(1*(z111^s))*((exp(0.01*m)))*m1*z111^n

Explicit equations

1 m = (0.0405*t – 0.01126*t^2 + 0.001462*t^3 – 0.00006868*t^4)

2 m1 = (0.0405 – 0.02252*t^1 + 0.004386*t^2 – 0.0002747*t^3)

3 m2 = 1.266*t + 2.12*t^2 – 3.077*t^3 + 1.085*t^4

4 fi = 0.1

5 k = 0.1

6 n = 1.0

7 s = 1.0

8 E = (0.625 – 0.375*(tanh(5*(t – 3))))*(1 – fi)+(1/k)* (0.625 – 0.375*(tanh(5*(t – 5))))*(fi)

9 A2 = 1

10 A3 = 1

11 p = –6
12 z = t*E–(z1 + z2)*((exp(–0.01*m)))–z3*(exp(–0.01*m2))
13 m21 = 1.266 + 4.24 *t – 9.231 *t^2 + 4.34 *t^3
14 f3 = if (t < 0.1) then (0) else (1)
15 f31 = (1/16)*t*(8 – t)
16 f32 = (exp(–0.1*t))
17 f33 = (sin(3.14*t/4))^2
18 f34 = if (t < 8) then (0) else (1)
19 f35 = (1/8)*t^1
20 E1 = (0.625 – 0.375*(tanh(5*(t – 3))))*1.0
21 z111 = t*E1–(z11+z22)*((exp(–0.01*m)))
22 Y1 = t*E

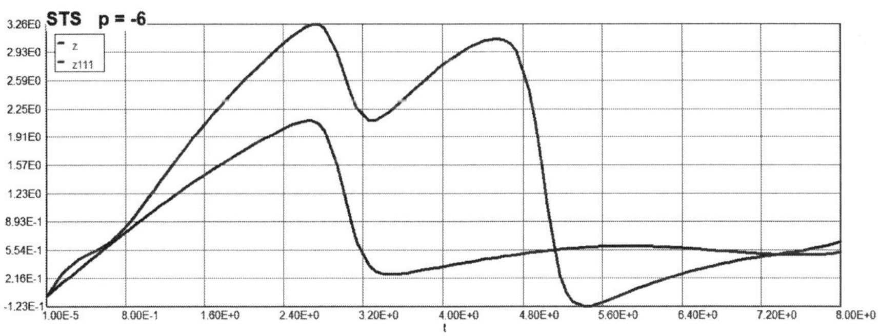

Figure 4.10: Stress-temperature-strain diagram A3 = $f_3(t)$; p = –6.

Approximation by polynomial (STS with nanomaterials creep)–upper curve

Model: z = a1*t + a2*t^2 + a3*t^3 + a4*t^4 + a5*t^5

Variable	Value
a1	0.3531991
a2	1.275595
a3	–0.5349269
a4	0.0700345
a5	–0.0029447

$$\sigma = 0.353\theta + 1.276\theta^2 - 0.535\theta^3 + 0.070\theta^4 - 0.00295\theta^5 \quad (4.27)$$

Approximation by polynomial (STS without nanomaterials creep)–lower curve

Model: z111 = a1*t + a2*t^2 + a3*t^3 + a4*t^4 + a5*t^5

Variable	Value
a1	1.534038
a2	–0.2666317
a3	–0.1153148
a4	0.0326275
a5	–0.0021258

$$\sigma = 1.527\theta - 0.261\theta^2 - 0.117\theta^3 + 0.0328\theta^4 - 0.00214\theta^5 \quad (4.28)$$

The same as in Example 4.5

Example 4.6:

Data: see previous Example 4.5, but p = –8

Calculated values of DEQ variables

	Variable	Initial value	Minimal value	Maximal value	Final value
1	A2	1.	1.	1.	1.
2	A3	1.	1.	1.	1.
3	E	1.9	0.475	1.9	0.475
4	E1	1.	0.25	1.	0.25
5	f3	0	0	1.	1.
6	f31	5.0E-06	0	0.9999442	0
7	f32	0.999999	0.449329	0.999999	0.449329
8	f33	6.162E-11	6.162E-11	0.9999681	1.015E-05
9	f34	0	0	1.	1.
10	f35	1.25E-06	1.25E-06	1.	1.
11	fi	0.1	0.1	0.1	0.1
12	k	0.1	0.1	0.1	0.1
13	m	4.05E-07	4.05E-07	0.0705907	0.0705907
14	m1	0.0404998	0.0003976	0.0404998	0.0003976
15	m2	1.266E-05	1.266E-05	3014.544	3014.544
16	m21	1.266042	0.5345736	1666.482	1666.482

17	n	1.	1.	1.	1.
18	p	–8.	–8.	–8.	–8.
19	s	1.	1.	1.	1.
20	t	1.0E-05	1.0E-05	8.	8.
21	Y1	1.9E-05	1.9E-05	5.596162	3.8
22	z	1.9E-05	–0.1902135	3.376297	0.6371068
23	z1	0	0	1.01189	1.01189
24	z11	0	0	0.79102	0.79102
25	z111	1.0E-05	1.0E-05	2.108427	0.5333333
26	z2	0	0	2.153167	2.153167
27	z22	0	0	0.6766824	0.6766824
28	z3	0	0	8.583E+08	8.583E+08

Differential equations

1 d(z1)/d(t) = (exp(t/(1 + 0.067*t)))*((exp(0.01*m)))*m1*z^n

2 d(z2)/d(t) = (exp(t/(1 + 0.067*t)))*(A2*(z^s))*((exp(0.01*m)))*m1*z^n

3 d(z3)/d(t) = k*(exp(t/(1 + 0.067*t)))*A3*(f3*(t^p/(1 + 0.1*t^p)))*((exp (0.01*m2)))*m21*z^n

4 d(z11)/d(t) = (exp(t/(1 + 0.067*t)))*((exp(0.01*m)))*m1*z111^n

5 d(z22)/d(t) = (exp(t/(1 + 0.067*t)))*(1*(z111^s))*((exp(0.01*m)))*m1* z111^n

Explicit equations

1 m = (0.0405*t – 0.01126*t^2 + 0.001462*t^3 – 0.00006868*t^4)

2 m1 = (0.0405 – 0.02252*t^1 + 0.004386*t^2 – 0.0002747*t^3)

3 m2 = 1.266*t + 2.12*t^2 – 3.077*t^3 + 1.085*t^4

4 fi = 0.1

5 k = 0.1

6 n = 1.0

7 s = 1.0

8 E = (0.625 – 0.375*(tanh(5*(t – 3))))*(1 – fi)+(1/k)* (0.625 – 0.375*(tanh(5*(t – 5))))*(fi)

9 A2 = 1

10 A3 = 1

11 p = –8

12 z = t*E–(z1+z2)*((exp(–0.01*m)))–z3*(exp(–0.01*m2))
13 m21 = 1.266 + 4.24 *t – 9.231 *t^2 + 4.34 *t^3
14 f3 = if (t < 0.1) then (0) else (1)
15 f31 = (1/16)*t*(8 – t)
16 f32 = (exp(–0.1*t))
17 f33 = (sin(3.14*t/4))^2
18 f34 = if (t < 8) then (0) else (1)
19 f35 = (1/8)*t^1
20 E1 = (0.625 – 0.375*(tanh(5*(t – 3))))*1.0
21 z111 = t*E1–(z11 + z22)*((exp(–0.01*m)))
22 Y1 = t*E

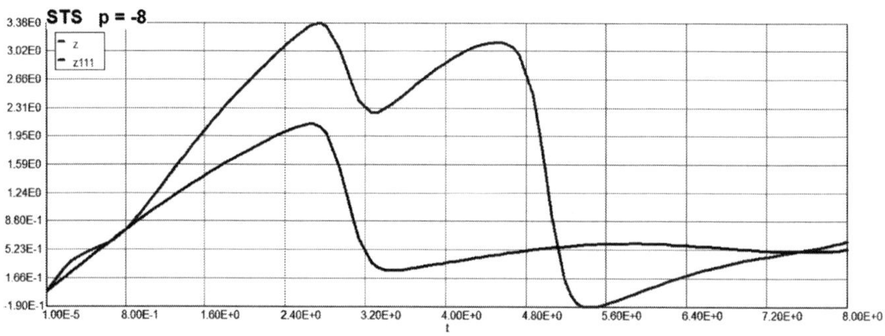

Figure 4.11: Stress-temperature-strain diagram A3 = $f_3(t)$; p = –8.

Approximation by polynomial (STS with nanomaterials creep)–upper curve

Model: z = a1*t + a2*t^2 + a3*t^3 + a4*t^4 + a5*t^5

Variable	Value
a1	0.054053
a2	1.589133
a3	–0.6343206
a4	0.0824429
a5	–0.0034822

$$\sigma = 0.054\theta + 1.5896\theta^2 - 0.634\theta^3 + 0.0824\theta^4 - 0.00348\theta^5 \quad (4.29)$$

Approximation by polynomial (STS without nanomaterials creep)–lower curve

The same as in Example 4.5

Example 4.7:

Data: see previous Example 4.5, but p = –8

Calculated values of DEQ variables

	Variable	Initial value	Minimal value	Maximal value	Final value
1	A2	1.	1.	1.	1.
2	A3	1.	1.	1.	1.
3	E	1.9	0.475	1.9	0.475
4	E1	1.	0.25	1.	0.25
5	f3	0	0	1.	1.
6	f31	5.0E-06	0	0.9999383	0
7	f32	0.999999	0.449329	0.999999	0.449329
8	f33	6.162E-11	6.162E-11	0.9999058	1.015E-05
9	f34	0	0	1.	1.
10	f35	1.25E-06	1.25E-06	1.	1.
11	fi	0.1	0.1	0.1	0.1
12	k	0.1	0.1	0.1	0.1
13	m	4.05E-07	4.05E-07	0.0705907	0.0705907
14	m1	0.0404998	0.0003976	0.0404998	0.0003976
15	m2	1.266E-05	1.266E-05	3014.544	3014.544
16	m21	1.266042	0.5349337	1666.482	1666.482
17	n	1.	1.	1.	1.
18	p	–10.	–10.	–10.	–10.
19	s	1.	1.	1.	1.
20	t	1.0E-05	1.0E-05	8.	8.
21	Y1	1.9E-05	1.9E-05	5.596146	3.8
22	z	1.9E-05	–0.1920421	3.387676	0.6364317
23	z1	0	0	1.009911	1.009911
24	z11	0	0	0.79102	0.79102
25	z111	1.0E-05	1.0E-05	2.108651	0.5333333
26	z2	0	0	2.15589	2.15589
27	z22	0	0	0.6766824	0.6766824
28	z3	0	0	1.361E+07	1.361E+07

Differential equations

```
1 d(z1)/d(t) = (exp(t/(1 + 0.067*t)))*((exp(0.01*m)))*m1*z^n
2 d(z2)/d(t) = (exp(t/(1 + 0.067*t)))*(A2*(z^s))*((exp(0.01*m)))*m1*z^n
3 d(z3)/d(t) = k*(exp(t/(1 + 0.067*t)))*A3*(f3*(t^p/(1 + 0.1*t^p)))*((exp
  (0.01*m2)))*m21*z^n
4 d(z11)/d(t) = (exp(t/(1 + 0.067*t)))*((exp(0.01*m)))*m1*z111^n
5 d(z22)/d(t) = (exp(t/(1 + 0.067*t)))*(1*(z111^s))*((exp(0.01*m)))*m1*
  z111^n
```

Explicit equations

```
1 m = (0.0405*t – 0.01126*t^2 + 0.001462*t^3 – 0.00006868*t^4)
2 m1 = (0.0405 – 0.02252*t^1 + 0.004386*t^2 – 0.0002747*t^3)
3 m2 = 1.266*t + 2.12*t^2 – 3.077*t^3 + 1.085*t^4
4 fi = 0.1
5 k = 0.1
6 n = 1.0
7 s = 1.0
8 E = (0.625 – 0.375*(tanh(5*(t – 3))))*(1 – fi)+(1/k)*
  (0.625 – 0.375*(tanh(5*(t – 5))))*(fi)
9 A2 = 1
10 A3 = 1
11 p = –10
12 z = t*E–(z1 + z2)*((exp(–0.01*m)))–z3*(exp(–0.01*m2))
13 m21 = 1.266 + 4.24 *t – 9.231 *t^2 + 4.34 *t^3
14 f3 = if (t < 0.1) then (0) else (1)
15 f31 = (1/16)*t*(8 – t)
16 f32 = (exp(–0.1*t))
17 f33 = (sin(3.14*t/4))^2
18 f34 = if (t < 8) then (0) else (1)
19 f35 = (1/8)*t^1
20 E1 = (0.625 – 0.375*(tanh(5*(t – 3))))*1.0
21 z111 = t*E1–(z11 + z22)*((exp(–0.01*m)))
22 Y1 = t*E
```

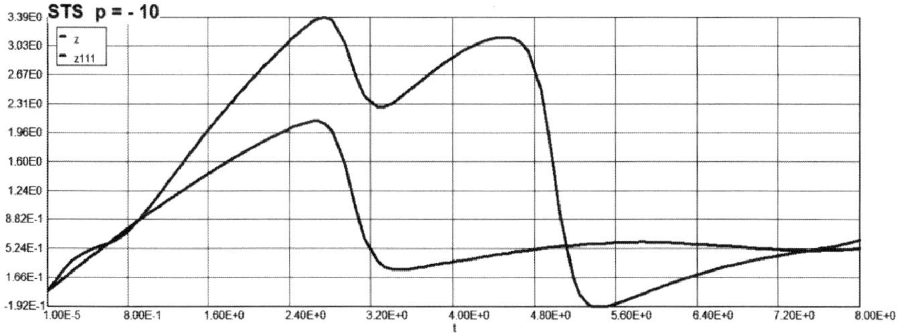

Figure 4.12: Stress-temperature-strain diagram A3 = $f_3(t)$; p = –10.

Approximation by polynomial (STS with nanomaterials creep)–upper curve

Model: z = a1*t + a2*t^2 + a3*t^3 + a4*t^4 + a5*t^5

Variable	Value
a1	–0.0383291
a2	1.667855
a3	–0.6571278
a4	0.0852049
a5	–0.0036022

$$\sigma = - 0.0383\theta + 1.668\theta^2 - 0.657\theta^3 + 0.08524\theta^4 - 0.0036\theta^5 \quad (4.30)$$

p	–1	–2	–4	–6	–8	–10
σ_{max}	2.214	2.264	2.732	3.364	3.376	3.388

Model: y = a0 + a1*x + a2*x^2 + a3*x^3 + a4*x^4

Variable	Value
a0	2.228993
a1	–0.2236186
a2	0.1635777
a3	–0.021152
a4	0.0008178

$$\sigma_{max} = 2.229 - 0.224p + 0.164p^2 - 0.0211p^3 + 0.000818p^4 \quad (4.31)$$

Finally, the effect of the inverse function of nanoparticles grows rate on maximum value of creep stresses are presented graphically in Fig. 4.13 below.

$0 < p < 10$

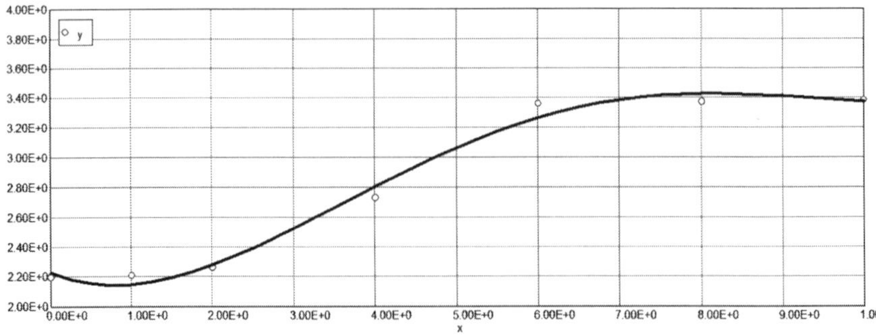

Figure 4.13: Maximum value of creep stresses vs. nanoparticles grows rate p.

Consider now the influence of other types of the first part of the function f3 (the transformation process of the reactant into final product of nanocomposite material) on the qualitative and quantitative changes in the creep stress-strain (STS) diagrams of nanocomposites. As examples, consider a linearly increasing, exponentially decreasing and step functions f3. In all the following examples, the parameter p will be fixed ($p = -4$), approximately the average of the interval $-10 < p < 0$.

Example 4.8:

Data: see previous Example 4.2, but $p = -4$ and f3 = f31

Calculated values of DEQ variables

	Variable	Initial value	Minimal value	Maximal value	Final value
1	A2	1.	1.	1.	1.
2	A3	1.	1.	1.	1.
3	E	1.9	0.475	1.9	0.475
4	E1	1.	0.25	1.	0.25
5	f3	0	0	1.	1.
6	f31	5.0E-06	0	0.9998138	0
7	f32	0.999999	0.449329	0.999999	0.449329
8	f33	6.162E-11	6.162E-11	0.9999808	1.015E-05
9	f34	0	0	1.	1.
10	f35	1.25E-06	1.25E-06	1.	1.
11	fi	0.1	0.1	0.1	0.1
12	k	0.1	0.1	0.1	0.1
13	m	4.05E-07	4.05E-07	0.0705907	0.0705907

14	m1	0.0404998	0.0003976	0.0404998	0.0003976
15	m2	1.266E-05	1.266E-05	3014.544	3014.544
16	m21	1.266042	0.5437351	1666.482	1666.482
17	n	1.	1.	1.	1.
18	p	–4.	–4.	–4.	–4.
19	s	1.	1.	1.	1.
20	t	1.0E-05	1.0E-05	8.	8.
21	Y1	1.9E-05	1.9E-05	5.592254	3.8
22	z	1.9E-05	1.9E-05	3.003057	0.8832258
23	z1	0	0	1.166819	1.166819
24	z11	0	0	0.79102	0.79102
25	z111	1.0E-05	1.0E-05	2.110648	0.5333333
26	z2	0	0	1.740674	1.740674
27	z22	0	0	0.6766824	0.6766824
28	z3	0	0	1.401E+11	1.401E+11

Differential equations

1 d(z1)/d(t) = (exp(t/(1 + 0.067*t)))*((exp(0.01*m)))*m1*z^n
2 d(z2)/d(t) = (exp(t/(1 + 0.067*t)))*(A2*(z^s))*((exp(0.01*m)))*m1*z^n
3 d(z3)/d(t) = k*(exp(t/(1 + 0.067*t)))*A3*(f31*(t^p/(1 + 0.1*t^p)))*((exp(0.01*m2)))*m21*z^n
4 d(z11)/d(t) = (exp(t/(1 + 0.067*t)))*((exp(0.01*m)))*m1*z111^n
5 d(z22)/d(t) = (exp(t/(1 + 0.067*t)))*(1*(z111^s))*((exp(0.01*m)))*m1*z111^n

Explicit equations

1 m = (0.0405*t – 0.01126*t^2 + 0.001462*t^3 – 0.00006868*t^4)
2 m1 = (0.0405 – 0.02252*t^1 + 0.004386*t^2 – 0.0002747*t^3)
3 m2 = 1.266*t + 2.12*t^2 – 3.077*t^3 + 1.085*t^4
4 fi = 0.1
5 k = 0.1
6 n = 1.0
7 s = 1.0
8 E = (0.625 – 0.375*(tanh(5*(t – 3))))*(1 – fi)+(1/k)* (0.625 – 0.375*(tanh(5*(t – 5))))*(fi)

9 A2 = 1
10 A3 = 1
11 p = –4
12 z = t*E–(z1 + z2)*((exp(–0.01*m)))–z3*(exp(–0.01*m2))
13 m21 = 1.266 + 4.24 *t – 9.231 *t^2 + 4.34 *t^3
14 f3 = if (t < 0.1) then (0) else (1)
15 f31 = (1/16)*t*(8 – t)
16 f32 = (exp(–0.1*t))
17 f33 = (sin(3.14*t/4))^2
18 f34 = if (t < 8) then (0) else (1)
19 f35 = (1/8)*t^1
20 E1 = (0.625 – 0.375*(tanh(5*(t – 3))))*1.0
21 z111 = t*E1–(z11 + z22)*((exp(–0.01*m)))
22 Y1 = t*E

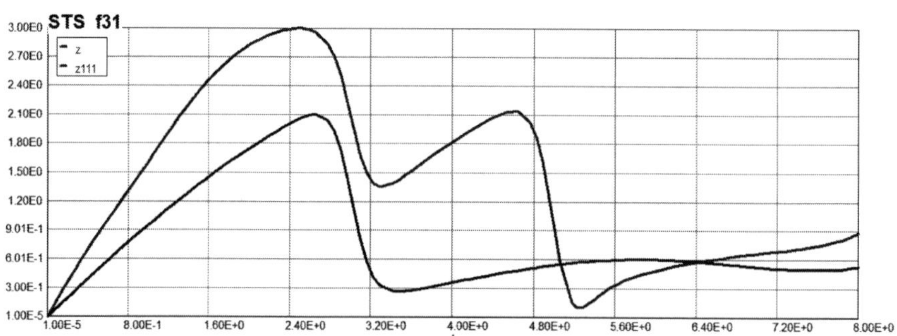

Figure 4.14: Stress-temperature-strain diagram A3 = f31(t); p = –4.

Approximation by polynomial (STS with nanomaterials creep)–upper curve

Model: z = a1*t + a2*t^2 + a3*t^3 + a4*t^4 + a5*t^5

Variable	Value
a1	2.09275
a2	–0.1992744
a3	–0.1610781
a4	0.034274
a5	–0.0018632

$$\sigma = 2.093\theta - 0.199\theta^2 - 0.161\theta^3 + 0.0343\theta^4 - 0.00186\theta^5 \quad (4.32)$$

Comparing maximum stress with similar Example 4.4 we have: σ_{max} = 2.731 and σ_{max} = 3.0.

Consider now the effect of functions f3 on maximum creep stress values. Let's assume now that p = –4. Each example in this series has comparison with the composite that doesn't have the nanomaterials enhancement (maximum allowable creep stresses are compared).

Example 4.9:

Data: see previous Example 4.2, but p = –4 and f3 = f32

Calculated values of DEQ variables

	Variable	Initial value	Minimal value	Maximal value	Final value
1	A2	1.	1.	1.	1.
2	A3	1.	1.	1.	1.
3	E	1.9	0.475	1.9	0.475
4	E1	1.	0.25	1.	0.25
5	f3	0	0	1.	1.
6	f31	5.0E-06	0	0.9999608	0
7	f32	0.999999	0.449329	0.999999	0.449329
8	f33	6.162E-11	6.162E-11	0.9996711	1.015E-05
9	f34	0	0	1.	1.
10	f35	1.25E-06	1.25E-06	1.	1.
11	fi	0.1	0.1	0.1	0.1
12	k	0.1	0.1	0.1	0.1
13	m	4.05E-07	4.05E-07	0.0705907	0.0705907
14	m1	0.0404998	0.0003976	0.0404998	0.0003976
15	m2	1.266E-05	1.266E-05	3014.544	3014.544
16	m21	1.266042	0.5371078	1666.482	1666.482
17	n	1.	1.	1.	1.
18	p	–4.	–4.	–4.	–4.
19	s	1.	1.	1.	1.
20	t	1.0E-05	1.0E-05	8.	8.
21	Y1	1.9E-05	1.9E-05	5.595651	3.8
22	z	1.9E-05	1.9E-05	2.858508	0.7609777
23	z1	0	0	1.157747	1.157747

24	z11	0	0	0.79102	0.79102
25	z111	1.0E-05	1.0E-05	2.10437	0.5333333
26	z2	0	0	1.732727	1.732727
27	z22	0	0	0.6766824	0.6766824
28	z3	0	0	1.861E+12	1.861E+12

Differential equations

1 d(z1)/d(t) = (exp(t/(1 + 0.067*t)))*((exp(0.01*m)))*m1*z^n

2 d(z2)/d(t) = (exp(t/(1 + 0.067*t)))*(A2*(z^s))*((exp(0.01*m)))*m1*z^n

3 d(z3)/d(t) = k*(exp(t/(1 + 0.067*t)))*A3*(f32*(t^p/(1 + 0.1*t^p)))*((exp (0.01*m2)))*m21*z^n

4 d(z11)/d(t) = (exp(t/(1 + 0.067*t)))*((exp(0.01*m)))*m1*z111^n

5 d(z22)/d(t) = (exp(t/(1 + 0.067*t)))*(1*(z111^s))*((exp(0.01*m)))*m1* z111^n

Explicit equations

1 m = (0.0405*t – 0.01126*t^2 + 0.001462*t^3 – 0.00006868*t^4)

2 m1 = (0.0405 – 0.02252*t^1 + 0.004386*t^2 – 0.0002747*t^3)

3 m2 = 1.266*t + 2.12*t^2 – 3.077*t^3 + 1.085*t^4

4 fi = 0.1

5 k = 0.1

6 n = 1.0

7 s = 1.0

8 E = (0.625 – 0.375*(tanh(5*(t – 3))))*(1 – fi)+(1/k)* (0.625 – 0.375*(tanh(5*(t – 5))))*(fi)

9 A2 = 1

10 A3 = 1

11 p = –4

12 z = t*E–(z1 + z2)*((exp(–0.01*m)))–z3*(exp(–0.01*m2))

13 m21 = 1.266 + 4.24 *t – 9.231 *t^2 + 4.34 *t^3

14 f3 = if (t < 0.1) then (0) else (1)

15 f31 = (1/16)*t*(8 – t)

16 f32 = (exp(–0.1*t))

17 f33 = (sin(3.14*t/4))^2

18 f34 = if (t < 8) then (0) else (1)

19 f35 = (1/8)*t^1

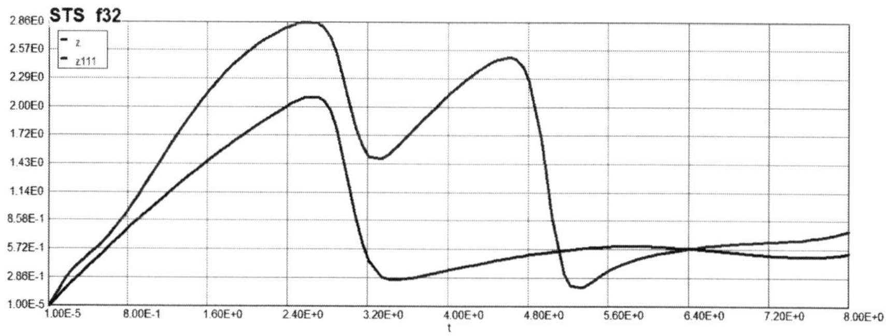

Figure 4.15: Stress-temperature-strain diagram A3 = f32(t); p = –4.

20 E1 = (0.625 – 0.375*(tanh(5*(t – 3))))*1.0

21 z111 = t*E1–(z11+z22)*((exp(–0.01*m)))

22 Y1 = t*E

Approximation by polynomial (STS with nanomaterials creep)–upper curve

Model: z = a1*t + a2*t^2 + a3*t^3 + a4*t^4 + a5*t^5

Variable	Value
a1	1.271564
a2	0.336782
a3	–0.2683125
a4	0.041755
a5	–0.0019698

$$\sigma = 1.272\theta + 0.336\theta^2 - 0.268\theta^3 + 0.0418\theta^4 - 0.00197\theta^5 \tag{4.33}$$

Comparing maximum stress with similar Example 4.4 we have: σ_{max} = 2.731 and σ_{max} = 2.8.

Example 4.10:

Data: see previous Example 4.2, but p = –4 and f3 = f33

Calculated values of DEQ variables

	Variable	Initial value	Minimal value	Maximal value	Final value
1	A2	1.	1.	1.	1.
2	A3	1.	1.	1.	1.
3	E	1.9	0.475	1.9	0.475

4	E1	1.	0.25	1.	0.25
5	f3	0	0	1.	1.
6	f31	5.0E-06	0	0.9999787	0
7	f32	0.999999	0.449329	0.999999	0.449329
8	f33	6.162E-11	6.162E-11	0.9999946	1.015E-05
9	f34	0	0	1.	1.
10	f35	1.25E-06	1.25E-06	1.	1.
11	fi	0.1	0.1	0.1	0.1
12	k	0.1	0.1	0.1	0.1
13	m	4.05E-07	4.05E-07	0.0705907	0.0705907
14	m1	0.0404998	0.0003976	0.0404998	0.0003976
15	m2	1.266E-05	1.266E-05	3014.544	3014.544
16	m21	1.266042	0.536555	1666.482	1666.482
17	n	1.	1.	1.	1.
18	p	–4.	–4.	–4.	–4.
19	s	1.	1.	1.	1.
20	t	1.0E-05	1.0E-05	8.	8.
21	Y1	1.9E-05	1.9E-05	5.594902	3.8
22	z	1.9E-05	–0.109324	3.049647	0.7464185
23	z1	0	0	1.055741	1.055741
24	z11	0	0	0.79102	0.79102
25	z111	1.0E-05	1.0E-05	2.102151	0.5333333
26	z2	0	0	1.998433	1.998433
27	z22	0	0	0.6766824	0.6766824
28	z3	0	0	1.931E+10	1.931E+10

Differential equations

1 d(z1)/d(t) = (exp(t/(1 + 0.067*t)))*((exp(0.01*m)))*m1*z^n

2 d(z2)/d(t) = (exp(t/(1 + 0.067*t)))*(A2*(z^s))*((exp(0.01*m)))*m1*z^n

3 d(z3)/d(t) = k*(exp(t/(1 + 0.067*t)))*A3*(f33*(t^p/(1 + 0.1*t^p)))*((exp (0.01*m2)))*m21*z^n

4 d(z11)/d(t) = (exp(t/(1 + 0.067*t)))*((exp(0.01*m)))*m1*z111^n

5 d(z22)/d(t) = (exp(t/(1 + 0.067*t)))*(1*(z111^s))*((exp(0.01*m)))*m1* z111^n

Explicit equations

1 m = (0.0405*t – 0.01126*t^2 + 0.001462*t^3 – 0.00006868*t^4)

2 m1 = (0.0405 – 0.02252*t^1 + 0.004386*t^2 – 0.0002747*t^3)

3 m2 = 1.266*t + 2.12*t^2 – 3.077*t^3 + 1.085*t^4

4 fi = 0.1

5 k = 0.1

6 n = 1.0

7 s = 1.0

8 E = (0.625 – 0.375*(tanh(5*(t – 3))))*(1 – fi)+(1/k)*
(0.625 – 0.375*(tanh(5*(t – 5))))*(fi)

9 A2 = 1

10 A3 = 1

11 p = –4

12 z = t*E–(z1+z2)*((exp(–0.01*m)))–z3*(exp(–0.01*m2))

13 m21 = 1.266 + 4.24 *t – 9.231 *t^2 + 4.34 *t^3

14 f3 = if (t < 0.1) then (0) else (1)

15 f31 = (1/16)*t*(8 – t)

16 f32 = (exp(–0.1*t))

17 f33 = (sin(3.14*t/4))^2

18 f34 = if (t < 8) then (0) else (1)

19 f35 = (1/8)*t^1

20 E1 = (0.625 – 0.375*(tanh(5*(t – 3))))*1.0

21 z111 = t*E1–(z11+z22)*((exp(–0.01*m)))

22 Y1 = t*E

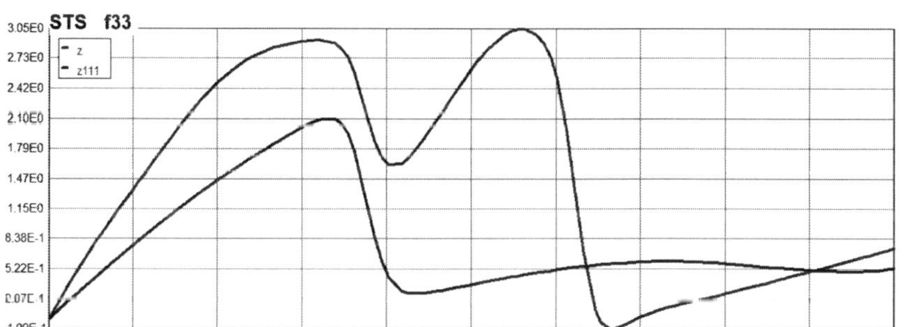

Figure 4.16: Stress-temperature-strain diagram A3 = f33(t); p = –4.

Approximation by polynomial (STS with nanomaterials creep)–upper curve

Model: z = a1*t + a2*t^2 + a3*t^3 + a4*t^4 + a5*t^5

Variable	Value
a1	1.975053
a2	–0.2244647
a3	–0.0754186
a4	0.0118681
a5	–0.0003151

$$\sigma = 1.975\theta - 0.224\theta^2 - 0.0754\theta^3 + 0.0119\theta^4 - 0.000315\theta^5 \quad (4.34)$$

Comparing maximum stress with similar Example 4.4 we have: σ_{max} = 2.731 and σ_{max} = 3.05.

Example 4.11:

Data: see previous Example 4.7, but p = –4 and f3 = f34

In this case obviously z ≡ z111, therefore it is more important to consider the case when the function f34 is a discontinuous function. For example: *f34 = f (θ) = if (θ* < 4) then (0) else (1)*, where $\theta_{gm} < \theta^* < \theta_{gf}$.

Data: see previous Example 4.7, but p = –4 and f3 = f34 = f (θ) = if (θ^* < 4) then (0) else (1). In this case:

$\theta < \theta^* = 4$ nanomaterials are not affected by high temperature load and the modulus of elasticity of the whole composition is computed based on the "mixture rule"; the third integrant of Equation 4.16 is ***not*** zero.

$\theta > \theta^*$ Total modulus of elasticity is computed based on the "mixture rule" and the effect of function f34; the third integrant of the third integral of Equation 4.16 is zero. The nanomaterials are suddenly affected by high temperature load and the system is moved to another viscous (semi liquid) phase.

θ_{gm} = 3-dimensional transitional temperature of the matrix; θ_{gf} = 5-dimensional transitional temperature of the nanomaterials.

Calculated values of DEQ variables

	Variable	Initial value	Minimal value	Maximal value	Final value
1	A2	1.	1.	1.	1.
2	A3	1.	1.	1.	1.
3	E	1.9	0.225	1.9	0.225

4	E1	1.	0.25	1.	0.25
5	f3	0	0	1.	1.
6	f31	5.0E-06	0	1.	0
7	f32	0.999999	0.449329	0.999999	0.449329
8	f33	6.162E-11	6.162E-11	0.9999427	1.015E-05
9	f34	1.	0	1.	0
10	f35	1.25E-06	1.25E-06	1.	1.
11	fi	0.1	0.1	0.1	0.1
12	k	0.1	0.1	0.1	0.1
13	m	4.05E-07	4.05E-07	0.0705907	0.0705907
14	m1	0.0404998	0.0003976	0.0404998	0.0003976
15	m2	1.266E-05	1.266E-05	3014.544	3014.544
16	m21	1.266042	0.5347826	1666.482	1666.482
17	n	1.	1.	1.	1.
18	p	–4.	–4.	–4.	–4.
19	s	1.	1.	1.	1.
20	t	1.0E-05	1.0E-05	8.	8.
21	Y1	1.9E-05	1.9E-05	5.090415	1.8
22	z	1.9E-05	–2.094176	2.726521	0.3475719
23	z1	0	0	0.4604933	0.4604933
24	z11	0	0	0.79102	0.79102
25	z111	1.0E-05	1.0E-05	2.10856	0.5333333
26	z2	0	0	0.9929605	0.9929605
27	z22	0	0	0.6766824	0.6766824
28	z3	0	0	5.774401	5.774401

Differential equations

1 d(z1)/d(t) = (exp(t/(1 + 0.067*t)))*((exp(0.01*m)))*m1*z^n

2 d(z2)/d(t) = (exp(t/(1 + 0.067*t)))*(A2*(z^s))*((exp(0.01*m)))*m1*z^n

3 d(z3)/d(t) = k*(exp(t/(1 + 0.067*t)))*A3*(f34*(t^p/(1 + 0.1*t^p)))*((exp (0.01*m2)))*m21*z^n

4 d(z11)/d(t) = (exp(t/(1 + 0.067*t)))*((exp(0.01*m)))*m1*z111^n

5 d(z22)/d(t) = (exp(t/(1 + 0.067*t)))*(1*(z111^s))*((exp(0.01*m)))*m1* z111^n

Explicit equations

1 m = (0.0405*t – 0.01126*t^2 + 0.001462*t^3 – 0.00006868*t^4)

2 m1 = (0.0405 – 0.02252*t^1 + 0.004386*t^2 – 0.0002747*t^3)

3 f34 = if (t < 4) then (1) else (0)

4 fi = 0.1

5 k = 0.1

6 n = 1.0

7 s = 1.0

8 E = (0.625 – 0.375*(tanh(5*(t – 3))))*(1 – fi)+(1/k)*
(0.625 – 0.375*(tanh(5*(t – 5))))*(fi)*f34

9 A2 = 1

10 A3 = 1

11 p = –4

12 m2 = 1.266*t + 2.12*t^2 – 3.077*t^3 + 1.085*t^4

13 m21 = 1.266 + 4.24 *t – 9.231 *t^2 + 4.34 *t^3

14 f3 = if (t < 0.1) then (0) else (1)

15 f31 = (1/16)*t*(8 – t)

16 f32 = (exp(–0.1*t))

17 f33 = (sin(3.14*t/4))^2

18 z = t*E–(z1 + z2)*((exp(–0.01*m)))–z3*(exp(–0.01*m2))

19 f35 = (1/8)*t^1

20 E1 = (0.625 – 0.375*(tanh(5*(t – 3))))*1.0

21 z111 = t*E1–(z11 + z22)*((exp(–0.01*m)))

22 Y1 = t*E

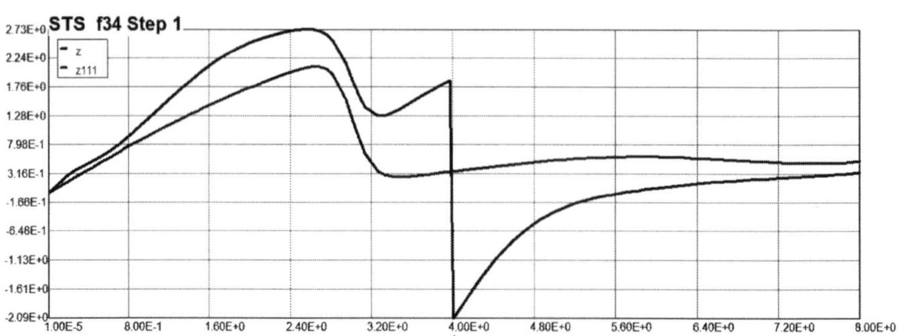

Figure 4.17: Stress-temperature-strain diagram A3 = f34(t); p = –4; $\theta^* = 4$.

Model: z = a1*t + a2*t^2 + a3*t^3 + a4*t^4 + a5*t^5

Variable	Value
a1	0.8935673
a2	1.257356
a3	–0.804119
a4	0.1416623
a5	–0.0078288

$$\sigma = 0.894\theta + 1.257\theta^2 - 0.804\theta^3 + 0.142\theta^4 - 0.00783\theta^5 \quad (4.35)$$

Or:

Case $\theta^* = 2$

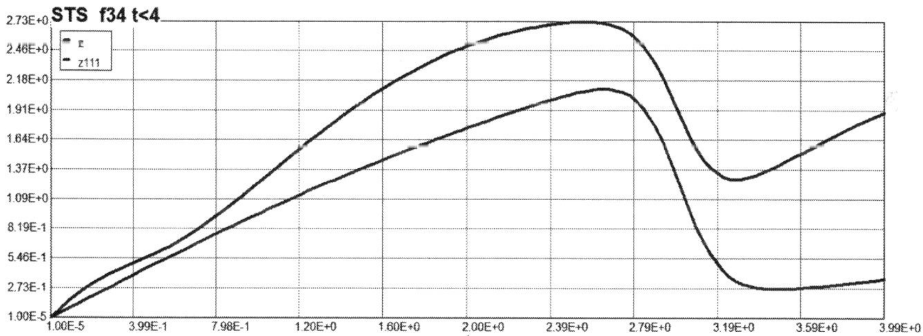

Figure 4.18: Stress-temperature-strain diagram A3 = f34(t); p = –4; $\theta^* = 2$.

Model: z = a1*t + a2*t^2 + a3*t^3 + a4*t^4 + a5*t^5

Variable	Value
a1	1.741305
a2	–1.880187
a3	2.099613
a4	–0.8454488
a5	0.1049543

$$\sigma = 1.741\theta - 1.88\theta^2 + 2.099\theta^3 - 0.845\theta^4 + 0.105\theta^5 \quad (4.36)$$

Comparing maximum stress with similar Example 4.2 we have: σ_{max} = 2.731 and σ_{max} = 2.73.

Similar computations are done for two other important cases: $\theta_{gm} > \theta^*$ and $\theta_{gf} < \theta^*$, where θ^* dimensionless transitional temperature

Calculated values of DEQ variables

	Variable	Initial value	Minimal value	Maximal value	Final value
1	A2	1.	1.	1.	1.
2	A3	1.	1.	1.	1.
3	E	1.9	0.225	1.9	0.225
4	E1	1.	0.25	1.	0.25
5	f3	0	0	1.	1.
6	f31	5.0E-06	0	0.9999733	0
7	f32	0.999999	0.449329	0.999999	0.449329
8	f33	6.162E-11	6.162E-11	0.9999995	1.015E-05
9	f34	1.	0	1.	0
10	f35	1.25E-06	1.25E-06	1.	1.
11	fi	0.1	0.1	0.1	0.1
12	k	0.1	0.1	0.1	0.1
13	m	4.05E-07	4.05E-07	0.0705907	0.0705907
14	m1	0.0404998	0.0003976	0.0404998	0.0003976
15	m2	1.266E-05	1.266E-05	3014.544	3014.544
16	m21	1.266042	0.5347826	1666.482	1666.482
17	n	1.	1.	1.	1.
18	p	–4.	–4.	–4.	–4.
19	s	1.	1.	1.	1.
20	t	1.0E-05	1.0E-05	8.	8.
21	Y1	1.9E-05	1.9E-05	3.767241	1.8
22	z	1.9E-05	–0.2514755	2.495185	0.5020923
23	z1	0	0	0.7107886	0.7107886
24	z11	0	0	0.79102	0.79102
25	z111	1.0E-05	1.0E-05	2.108678	0.5333333
26	z2	0	0	0.5880357	0.5880357
27	z22	0	0	0.6766824	0.6766824
28	z3	0	0	0.9062283	0.9062283

Differential equations

1 d(z1)/d(t) = (exp(t/(1 + 0.067*t)))*((exp(0.01*m)))*m1*z^n

2 d(z2)/d(t) = (exp(t/(1 + 0.067*t)))*(A2*(z^s))*((exp(0.01*m)))*m1*z^n

3 d(z3)/d(t) = k*(exp(t/(1 + 0.067*t)))*A3*(f34*(t^p/(1 + 0.1*t^p)))*((exp (0.01*m2)))*m21*z^n

4 d(z11)/d(t) = (exp(t/(1 + 0.067*t)))*((exp(0.01*m)))*m1*z111^n

5 d(z22)/d(t) = (exp(t/(1 + 0.067*t)))*(1*(z111^s))*((exp(0.01*m)))*m1* z111^n

Explicit equations

1 m = (0.0405*t – 0.01126*t^2 + 0.001462*t^3 – 0.00006868*t^4)

2 m1 = (0.0405 – 0.02252*t^1 + 0.004386*t^2 – 0.0002747*t^3)

3 f34 = if (t < 2) then (1) else (0)

4 fi = 0.1

5 k = 0.1

6 n = 1.0

7 s = 1.0

8 E = (0.625 – 0.375*(tanh(5*(t – 3))))*(1 – fi)+(1/k)* (0.625 – 0.375*(tanh(5*(t – 5))))*(fi)*f34

9 A2 = 1

10 A3 = 1

11 p = –4

12 m2 = 1.266*t + 2.12*t^2 – 3.077*t^3 + 1.085*t^4

13 m21 = 1.266 + 4.24 *t – 9.231 *t^2 + 4.34 *t^3

14 f3 = if (t < 0.1) then (0) else (1)

15 f31 = (1/16)*t*(8 – t)

16 f32 = (exp(–0.1*t))

17 f33 = (sin(3.14*t/4))^2

18 z = t*E–(z1 + z2)*((exp(–0.01*m)))–z3*(exp(–0.01*m2))

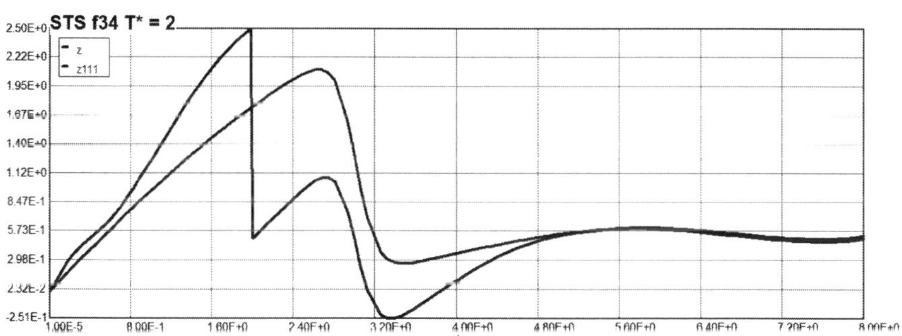

Figure 4.19: Stress-temperature-strain diagram A3 = f34(t); p = –4; θ* = 2.

19 f35 = (1/8)*t^1
20 E1 = (0.625 – 0.375*(tanh(5*(t – 3))))*1.0
21 z111 = t*E1–(z11+z22)*((exp(–0.01*m)))
22 Y1 = t*E

Model: z = a1*t + a2*t^2 + a3*t^3 + a4*t^4 + a5*t^5

Variable	Value
a1	3.269132
a2	–2.218342
a3	0.5506643
a4	–0.0579924
a5	0.0021897

$$\sigma = 3.269\theta - 2.218\theta^2 + 0.55\theta^3 - 0.058\theta^4 + 0.0022\theta^5 \quad (4.37)$$

Case $\theta^* = 7$

Calculated values of DEQ variables

	Variable	Initial value	Minimal value	Maximal value	Final value
1	A2	1.	1.	1.	1.
2	A3	1.	1.	1.	1.
3	E	1.9	0.225	1.9	0.225
4	E1	1.	0.25	1.	0.25
5	f3	0	0	1.	1.
6	f31	5.0E-06	0	0.9999408	0
7	f32	0.999999	0.449329	0.999999	0.449329
8	f33	6.162E-11	6.162E-11	0.9997952	1.015E-05
9	f34	1.	0	1.	0
10	f35	1.25E-06	1.25E-06	1.	1.
11	fi	0.1	0.1	0.1	0.1
12	k	0.1	0.1	0.1	0.1
13	m	4.05E-07	4.05E-07	0.0705907	0.0705907
14	m1	0.0404998	0.0003976	0.0404998	0.0003976
15	m2	1.266E-05	1.266E-05	3014.544	3014.544
16	m21	1.266042	0.5347826	1666.482	1666.482
17	n	1.	1.	1.	1.
18	p	–4.	–4.	–4.	–4.

19	s	1.	1.	1.	1.
20	t	1.0E-05	1.0E-05	8.	8.
21	Y1	1.9E-05	1.9E-05	5.596146	1.8
22	z	1.9E-05	–0.9363994	2.726521	–0.4830138
23	z1	0	0	0.915258	0.7017927
24	z11	0	0	0.79102	0.79102
25	z111	1.0E-05	1.0E-05	2.10856	0.5333333
26	z2	0	0	1.582833	1.582833
27	z22	0	0	0.6766824	0.6766824
28	z3	0	0	5.348E+06	5.348E+06

Differential equations

1 d(z1)/d(t) = (exp(t/(1 + 0.067*t)))*((exp(0.01*m)))*m1*z^n
2 d(z2)/d(t) = (exp(t/(1 + 0.067*t)))*(A2*(z^s))*((exp(0.01*m)))*m1*z^n
3 d(z3)/d(t) = k*(exp(t/(1 + 0.067*t)))*A3*(f34*(t^p/(1 + 0.1*t^p)))*((exp (0.01*m2)))*m21*z^n
4 d(z11)/d(t) = (exp(t/(1 + 0.067*t)))*((exp(0.01*m)))*m1*z111^n
5 d(z22)/d(t) = (exp(t/(1 + 0.067*t)))*(1*(z111^s))*((exp(0.01*m)))*m1* z111^n

Explicit equations

1 m = (0.0405*t – 0.01126*t^2 + 0.001462*t^3 – 0.00006868*t^4)
2 m1 = (0.0405 – 0.02252*t^1 + 0.004386*t^2 – 0.0002747*t^3)
3 f34 = if (t < 7) then (1) else (0)
4 fi = 0.1
5 k = 0.1
6 n = 1.0
7 s = 1.0
8 E = (0.625 – 0.375*(tanh(5*(t – 3))))*(1 – fi)+(1/k)* (0.625 – 0.375*(tanh(5*(t – 5))))*(fi)*f34
9 A2 = 1
10 A3 = 1
11 p = –4
12 m2 = 1.266*t + 2.12*t^2 – 3.077*t^3 + 1.085*t^4
13 m21 = 1.266 + 4.24 *t – 9.231 *t^2 + 4.34 *t^3
14 f3 = if (t < 0.1) then (0) else (1)

15 f31 = (1/16)*t*(8 – t)

16 f32 = (exp(–0.1*t))

17 f33 = (sin(3.14*t/4))^2

18 z = t*E–(z1+z2)*((exp(–0.01*m)))–z3*(exp(–0.01*m2))

19 f35 = (1/8)*t^1

20 E1 = (0.625 – 0.375*(tanh(5*(t – 3))))*1.0

21 z111 = t*E1–(z11+z22)*((exp(–0.01*m)))

22 Y1 = t*E

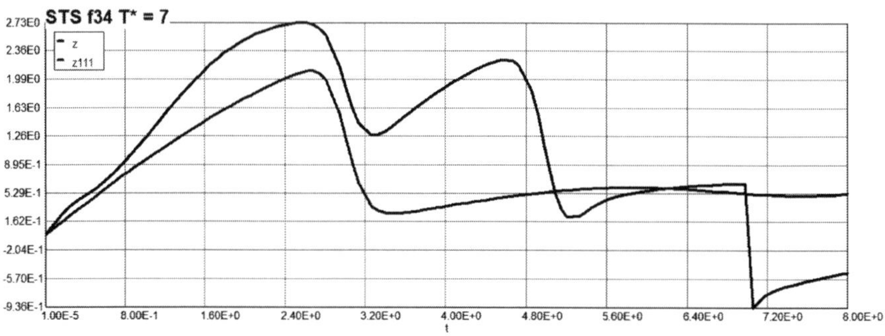

Figure 4.20: Stress-temperature-strain diagram A3 = f34(t); p = –4; $\theta^* = 7$.

Model: z = a1*t + a2*t^2 + a3*t^3 + a4*t^4 + a5*t^5

Variable	Value
a1	1.63061
a2	–0.0859712
a3	–0.1337692
a4	0.0261698
a5	–0.0014371

$$\sigma = 1.631\theta - 0.086\theta^2 - 0.134\theta^3 + 0.0262\theta^4 - 0.00144\theta^5 \quad (4.38)$$

Example 4.12

Data: see previous Example 4.11, but p = –4 and f3 = f35 = f (**θ**) = (1/8)*t^1

Calculated values of DEQ variables

	Variable	Initial value	Minimal value	Maximal value	Final value
1	A2	1.	1.	1.	1.
2	A3	1.	1.	1.	1.

3	E	1.9	0.475	1.9	0.475
4	E1	1.	0.25	1.	0.25
5	f3	0	0	1.	1.
6	f31	5.0E-06	0	0.9999822	0
7	f32	0.999999	0.449329	0.999999	0.449329
8	f33	6.162E-11	6.162E-11	0.9999838	1.015E-05
9	f34	1.	0	1.	0
10	f35	1.25E-06	1.25E-06	1.	1.
11	fi	0.1	0.1	0.1	0.1
12	k	0.1	0.1	0.1	0.1
13	m	4.05E-07	4.05E-07	0.0705907	0.0705907
14	m1	0.0404998	0.0003976	0.0404998	0.0003976
15	m2	1.266E-05	1.266E-05	3014.544	3014.544
16	m21	1.266042	0.5360993	1666.482	1666.482
17	n	1.	1.	1.	1.
18	p	–4.	–4.	–4.	–4.
19	s	1.	1.	1.	1.
20	t	1.0E-05	1.0E-05	8.	8.
21	Y1	1.9E-05	1.9E-05	5.594308	3.8
22	z	1.9E-05	–0.1223136	3.418811	0.5893658
23	z1	0	0	1.030128	1.030128
24	z11	0	0	0.79102	0.79102
25	z111	1.0E-05	1.0E-05	2.101565	0.5333333
26	z2	0	0	1.927205	1.927205
27	z22	0	0	0.6766824	0.6766824
28	z3	0	0	3.156E+12	3.156E+12

Differential equations

1 d(z1)/d(t) = (exp(t/(1 + 0.067*t)))*((exp(0.01*m)))*m1*z^n

2 d(z2)/d(t) = (exp(t/(1 + 0.067*t)))*(A2*(z^s))*((exp(0.01*m)))*m1*z^n

3 d(z3)/d(t) = k*(exp(t/(1 + 0.067*t)))*A3*(f35*(t^p/(1 + 0.1*t^p)))*((exp(0.01*m2)))*m21*z^n

4 d(z11)/d(t) = (exp(t/(1 + 0.067*t)))*((exp(0.01*m)))*m1*z111^n

5 d(z22)/d(t) = (exp(t/(1 + 0.067*t)))*(1*(z111^s))*((exp(0.01*m)))*m1*z111^n

Explicit equations

1 m = (0.0405*t – 0.01126*t^2 + 0.001462*t^3 – 0.00006868*t^4)
2 m1 = (0.0405 – 0.02252*t^1 + 0.004386*t^2 – 0.0002747*t^3)
3 m2 = 1.266*t + 2.12*t^2 – 3.077*t^3 + 1.085*t^4
4 fi = 0.1
5 k = 0.1
6 n = 1.0
7 s = 1.0
8 E = (0.625 – 0.375*(tanh(5*(t – 3))))*(1 – fi)+(1/k)* (0.625 – 0.375*(tanh(5*(t – 5))))*(fi)
9 A2 = 1
10 A3 = 1
11 p = –4
12 z = t*E–(z1 + z2)*((exp(–0.01*m)))–z3*(exp(–0.01*m2))
13 m21 = 1.266 + 4.24 *t – 9.231 *t^2 + 4.34 *t^3
14 f3 = if (t < 0.1) then (0) else (1)
15 f31 = (1/16)*t*(8 – t)
16 f32 = (exp(–0.1*t))
17 f33 = (sin(3.14*t/4))^2
18 f34 = if (t < 7) then (1) else (0)
19 f35 = (1/8)*t^1
20 E1 = (0.625 – 0.375*(tanh(5*(t – 3))))*1.0
21 z111 = t*E1–(z11 + z22)*((exp(–0.01*m)))
22 Y1 = t*E

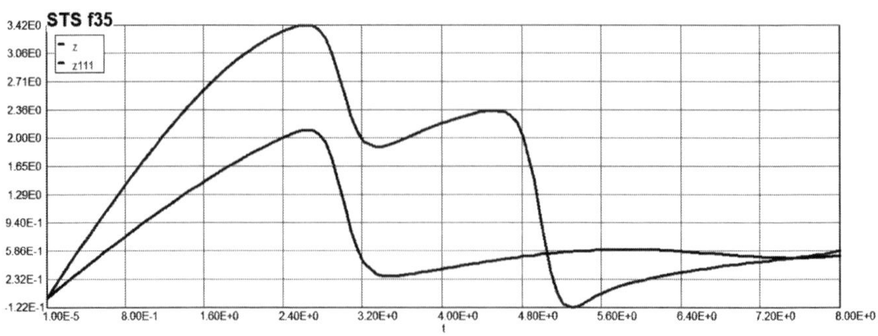

Figure 4.21: Stress-temperature-strain diagram f3 = f35(t) (with dawning portion).

Model: z = a1*t + a2*t^2 + a3*t^3 + a4*t^4 + a5*t^5

Variable	Value
a1	1.713601
a2	0.4042426
a3	–0.3746404
a4	0.0619847
a5	–0.0030829

$$\sigma = 1.714\theta + 0.404\theta^2 - 0.375\theta^3 + 0.062\theta^4 - 0.00308\theta^5 \quad (4.39)$$

Example 4.13

If the nanocrystalline growth process is described by not linear but for example parabolic function (e.g., f3 = f35 = (1/64)(θ^2), then the engineering creep allowable stress also is increasing (comparable to the linear growth process) and an example below illustrates this.

Data: see previous Example 4.11, but p = –4 and f3 = f35 = f (θ) = (1/64)* θ^2

Calculated values of DEQ variables

	Variable	Initial value	Minimal value	Maximal value	Final value
1	A2	1.	1.	1.	1.
2	A3	1.	1.	1.	1.
3	E	1.9	0.475	1.9	0.475
4	E1	1.	0.25	1.	0.25
5	f3	0	0	1.	1.
6	f31	5.0E-06	0	0.9999807	0
7	f32	0.999999	0.449329	0.999999	0.449329
8	f33	6.162E-11	6.162E-11	0.9999749	1.015E-05
9	f34	1.	0	1.	0
10	f35	1.563E-12	1.563E-12	1.	1.
11	fi	0.1	0.1	0.1	0.1
12	k	0.1	0.1	0.1	0.1
13	m	4.05E-07	4.05E-07	0.0705907	0.0705907
14	m1	0.0404998	0.0003976	0.0404998	0.0003976
15	m2	1.266E-05	1.266E-05	3014.544	3014.544
16	m21	1.266042	0.5363002	1666.482	1666.482

17	n	1.	1.	1.	1.
18	p	–4.	–4.	–4.	–4.
19	s	1.	1.	1.	1.
20	t	1.0E-05	1.0E-05	8.	8.
21	Y1	1.9E-05	1.9E-05	5.594458	3.8
22	z	1.9E-05	–0.2750274	3.620656	0.5075764
23	z1	0	0	0.9646977	0.9646977
24	z11	0	0	0.79102	0.79102
25	z111	1.0E-05	1.0E-05	2.10183	0.5333333
26	z2	0	0	2.11242	2.11242
27	z22	0	0	0.6766824	0.6766824
28	z3	0	–21.4011	2.688E+12	2.688E+12

Differential equations

1 d(z1)/d(t) = (exp(t/(1 + 0.067*t)))*((exp(0.01*m)))*m1*z^n

2 d(z2)/d(t) = (exp(t/(1 + 0.067*t)))*(A2*(z^s))*((exp(0.01*m)))*m1*z^n

3 d(z3)/d(t) = k*(exp(t/(1 + 0.067*t)))*A3*(f35*(t^p/(1 + 0.1*t^p)))*((exp (0.01*m2)))*m21*z^n

4 d(z11)/d(t) = (exp(t/(1 + 0.067*t)))*((exp(0.01*m)))*m1*z111^n

5 d(z22)/d(t) = (exp(t/(1 + 0.067*t)))*(1*(z111^s))*((exp(0.01*m)))*m1* z111^n

Explicit equations

1 m = (0.0405*t – 0.01126*t^2 + 0.001462*t^3 – 0.00006868*t^4)

2 m1 = (0.0405 – 0.02252*t^1 + 0.004386*t^2 – 0.0002747*t^3)

3 m2 = 1.266*t + 2.12*t^2 – 3.077*t^3 + 1.085*t^4

4 fi = 0.1

5 k = 0.1

6 n = 1.0

7 s = 1.0

8 E = (0.625 – 0.375*(tanh(5*(t – 3))))*(1 – fi)+(1/k)* (0.625 – 0.375*(tanh(5*(t – 5))))*(fi)

9 A2 = 1

10 A3 = 1

11 p = –4

12 z = t*E–(z1 + z2)*((exp(–0.01*m)))–z3*(exp(–0.01*m2))

```
13 m21 = 1.266 + 4.24 *t – 9.231 *t^2 + 4.34 *t^3
14 f3 = if (t < 0.1) then (0) else (1)
15 f31 = (1/16)*t*(8 – t)
16 f32 = (exp(–0.1*t))
17 f33 = (sin(3.14*t/4))^2
18 f34 = if (t < 7) then (1) else (0)
19 f35 = (1/64)*t^2
20 E1 = (0.625 – 0.375*(tanh(5*(t – 3))))*1.0
21 z111 = t*E1–(z11+z22)*((exp(–0.01*m)))
22 Y1 = t*E
```

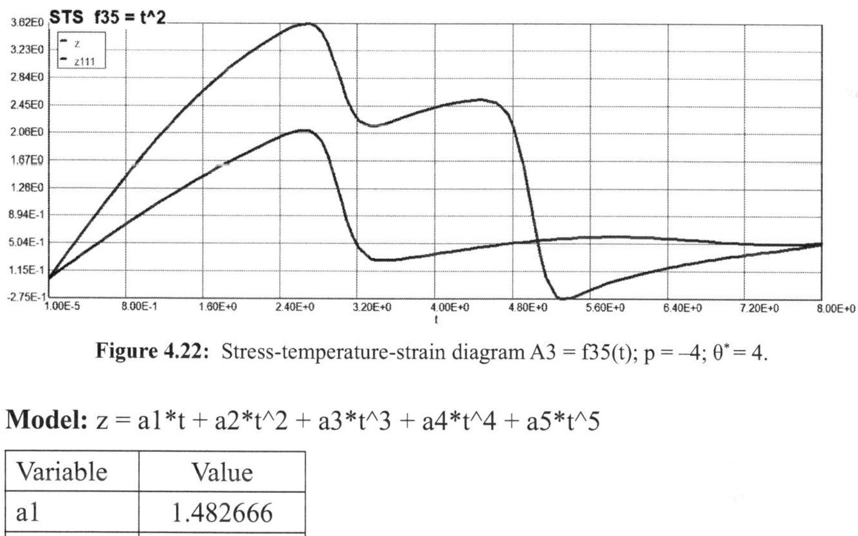

Figure 4.22: Stress-temperature-strain diagram A3 = f35(t); p = –4; $\theta^* = 4$.

Model: z = a1*t + a2*t^2 + a3*t^3 + a4*t^4 + a5*t^5

Variable	Value
a1	1.482666
a2	0.7150971
a3	–0.4739853
a4	0.0735177
a5	0.0035231

$$\sigma = 1.483\theta + 0.715\theta^2 - 0.474\theta^3 + 0.0735\theta^4 \quad 0.00352\theta^5 \qquad (4.40)$$

Comparing maximum stress with similar example we have: $\sigma_{max} = 3.4188$ and $\sigma_{max} = 3.6206$ (5.9%).

4.6.2 *Effect of E_0/E_2 ratio on creep process*

Consider now the case when modulus of elasticity E_0 of matrix material is much smaller then the corresponding modulus of elasticity E_2 of nanomaterials

$-E_0/E_2 = 0.05$. It is also assumed in the following example that function f3 is a continuous function: f3 = f35 = (1/64)(θ^2).

Example 4.14

Data: see previous Example 4.7, but p = –4 and f3 = f35 = f (θ) = (1/64)(θ^2); $E_0/E_2 = 0.05$

Calculated values of DEQ variables

	Variable	Initial value	Minimal value	Maximal value	Final value
1	A2	1.	1.	1.	1.
2	A3	1.	1.	1.	1.
3	E	2.9	0.725	2.9	0.725
4	E1	1.	0.25	1.	0.25
5	f3	0	0	1.	1.
6	f31	5.0E-06	0	0.9999738	0
7	f32	0.999999	0.449329	0.999999	0.449329
8	f33	6.162E-11	6.162E-11	0.9999977	1.015E-05
9	f34	1.	0	1.	0
10	f35	1.563E-12	1.563E-12	1.	1.
11	fi	0.1	0.1	0.1	0.1
12	k	0.05	0.05	0.05	0.05
13	m	4.05E-07	4.05E-07	0.0705907	0.0705907
14	m1	0.0404998	0.0003976	0.0404998	0.0003976
15	m2	1.266E-05	1.266E-05	3014.544	3014.544
16	m21	1.266042	0.5329491	1666.482	1666.482
17	n	1.	1.	1.	1.
18	p	–4.	–4.	–4.	–4.
19	s	1.	1.	1.	1.
20	t	1.0E-05	1.0E-05	8.	8.
21	Y1	2.9E-05	2.9E-05	10.14275	5.8
22	z	2.9E-05	–2.020004	5.130823	–0.8482906
23	z1	0	–0.1012231	1.21402	–0.1012231
24	z11	0	0	0.79102	0.79102
25	z111	1.0E-05	1.0E-05	2.10871	0.5333333
26	z2	0	0	6.950089	6.950089

27	z22	0	0	0.6766824	0.6766824
28	z3	0	–2.419E+12	13.2692	–2.419E+12

Differential equations

1 d(z1)/d(t) = (exp(t/(1 + 0.067*t)))*((exp(0.01*m)))*m1*z^n

2 d(z2)/d(t) = (exp(t/(1 + 0.067*t)))*(A2*(z^s))*((exp(0.01*m)))*m1*z^n

3 d(z3)/d(t) = k*(exp(t/(1 + 0.067*t)))*A3*(f35*(t^p/(1 + 0.1*t^p)))*((exp (0.01*m2)))*m21*z^n

4 d(z11)/d(t) = (exp(t/(1 + 0.067*t)))*((exp(0.01*m)))*m1*z111^n

5 d(z22)/d(t) = (exp(t/(1 + 0.067*t)))*(1*(z111^s))*((exp(0.01*m)))*m1* z111^n

Explicit equations

1 m = (0.0405*t – 0.01126*t^2 + 0.001462*t^3 – 0.00006868*t^4)

2 m1 = (0.0405 – 0.02252*t^1 + 0.004386*t^2 – 0.0002747*t^3)

3 m2 = 1.266*t + 2.12*t^2 – 3.077*t^3 + 1.085*t^4

4 fi = 0.1

5 k = 0.05

6 n = 1.0

7 s = 1.0

8 E = (0.625 – 0.375*(tanh(5*(t – 3))))*(1 – fi)+(1/k)* (0.625 – 0.375*(tanh(5*(t – 5))))*(fi)

9 A2 = 1

10 A3 = 1

11 p = –4

12 z = t*E–(z1 + z2)*((exp(–0.01*m)))–z3*(exp(–0.01*m2))

13 m21 = 1.266 + 4.24 *t – 9.231 *t^2 + 4.34 *t^3

14 f3 = if (t < 0.1) then (0) else (1)

15 f31 = (1/16)*t*(8 – t)

16 f32 = (exp(–0.1*t))

17 f33 = (sin(3.14*t/4))^2

18 f34 = if (t < 7) then (1) else (0)

19 f35 = (1/64)*t^2

20 E1 = (0.625 – 0.375*(tanh(5*(t – 3))))*1.0

21 z111 = t*E1–(z11 + z22)*((exp(–0.01*m)))

22 Y1 = t*E

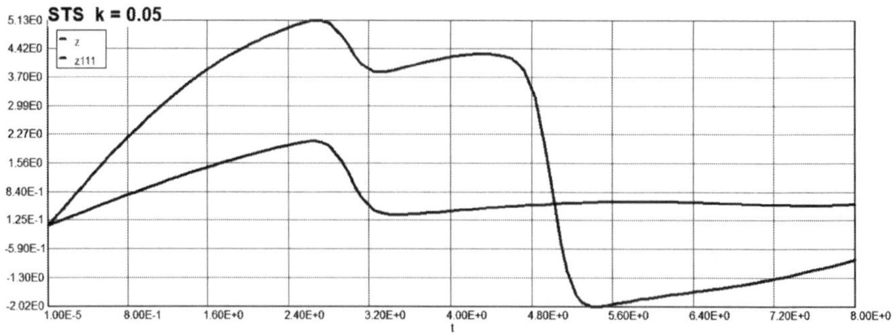

Figure 4.23: Stress-temperature-strain diagram $E_0/E_2 = 0.05$ (with dawning portion).

z111—Stress-Temperature-Strain (STS) curve without creep

z—Stress-Temperature-Strain (STS) curve with creep

Model: z = a1*t + a2*t^2 + a3*t^3 + a4*t^4 + a5*t^5

Variable	Value
a1	1.581664
a2	1.557429
a3	–0.7777135
a4	0.1026803
a5	–0.0041149

$$\sigma = 1.582\theta + 1.557\theta^2 - 0.778\theta^3 + 0.103\theta^4 - 0.00411\theta^5 \qquad (4.41)$$

Comparing results of Examples 4.5 and 4.7 one can see that the maximum allowable stress increases when the ratio E_0/E_2 decreases ($\sigma_{max} = 5.131$ vs. $\sigma_{max} = 3.621$). Uniformly dispersed very strong nanoparticles in this case increase the overall strength of a composite material. Comparing results from Examples 4.7, 4.8 and 4.9 one can see that maximum stresses and the shapes of these curves are qualitatively different. Therefore the correct choice of functions f_3 must be determined by using the "best-to-fit" regression method with experimental data.

4.6.3 *Effect of volumetric fillers ratio φ_f on creep process*

The effect of fillers amount on properties of nanocomposite materials is considered in this section. It is shown that the properties of the filled polymer composites are determined by the characteristics of the polymer matrix; disperse filler and their interaction at the interface. It was noted that the content of fillers in the polymer composite should be optimal both from the point of view of the possibility of its processing and from the point of view of its effect on performance characteristics. When the filler content is above the optimum, many properties of the composite deteriorate.

Example 4.15

Data: see previous Example 4.7, but p = –4 and f3 = f35 = f (θ) = (1/64)(θ^2); E_0/E_2 = 0.1; fi = 0.01

Calculated values of DEQ variables

	Variable	Initial value	Minimal value	Maximal value	Final value
1	A2	1.	1.	1.	1.
2	A3	1.	1.	1.	1.
3	E	1.09	0.2725	1.09	0.2725
4	E1	1.	0.25	1.	0.25
5	f3	0	0	1.	1.
6	f31	5.0E-06	0	0.9999705	0
7	f32	0.999999	0.449329	0.999999	0.449329
8	f33	6.162E-11	6.162E-11	0.9999995	1.015E-05
9	f34	1.	0	1.	0
10	f35	1.563E-12	1.563E-12	1.	1.
11	fi	0.01	0.01	0.01	0.01
12	k	0.1	0.1	0.1	0.1
13	m	4.05E-07	4.05E-07	0.0705907	0.0705907
14	m1	0.0404998	0.0003976	0.0404998	0.0003976
15	m2	1.266E-05	1.266E-05	3014.544	3014.544
16	m21	1.266042	0.5414558	1666.482	1666.482
17	n	1.	1.	1.	1.
18	p	–4.	–4.	–4.	–4.
19	з	1.	1.	1.	1.
20	t	1.0E-05	1.0E-05	8.	8.
21	Y1	1.09E-05	1.09E-05	2.831929	2.18
22	z	1.09E-05	1.09E-05	2.23294	0.4941499
23	z1	0	0	0.775332	0.775332
24	z11	0	0	0.79102	0.79102
25	z111	1.0E-05	1.0E-05	2.110468	0.5333333

26	z2	0	0	0.6962726	0.6962726
27	z22	0	0	0.6766824	0.6766824
28	z3	0	0	2.661E+12	2.661E+12

Differential equations

1 d(z1)/d(t) = (exp(t/(1 + 0.067*t)))*((exp(0.01*m)))*m1*z^n

2 d(z2)/d(t) = (exp(t/(1+ 0.067*t)))*(A2*(z^s))*((exp(0.01*m)))*m1*z^n

3 d(z3)/d(t) = k*(exp(t/(1 + 0.067*t)))*A3*(f35*(t^p/(1 + 0.1*t^p)))*((exp(0.01*m2)))*m21*z^n

4 d(z11)/d(t) = (exp(t/(1 + 0.067*t)))*((exp(0.01*m)))*m1*z111^n

5 d(z22)/d(t) = (exp(t/(1 + 0.067*t)))*(1*(z111^s))*((exp(0.01*m)))*m1*z111^n

Explicit equations

1 m = (0.0405*t – 0.01126*t^2 + 0.001462*t^3 – 0.00006868*t^4)

2 m1 = (0.0405 – 0.02252*t^1 + 0.004386*t^2 – 0.0002747*t^3)

3 m2 = 1.266*t + 2.12*t^2 – 3.077*t^3 + 1.085*t^4

4 fi = 0.01

5 k = 0.1

6 n = 1.0

7 s = 1.0

8 E = (0.625 – 0.375*(tanh(5*(t – 3))))*(1 – fi)+(1/k)*(0.625 – 0.375*(tanh(5*(t – 5))))*(fi)

9 A2 = 1

10 A3 = 1

11 p = –4

12 z = t*E–(z1 + z2)*((exp(–0.01*m)))–z3*(exp(–0.01*m2))

13 m21 = 1.266 + 4.24 *t – 9.231 *t^2 + 4.34 *t^3

14 f3 = if (t < 0.1) then (0) else (1)

15 f31 = (1/16)*t*(8 – t)

16 f32 = (exp(–0.1*t))

17 f33 = (sin(3.14*t/4))^2

18 f34 = if (t < 7) then (1) else (0)

19 f35 = (1/64)*t^2

20 E1 = (0.625 – 0.375*(tanh(5*(t – 3))))*1.0

21 z111 = t*E1–(z11 + z22)*((exp(–0.01*m)))

22 Y1 = t*E

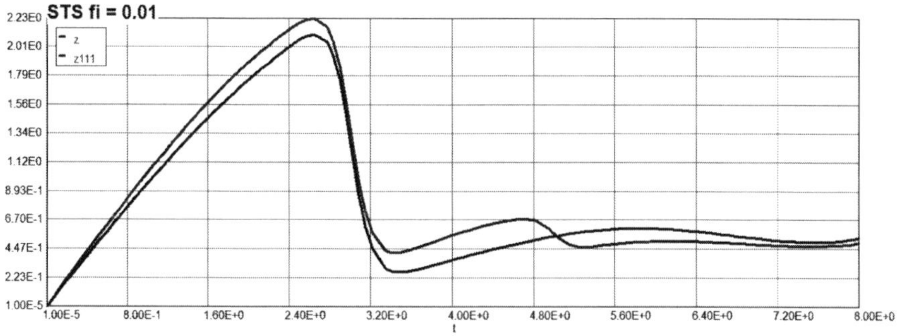

Figure 4.24: Stress-temperature-strain diagram A3 = f35(t); p = –4; θ^* = 7; φ = 0.01.

Model: z = a1*t + a2*t^2 + a3*t^3 + a4*t^4 + a5*t^5

Variable	Value
a1	1.54119
a2	–0.1862496
a3	–0.1454993
a4	0.0359771
a5	–0.002231

$$\sigma = 1.541\theta - 0.186\theta^2 - 0.146\theta^3 + 0.036\theta^4 - 0.00223\theta^5 \quad (4.42)$$

Example 4.16

Data: see previous Example 4.7, but p = –4 and f3 = f35 = f (θ) = (1/64)(θ^2); E_0/E_2 = 0.1; fi = 0.02

Calculated values of DEQ variables

	Variable	Initial value	Minimal value	Maximal value	Final value
1	A2	1.	1.	1.	1.
2	A3	1.	1.	1.	1.
3	E	1.18	0.295	1.18	0.295
4	E1	1.	0.25	1.	0.25
5	f3	0	0	1.	1.
6	f31	5.0E-06	0	0.9999716	0
7	f32	0.999999	0.449329	0.999999	0.449329
8	f33	6.162E-11	6.162E-11	0.9997208	1.015E-05
9	f34	1.	0	1.	0
10	f35	1.563E-12	1.563E-12	1.	1.

11	fi	0.02	0.02	0.02	0.02
12	k	0.1	0.1	0.1	0.1
13	m	4.05E-07	4.05E-07	0.0705907	0.0705907
14	m1	0.0404998	0.0003976	0.0404998	0.0003976
15	m2	1.266E-05	1.266E-05	3014.544	3014.544
16	m21	1.266042	0.5341124	1666.482	1666.482
17	n	1.	1.	1.	1.
18	p	–4.	–4.	–4.	–4.
19	s	1.	1.	1.	1.
20	t	1.0E-05	1.0E-05	8.	8.
21	Y1	1.18E-05	1.18E-05	3.092304	2.36
22	z	1.18E-05	1.18E-05	2.396634	0.5078425
23	z1	0	0	0.8232942	0.8232942
24	z11	0	0	0.79102	0.79102
25	z111	1.0E-05	1.0E-05	2.108096	0.5333333
26	z2	0	0	0.8090332	0.8090332
27	z22	0	0	0.6766824	0.6766824
28	z3	0	0	2.731E+12	2.731E+12

Differential equations

1 d(z1)/d(t) = (exp(t/(1 + 0.067*t)))*((exp(0.01*m)))*m1*z^n

2 d(z2)/d(t) = (exp(t/(1 + 0.067*t)))*(A2*(z^s))*((exp(0.01*m)))*m1*z^n

3 d(z3)/d(t) = k*(exp(t/(1 + 0.067*t)))*A3*(f35*(t^p/(1 + 0.1*t^p)))*((exp (0.01*m2)))*m21*z^n

4 d(z11)/d(t) = (exp(t/(1 + 0.067*t)))*((exp(0.01*m)))*m1*z111^n

5 d(z22)/d(t) = (exp(t/(1 + 0.067*t)))*(1*(z111^s))*((exp(0.01*m)))*m1* z111^n

Explicit equations

1 m = (0.0405*t – 0.01126*t^2 + 0.001462*t^3 – 0.00006868*t^4)

2 m1 = (0.0405 – 0.02252*t^1 + 0.004386*t^2 – 0.0002747*t^3)

3 m2 = 1.266*t + 2.12*t^2 – 3.077*t^3 + 1.085*t^4

4 fi = 0.02

5 k = 0.1

6 n = 1.0

7 s = 1.0

```
8 E = (0.625 – 0.375*(tanh(5*(t – 3))))*(1 – fi)+(1/k)*
  (0.625 – 0.375*(tanh(5*(t – 5))))*(fi)
9 A2 = 1
10 A3 = 1
11 p = –4
12 z = t*E–(z1 + z2)*((exp(–0.01*m)))–z3*(exp(–0.01*m2))
13 m21 = 1.266 + 4.24 *t – 9.231 *t^2 + 4.34 *t^3
14 f3 = if (t < 0.1) then (0) else (1)
15 f31 = (1/16)*t*(8 – t)
16 f32 = (exp(–0.1*t))
17 f33 = (sin(3.14*t/4))^2
18 f34 = if (t < 7) then (1) else (0)
19 f35 = (1/64)*t^2
20 E1 = (0.625 – 0.375*(tanh(5*(t – 3))))*1.0
21 z111 = t*E1–(z11 + z22)*((exp(–0.01*m)))
22 Y1 – t*E
```

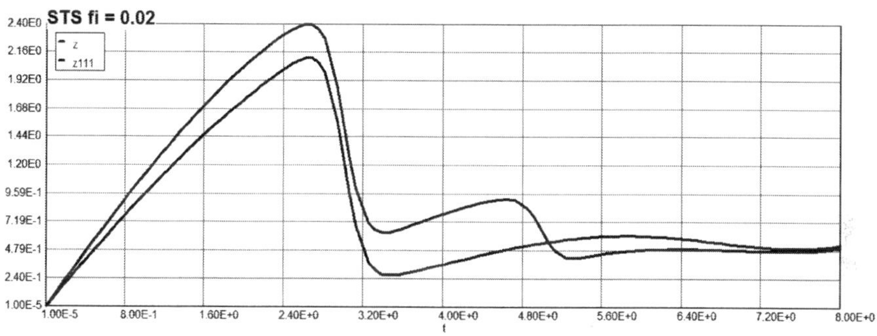

Figure 4.25: Stress-temperature-strain diagram A3 = $f_3(t)$; p = –4; $\theta^* = 4$; $\varphi = 0.02$.

Model: z = a1*t + a2*t^2 + a3*t^3 + a4*t^4 + a5*t^5

Variable	Value
a1	1.552037
a2	–0.1081338
a3	–0.17236
a4	0.0386226
a5	–0.0022951

$$\sigma = 1.541\theta - 0.186\theta^2 - 0.146\theta^3 + 0.036\theta^4 - 0.00223\theta^5 \tag{4.43}$$

Example 4.17

Data: see previous Example 4.7, but p = –4 and f3 = f35 = f (θ) = (1/64)(θ^2); E_0/E_2 = 0.1; fi = 0.05

Calculated values of DEQ variables

	Variable	Initial value	Minimal value	Maximal value	Final value
1	A2	1.	1.	1.	1.
2	A3	1.	1.	1.	1.
3	E	1.45	0.3625	1.45	0.3625
4	E1	1.	0.25	1.	0.25
5	f3	0	0	1.	1.
6	f31	5.0E-06	0	0.9999808	0
7	f32	0.999999	0.449329	0.999999	0.449329
8	f33	6.162E-11	6.162E-11	0.9998469	1.015E-05
9	f34	1.	0	1.	0
10	f35	1.563E-12	1.563E-12	1.	1.
11	fi	0.05	0.05	0.05	0.05
12	k	0.1	0.1	0.1	0.1
13	m	4.05E-07	4.05E-07	0.0705907	0.0705907
14	m1	0.0404998	0.0003976	0.0404998	0.0003976
15	m2	1.266E-05	1.266E-05	3014.544	3014.544
16	m21	1.266042	0.5328301	1666.482	1666.482
17	n	1.	1.	1.	1.
18	p	–4.	–4.	–4.	–4.
19	s	1.	1.	1.	1.
20	t	1.0E-05	1.0E-05	8.	8.
21	Y1	1.45E-05	1.45E-05	3.83794	2.9
22	z	1.45E-05	1.45E-05	2.876543	0.5309334
23	z1	0	0	0.9244367	0.9244367
24	z11	0	0	0.79102	0.79102
25	z111	1.0E-05	1.0E-05	2.108878	0.5333333
26	z2	0	0	1.216104	1.216104
27	z22	0	0	0.6766824	0.6766824
28	z3	0	0	2.843E+12	2.843E+12

Differential equations

```
1 d(z1)/d(t) = (exp(t/(1 + 0.067*t)))*((exp(0.01*m)))*m1*z^n
2 d(z2)/d(t) = (exp(t/(1 + 0.067*t)))*(A2*(z^s))*((exp(0.01*m)))*m1*z^n
3 d(z3)/d(t) = k*(exp(t/(1 + 0.067*t)))*A3*(f35*(t^p/(1 + 0.1*t^p)))*((exp
  (0.01*m2)))*m21*z^n
4 d(z11)/d(t) = (exp(t/(1 + 0.067*t)))*((exp(0.01*m)))*m1*z111^n
5 d(z22)/d(t) = (exp(t/(1 + 0.067*t)))*(1*(z111^s))*((exp(0.01*m)))*m1*z
  111^n
```

Explicit equations

```
1 m = (0.0405*t – 0.01126*t^2 + 0.001462*t^3 – 0.00006868*t^4)
2 m1 = (0.0405 – 0.02252*t^1 + 0.004386*t^2 – 0.0002747*t^3)
3 m2 = 1.266*t + 2.12*t^2 – 3.077*t^3 + 1.085*t^4
4 fi = 0.05
5 k = 0.1
6 n = 1.0
7 s = 1.0
8 E = (0.625 – 0.375*(tanh(5*(t – 3))))*(1 – fi)+(1/k)*
  (0.625 – 0.375*(tanh(5*(t – 5))))*(fi)
9 A2 = 1
10 A3 = 1
11 p = –4
12 z = t*E–(z1 + z2)*((exp(–0.01*m)))–z3*(exp(–0.01*m2))
13 m21 = 1.266 + 4.24 *t – 9.231 *t^2 + 4.34 *t^3
14 f3 = if (t < 0.1) then (0) else (1)
15 f31 = (1/16)*t*(8 – t)
16 f32 = (exp(–0.1*t))
17 f33 = (sin(3.14*t/4))^2
18 f34 = if (t < 7) then (1) else (0)
19 f35 = (1/64)*t^2
20 E1 = (0.625 – 0.375*(tanh(5*(t – 3))))*1.0
21 z111 = t*E1–(z11 + z22)*((exp(–0.01*m)))
22 Y1 = t*E
```

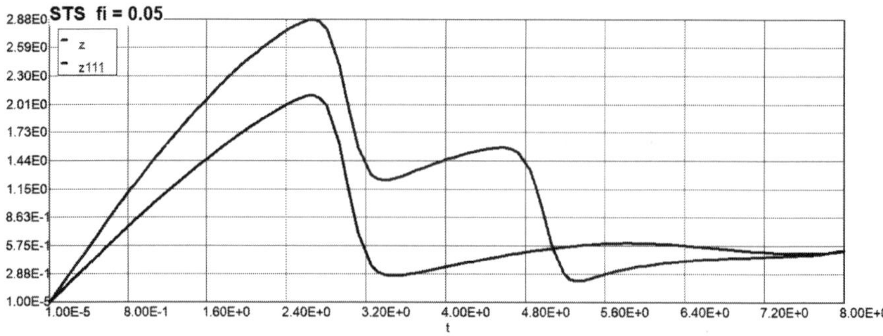

Figure 4.26: Stress-temperature-strain diagram $A3 = f_3(t)$; $p = -4$; $\theta^* = 4$; $\varphi = 0.05$.

Model: z = a1*t + a2*t^2 + a3*t^3 + a4*t^4 + a5*t^5

Variable	Value
a1	1.531793
a2	0.1920094
a3	–0.2794209
a4	0.0505784
a5	–0.0026929

$$\sigma = 1.532\theta + 0.192\theta^2 - 0.279\theta^3 + 0.0506\theta^4 - 0.00269\theta^5 \quad (4.44)$$

Example 4.18

Data: see previous Example 4.7, but $p = -4$ and $f3 = f35 = f(\theta) = (1/64)(\theta^2)$; $E_0/E_2 = 0.1$; fi = 0.08

Calculated values of DEQ variables

	Variable	Initial value	Minimal value	Maximal value	Final value
1	A2	1.	1.	1.	1.
2	A3	1.	1.	1.	1.
3	E	1.72	0.43	1.72	0.43
4	E1	1.	0.25	1.	0.25
5	f3	0	0	1.	1.
6	f31	5.0E-06	0	0.999823	0
7	f32	0.999999	0.449329	0.999999	0.449329
8	f33	6.162E-11	6.162E-11	0.9999891	1.015E-05
9	f34	1.	0	1.	0
10	f35	1.563E-12	1.563E-12	1.	1.

11	fi	0.08	0.08	0.08	0.08
12	k	0.1	0.1	0.1	0.1
13	m	4.05E-07	4.05E-07	0.0705907	0.0705907
14	m1	0.0404998	0.0003976	0.0404998	0.0003976
15	m2	1.266E-05	1.266E-05	3014.544	3014.544
16	m21	1.266042	0.5430382	1666.482	1666.482
17	n	1.	1.	1.	1.
18	p	–4.	–4.	–4.	–4.
19	s	1.	1.	1.	1.
20	t	1.0E-05	1.0E-05	8.	8.
21	Y1	1.72E-05	1.72E-05	4.708834	3.44
22	z	1.72E-05	–0.060784	3.330802	0.5272795
23	z1	0	0	0.9675236	0.9675236
24	z11	0	0	0.79102	0.79102
25	z111	1.0E-05	1.0E-05	2.110607	0.5333333
26	z2	0	0	1.719961	1.719961
27	z22	0	0	0.6766824	0.6766824
28	z3	0	0	2.807E+12	2.807E+12

Differential equations

1 d(z1)/d(t) = (exp(t/(1 + 0.067*t)))*((exp(0.01*m)))*m1*z^n

2 d(z2)/d(t) = (exp(t/(1 + 0.067*t)))*(A2*(z^s))*((exp(0.01*m)))*m1*z^n

3 d(z3)/d(t) = k*(exp(t/(1 + 0.067*t)))*A3*(f35*(t^p/(1 + 0.1*t^p)))*((exp (0.01*m2)))*m21*z^n

4 d(z11)/d(t) = (exp(t/(1 + 0.067*t)))*((exp(0.01*m)))*m1*z111^n

5 d(z22)/d(t) = (exp(t/(1 + 0.067*t)))*(1*(z111^s))*((exp(0.01*m)))*m1* z111^n

Explicit equations

1 m = (0.0405*t – 0.01126*t^2 + 0.001462*t^3 – 0.00006868*t^4)

2 m1 = (0.0405 – 0.02252*t^1 + 0.004386*t^2 – 0.0002747*t^3)

3 m2 = 1.266*t + 2.12*t^2 – 3.077*t^3 + 1.085*t^4

4 fi = 0.08

5 k = 0.1

6 n = 1.0

7 s = 1.0

```
8  E = (0.625 – 0.375*(tanh(5*(t – 3))))*(1 – fi)+(1/k)* (0.625 –
   0.375*(tanh(5*(t – 5))))*(fi)
9  A2 = 1
10 A3 = 1
11 p = –4
12 z = t*E–(z1 + z2)*((exp(–0.01*m)))–z3*(exp(–0.01*m2))
13 m21 = 1.266 + 4.24 *t – 9.231 *t^2 + 4.34 *t^3
14 f3 = if (t < 0.1) then (0) else (1)
15 f31 = (1/16)*t*(8 – t)
16 f32 = (exp(–0.1*t))
17 f33 = (sin(3.14*t/4))^2
18 f34 = if (t < 7) then (1) else (0)
19 f35 = (1/64)*t^2
20 E1 = (0.625 – 0.375*(tanh(5*(t – 3))))*1.0
21 z111 = t*E1–(z11 + z22)*((exp(–0.01*m)))
22 Y1 = t*E
```

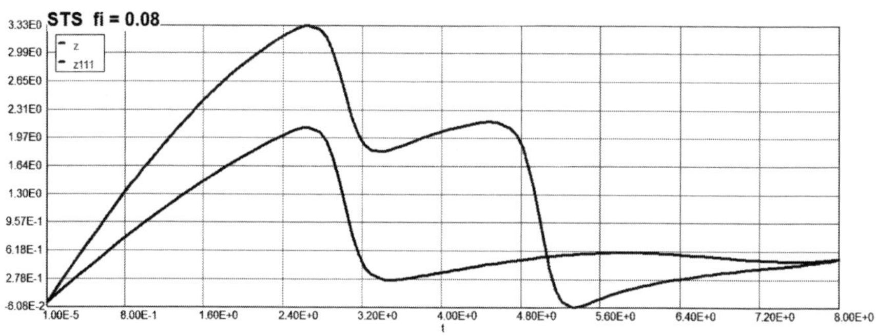

Figure 4.27: Stress-temperature-strain diagram $A3 = f_3(t)$; $p = -4$; $\theta^* = 4$; $\varphi = 0.08$.

Model: z = a1*t + a2*t^2 + a3*t^3 + a4*t^4 + a5*t^5

Variable	Value
a1	1.49415
a2	0.5190015
a3	–0.4013184
a4	0.0651564
a5	–0.0032358

$$\sigma = 1.532\theta + 0.192\theta^2 - 0.279\theta^3 + 0.0506\theta^4 - 0.00269\theta^5 \quad (4.45)$$

Example 4.19

Data: see previous Example 4.7, but p = –4 and f3 = f35 = f (θ) = (1/64)($θ^2$); E_0/E_2 = 0.1; fi = 0.1

Calculated values of DEQ variables

	Variable	Initial value	Minimal value	Maximal value	Final value
1	A2	1.	1.	1.	1.
2	A3	1.	1.	1.	1.
3	E	1.9	0.475	1.9	0.475
4	E1	1.	0.25	1.	0.25
5	f3	0	0	1.	1.
6	f31	5.0E-06	0	0.9999807	0
7	f32	0.999999	0.449329	0.999999	0.449329
8	f33	6.162E-11	6.162E-11	0.9999749	1.015E-05
9	f34	1.	0	1.	0
10	f35	1.563E-12	1.563E-12	1.	1.
11	fi	0.1	0.1	0.1	0.1
12	k	0.1	0.1	0.1	0.1
13	m	4.05E-07	4.05E-07	0.0705907	0.0705907
14	m1	0.0404998	0.0003976	0.0404998	0.0003976
15	m2	1.266E-05	1.266E-05	3014.544	3014.544
16	m21	1.266042	0.5363002	1666.482	1666.482
17	n	1.	1.	1.	1.
18	p	–4.	–4.	–4.	–4.
19	s	1.	1.	1.	1.
20	t	1.0E-05	1.0E-05	8.	8.
21	Y1	1.9E-05	1.9E-05	5.594458	3.8
22	z	1.9E-05	–0.2750274	3.620656	0.5075764
23	z1	0	0	0.9646977	0.9646977
24	z11	0	0	0.79102	0.79102
25	z111	1.0E-05	1.0E-05	2.10183	0.5333333
26	z2	0	0	2.11242	2.11242
27	z22	0	0	0.6766824	0.6766824
28	z3	0	–21.4011	2.688E+12	2.688E+12

Differential equations

1 d(z1)/d(t) = (exp(t/(1 + 0.067*t)))*((exp(0.01*m)))*m1*z^n

2 d(z2)/d(t) = (exp(t/(1 + 0.067*t)))*(A2*(z^s))*((exp(0.01*m)))*m1*z^n

3 d(z3)/d(t) = k*(exp(t/(1 + 0.067*t)))*A3*(f35*(t^p/(1 + 0.1*t^p)))*((exp (0.01*m2)))*m21*z^n

4 d(z11)/d(t) = (exp(t/(1 + 0.067*t)))*((exp(0.01*m)))*m1*z111^n

5 d(z22)/d(t) = (exp(t/(1 + 0.067*t)))*(1*(z111^s))*((exp(0.01*m)))*m1* z111^n

Explicit equations

1 m = (0.0405*t – 0.01126*t^2 + 0.001462*t^3 – 0.00006868*t^4)

2 m1 = (0.0405 – 0.02252*t^1 + 0.004386*t^2 – 0.0002747*t^3)

3 m2 = 1.266*t + 2.12*t^2 – 3.077*t^3 + 1.085*t^4

4 fi = 0.1

5 k = 0.1

6 n = 1.0

7 s = 1.0

8 E = (0.625 – 0.375*(tanh(5*(t – 3))))*(1 – fi)+(1/k)* (0.625 – 0.375*(tanh(5*(t – 5))))*(fi)

9 A2 = 1

10 A3 = 1

11 p = –4

12 z = t*E–(z1 + z2)*((exp(–0.01*m)))–z3*(exp(–0.01*m2))

13 m21 = 1.266 + 4.24 *t – 9.231 *t^2 + 4.34 *t^3

14 f3 = if (t < 0.1) then (0) else (1)

15 f31 = (1/16)*t*(8 – t)

16 f32 = (exp(–0.1*t))

17 f33 = (sin(3.14*t/4))^2

18 f34 = if (t < 7) then (1) else (0)

19 f35 = (1/64)*t^2

20 E1 = (0.625 – 0.375*(tanh(5*(t – 3))))*1.0

21 z111 = t*E1–(z11 + z22)*((exp(–0.01*m)))

22 Y1 = t*E

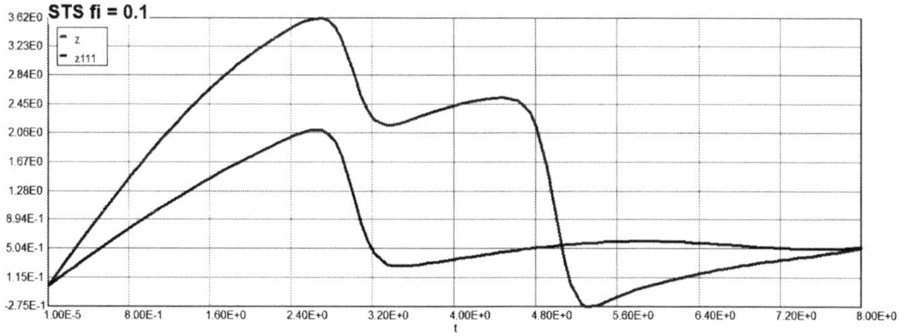

Figure 4.28: Stress-temperature-strain diagram A3 = $f_3(t)$; p = –4; $\theta^* = 4$; $\varphi = 0.1$.

Model: z = a1*t + a2*t^2 + a3*t^3 + a4*t^4 + a5*t^5

Variable	Value
a1	1.482666
a2	0.7150971
a3	0.1739853
a4	0.0735177
a5	–0.0035231

$$\sigma = 1.532\theta + 0.192\theta^2 - 0.279\theta^3 + 0.0506\theta^4 - 0.00269\theta^5 \qquad (4.46)$$

Example 4.20

Data: see previous Example 4.7, but p = –4 and f3 = f35 = f (θ) = (1/64)(θ^2); $E_0/E_2 = 0.1$; fi = 0.15

Calculated values of DEQ variables

	Variable	Initial value	Minimal value	Maximal value	Final value
1	A2	1.	1.	1.	1.
2	A3	1.	1.	1.	1.
3	E	2.35	0.5875	2.35	0.5875
4	E1	1.	0.25	1.	0.25
5	f3	0	0	1.	1.
6	f31	5.0E-06	0	0.999853	0
7	f32	0.999999	0.449329	0.999999	0.449329
8	f33	6.162E-11	6.162E-11	0.9999998	1.015E-05
9	f34	1.	0	1.	0
10	f35	1.563E-12	1.563E-12	1.	1.

11	fi	0.15	0.15	0.15	0.15
12	k	0.1	0.1	0.1	0.1
13	m	4.05E-07	4.05E-07	0.0705907	0.0705907
14	m1	0.0404998	0.0003976	0.0404998	0.0003976
15	m2	1.266E-05	1.266E-05	3014.544	3014.544
16	m21	1.266042	0.5408078	1666.482	1666.482
17	n	1.	1.	1.	1.
18	p	–4.	–4.	–4.	–4.
19	s	1.	1.	1.	1.
20	t	1.0E-05	1.0E-05	8.	8.
21	Y1	2.35E-05	2.35E-05	7.804924	4.7
22	z	2.35E-05	–0.9265854	4.305684	0.3642367
23	z1	0	0	0.9878946	0.830322
24	z11	0	0	0.79102	0.79102
25	z111	1.0E-05	1.0E-05	2.110389	0.5333333
26	z2	0	0	3.357069	3.357069
27	z22	0	0	0.6766824	0.6766824
28	z3	0	–2.303E+05	1.87E+12	1.87E+12

Differential equations

1 d(z1)/d(t) = (exp(t/(1 + 0.067*t)))*((exp(0.01*m)))*m1*z^n

2 d(z2)/d(t) = (exp(t/(1 + 0.067*t)))*(A2*(z^s))*((exp(0.01*m)))*m1*z^n

3 d(z3)/d(t) = k*(exp(t/(1 + 0.067*t)))*A3*(f35*(t^p/(1 + 0.1*t^p)))*((exp (0.01*m2)))*m21*z^n

4 d(z11)/d(t) = (exp(t/(1 + 0.067*t)))*((exp(0.01*m)))*m1*z111^n

5 d(z22)/d(t) = (exp(t/(1 + 0.067*t)))*(1*(z111^s))*((exp(0.01*m)))*m1* z111^n

Explicit equations

1 m = (0.0405*t – 0.01126*t^2 + 0.001462*t^3 – 0.00006868*t^4)

2 m1 = (0.0405 – 0.02252*t^1 + 0.004386*t^2 – 0.0002747*t^3)

3 m2 = 1.266*t + 2.12*t^2 – 3.077*t^3 + 1.085*t^4

4 fi = 0.15

5 k = 0.1

6 n = 1.0

7 s = 1.0

```
E = (0.625 – 0.375*(tanh(5*(t – 3))))*(1 – fi)+(1/k)*
  (0.625 – 0.375*(tanh(5*(t – 5))))*(fi)
A2 = 1
A3 = 1
p = –4
z = t*E–(z1 + z2)*((exp(–0.01*m)))–z3*(exp(–0.01*m2))
m21 = 1.266 + 4.24 *t – 9.231 *t^2 + 4.34 *t^3
f3 = if (t < 0.1) then (0) else (1)
f31 = (1/16)*t*(8 – t)
f32 = (exp(–0.1*t))
f33 = (sin(3.14*t/4))^2
f34 = if (t < 7) then (1) else (0)
f35 = (1/64)*t^2
E1 = (0.625 – 0.375*(tanh(5*(t – 3))))*1.0
z111 = t*E1–(z11 + z22)*((exp(–0.01*m)))
Y1 = t*E
```

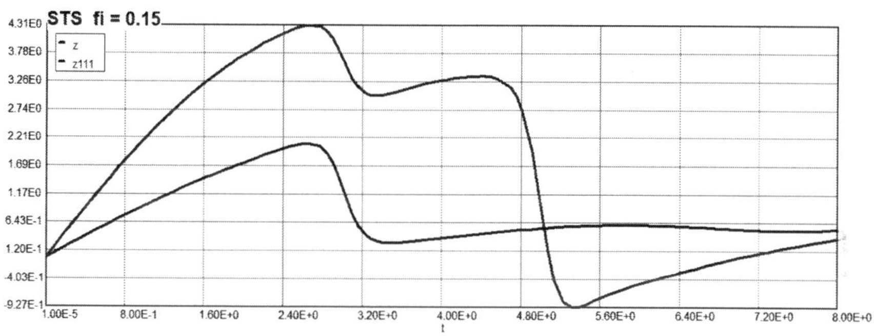

Figure 4.29: Stress-temperature-strain diagram A3 = $f_3(t)$; p = –4; $\theta^* = 4$; $\varphi = 0.15$.

Model: z = a1*t + a2*t^2 + a3*t^3 + a4*t^4 + a5*t^5

Variable	Value
a1	1.435905
a2	1.230572
a3	–0.6703235
a4	0.0969525
a5	–0.0043767

$$\sigma = 1.532\theta + 0.192\theta^2 - 0.279\theta^3 + 0.0506\theta^4 - 0.00269\theta^5 \quad (4.47)$$

Conclusions

1. Phenomenological creep deformation models, suitable in its various crystalline composites and nanocomposites structural materials are presented.
2. The single integral constitutive models are presented in terms of *creep compliance dimensionless non-linear equations* that are solved using approximate numerical methods.
3. Matrix degradation of composite materials often follows first-order kinetics as validated by experimental studies.
4. The entropy of a mass element changes with time for two reasons. First, because entropy flows into the volume element occupied by this mass element, second because there is an entropy source due to irreversible phenomena inside the volume element. Therefore the entropy of the system is the sum of the entropies of the subsystems.
5. The rate constants for many simple chemical reactions were found to be accurately described by the Arrhenius law, and in case of kinetics of nanoparticles synthesis processes can be described as a first order chemical reaction or as autocatalytic reactions: they are chemical reactions in which at least one of the products is also a reactant. The rate of these equations for autocatalytic reactions is fundamentally nonlinear and they should be presented by the conservation of mass differential equation.
6. The assumptions are made in order to simplify the corresponding heat and mass conservation equations.

References

[1] Kassing, R., Petkov, P., Kulisch, W. and Popov, C. (eds.). 2006. Functional Properties of Nanostructured Materials. Springer, N.Y.
[2] Kumar, S.K. and Krishnamoorti, R. 2010. Nanocomposites: Structure, phase behavior, and properties. Annual Review of Chemical and Biomolecular Engineering 1: 37–58.
[3] Milburn, C. 2008. Nanovision: Engineering the Future. Duke University Press, Durham, NC.
[4] Toumey, C. 2008. Reading feynman into nanotechnology, a text for a new science (PDF). Techné 13(3): 133–168.
[5] Silvestre, J., Silvestre, N. and de Brito, J. 2016. Polymer nanocomposites for structural applications: Recent trends and new perspectives. Journal Mechanics of Advanced Materials and Structures 23(11): 1263–1277.
[6] Davim, J.P. and Charitidis, C.A. (eds.). 2013. Nanocomposites Materials, Manufacturing and Engineering. De Gruyter, Berlin.
[7] Chen, F.F. 1990. Introduction to Plasma Physics and Controlled Fusion. Plenum, New York.
[8] Blackle, D.C. 1997. Polymer Latices: Science and Technology Volume 3: Applications of lattices. Chapman & Hall, London, UK.
[9] Iakoubovskii, K. Mitsuishi, K. and Furuya, K. 2008. High-resolution electron microscopy of detonation nano-diamond. Nanotechnology 19(15): 155705.

[10] Hanemann, T. and Szabó, D.V. 2010. Polymer-nanoparticle composites: From synthesis to modern applications. Materials 3: 3468–3517.
[11] Xu, H.X. and Suslick, K.S. 2010. Water-soluble fluorescent silver nanoclusters. Advanced Materials 22: 1078–1082.
[12] Bang, J.H., Helmich, R.J. and Suslick, K.S. 2008. Nanostructured ZnS:Ni2+ photocatalysts prepared by ultrasonic spray pyrolysis. Advanced Materials 20: 2599–2603.
[13] Mansour, A.F., Mansour, S.F. and Abdo, M.A. 2015. Improvement structural and optical properties of nanocomposites. OSR Journal of Applied Physics (IOSR-JAP) 7(2): Ver. II (Mar.–Apr. 2015) 60–69.
[14] Wulff, G. 1902. Zur Frage der Geschwindigkeit des Wachsthums und der Aufl¨osung der Krystallfl¨achen. Z. Kristallogr. 34: 449–530.
[15] Curie, M.P. 1885. Sur la formation des cristaux et sur les constantes capillaires deleurs diverse faces. Bull. de la Soci´et´e Min´eralogique de France 6: 145–150.
[16] Fonseca, I., Fusco, N., Leoni, G. and Millot, V. 2011. Material voids in elastic solids with anisotropic surface energies. J. Math. Pures Appl. 96(6): 591–639.
[17] Bailey, A. 1957. The adhesion of mica crystals surfaces. 2nd Int. Congr. Surf. Act 3: 406–417.
[18] Aufray, M., Menuel, S., Fort, Y., Eschbach, J., Rouxel, D. and Vincent, B. 2009. New synthesis of nanosized niobium oxides and lithium niobate particles and their characterization by XPS analysis. Journal of Nanoscience and Nanotechnology 9(8): 4780–4789.
[19] Russel, W.B., Saville, D.A. and Schowalter, W.R. 1989. Colloidal Dispersions. Cambridge University Press, Cambridge, UK.
[20] Dascri, A., Yu, Z.Z. and Nai, Y.W. 2009. Fundamental aspects and recent progress on wear/scratch damage in polymer nanocomposites. Mater. Sci. Eng. R 63: 31–80.
[21] Zhang, Z., Yang, J.L. and Friedrich, K. 2004. Creep resistant polymeric nanocomposites. Polymer 45: 3481–3485.
[22] Ritzhaupt-Kleissl, E., Haußelt, J. and Hanemann, T. 2005. Thermo-mechanical properties of thermoplastic polymer-nanofiller composites. pp. 87–90. *In*: Proceedings of the 4M 2005 Conference (Multi-MaterialMicro-Manufacture). Elsevier Publisher: Oxford, UK.
[23] Jordan, J., Jacob, K.I., Tannenbaum, R., Sharaf, M.A. and Jasiuk, I. 2005. Experimental trends in polymer nanocomposites—a review. Mater. Sci. Eng. A 393: 1–11.
[24] Razdolsky, L. 2016. Reliability Index and structural fire resistance of spacecraft and aircraft framing systems. AIAA SPACE 2016, AIAA SPACE Forum, AIAA 2016–5413.
[25] Wachtman, J.B. 1963. Chapter 6. pp. 139–169. *In*: J.B. Wachtman (ed.). Mechanical and Thermal Properties of Ceramics. NBS Washington: NBS Special Publication.
[26] Mackenzie, J.K. 1950. The elastic constants of a solid containing spherical holes. Proceedings of the Physical Society, Section B, Volume 63, Number 1.
[27] Hall, E.O. 1951. The deformation and ageing of mild steel: III discussion of results. Proc. Phys. Soc. London. 64: 747–753.
[28] Petch, N.J. 1953. The cleavage strength of polycristals. J. Iron Steel Inst. London 173: 25–28.
[29] Razdolsky, L. 2017. Probability Based High Temperature Engineering. Springer Nature Publishing Co., AG Switzerland.
[30] Razdolsky, L. 2017. Phenomenological high temperature creep models of composites and nanomaterials. AIAA SPACE 2017, AIAA SPACE 2017 Forum, AIAA 2017–5413.
[31] Frenkel, J. 1955. Kinetic Theory of Liquids. Dover, New York.
[32] Zeldovich, J.B. 1943. On the theory of new phase formation: cavitations. Acta Physicochim. (In English) URSS 18: 1–22.
[33] Brennen, C.E. 1995. Cavitations and Bubble Dynamics. Oxford University Press, Oxford, UK.

5

Physical Chemistry of Nanoparticles

5.1 Introduction

The objective of this chapter is to obtain a theoretical description of the rates of chemical reactions on a macroscopic level and to relate corresponding laws to mechanisms for reactions on a microscopic level. The rate of a reaction depends on a variety of factors: on the temperature, pressure; concentrations of the reactants and products; and on whether or not a catalyst is present. The *rate law* for a reaction is defined as the change in the concentration of one of the reactants or products with respect to time. In general, the rate of change of the chosen species will be a function of the concentrations of the reactant and product species as well as the external parameters such as temperature. The stochiometry of the reaction determines its proportionality constant. Consider the general reaction a*A + b*B = c*C + d*D, where a; b; c; d are the stochiometric coefficients and the rate of concentration [C] change is defined as: rate r = [(1/c)]d[C]/dt. This rate varies with time and is equal to some functions of the concentrations: [(1/c)]d[C]dt = f([A],[B],[C],[D]). Of course, the rates of change for the concentrations of the other species in the reaction are related to that of the first species, by the stochiometry of the reaction. For the example presented above, it is as follows:

$$\frac{d[C]}{cdt} = \frac{dD}{ddt} = -\frac{d[A]}{adt} = -\frac{d[B]}{bdt} \tag{5.1}$$

Let's agree that the rate will be called positive if the reaction proceeds from left to right, and positive derivatives are called for the products and negative ones for the reactants. The equation is:

$$(1/c)d[C]/dt = f([A],[B],[C],[D]) \tag{5.2}$$

and it called *the rate law for the reaction.* While f([A],[B],[C],[D]) might in general be a complicated function of the concentrations, it often occurs that f(x) can be expressed as a simple product of rates. The rate law in its differential form describes in the simplest of terms how the rate of the reaction depends on the concentrations, however it is useful to determine how the concentrations themselves vary in time. Of course, if we know d[C]/dt, in principle we can find [C] as a function of time by integration. In practice it is useful to consider both, the *differential and integrated rate laws* [1]. While the *differential rate law* describes the rate of the reaction, the *integrated rate law* describes the concentrations. In other words, the *differential rate law* is describing the chemical reaction process on a micro level and the *integrated rate law* describes this process from a macro kinetics point of view. Macro kinetics is the kinetics of macroscopic processes, describing the course of chemical transformations in their interrelation with the physical processes of matter (mass), heat transfer and electric charge transfer. The term macroscopic began to be used in the early 40's of the 20th century (in particular, in the works of D.A. Frank-Kamenetsky and M. Boudart [2–4]) and covers all phenomena that arise as a result of the influence of the processes of transport, of matter and heat on the rate of chemical transformation. Chemical kinetics considers the rate of only the chemical reaction itself. Under actual conditions, chemical transformation is often accompanied by processes of mass and heat transfer, depending on the hydrodynamic conditions of motion of gas, liquid or solid particles, and the rates of these latter processes often limit the overall process speed. A complex chemical-technological process is divided into chemical and physical components, and a separate study of them is carried out, after which their mutual influence is determined by mathematical methods using a computer. This is due to the impossibility of reproducing in the laboratory settings all the features of a real process, accompanied by the transfer of matter and heat, as is the case in industrial conditions.

Chemical reactions are always associated with a variety of physical processes: heat transfer, absorption or emission of electromagnetic waves, electrical phenomena and others. Thus, a mixture of substances which produces a chemical reaction, releases energy into the external environment in the form of heat or absorbs it from the outside. As the temperature of the substance rises, the intensity of the vibration motions inside the molecules increases and the bond between the atoms in the molecules weakens. After reaching a known critical energy value, molecules can dissociate or interact with other molecules in a collision, that is, a chemical process occurs. In all cases, there is a close connection between the physical and chemical phenomena and their interaction. Physical chemistry is the study of the relationship between physical and chemical phenomena. It deals with the multilateral studies of chemical reactions and their corresponding physical processes using theoretical and experimental methods of physics and chemistry. Physical chemistry as a boundary science covers the phenomena studied from several angles. It takes into account the nature of the interaction of atoms and molecules and thus examines complex and interconnected phenomena of the material world. Physical chemistry pays the most attention to the study of the laws of the course of chemical processes, *in time* and the laws of chemical equilibrium. The main task of physical chemistry is prediction—the chemical process over time and the state of equilibrium in various conditions, on the basis of the data on the structure and properties of molecules that comprise the system. Knowledge of the conditions of the chemical reaction leads to the possibility of controlling the chemical processes of the technological production of nanomaterials.

5.2 Disperse systems

A substance that is in state of strong grinding (high degree of dispersion) is called a dispersed phase. The medium in which the disintegrated substance is placed is called the dispersion medium. There are several ways to classify disperse systems [5, 6].

5.2.1 *Classification by degree of dispersion*

Systems in which the dispersed phase has particles with a size $r > 10^{-5}$ cm are called coarse dispersions. If r is $10^{-7} \leq r \leq 10^{-5}$ cm, the system is called a colloidal solution, or a colloidal dispersion. When $r < 10^{-7}$ cm, we have a molecular dispersion or a true solution. The concept of "colloid" was introduced by the English scientist Thomas Graham [7], he studied the behavior of glue solutions in water and found that through the parchment, molecules of water pass, and the molecules glue-no. Parchment was selectively permeable or a semi-permeable septum. The separation of molecules is called dialysis. Glue in Greek language is καλα, hence the name—colloids. Graham suggested using the value of the rate of diffusion of the molecules through a parchment partition as a criterion for dividing all substances into crystalloids and colloids. He believed that if substances diffuse rapidly and are capable of crystallization, they are crystalloids.

5.2.2 *Classification of disperse systems*

The concepts introduced above are rather arbitrary, since many polymer molecules consist of millions of atoms and have a size $r > 10^{-5}$ cm. According to their chemical nature, they should be attributed to true solutions, but are included with colloids in particle sizes. All disperse systems are thermodynamically unstable, since they have a highly developed surface. As a consequence, such systems are characterized by the high value of free surface

Gibbs energies G_ω. Many of the finely dispersed systems grow so slowly that this process cannot be detected. Thus, a suspension of gold particles in water, cooked more than two centuries ago by Faraday, and stored in England at the Royal Museum has practically no sludge. The reason for this stability lies in the kinetic factors—the Brownian motion pushes out tiniest particles and prevents them from connecting. Many regularities of the behavior of disperse systems can be explained by the surface properties of the phases and their dependencies on various factors. The surface phenomena can be ignored if the relative phase surface is small: $W/V \rightarrow 0$—the ratio of the surface area to the volume is small. On the other hand, this ratio characterizes the ratio of the number of atoms (molecules) on the surface N_ω and their number in the volume N_0 of solid particles. If the ratio W/V is small, then the ratio N_ω/N_0 is also small and the bulk of the atoms (molecules) of the substance are in the bulk phase. However, as the substance is fragmented, its specific surface, and the fraction of atoms located on the surface also grows.

For instance take a cube of matter with a side of 1 cm and split it, dividing each side of the cube into 100 pieces. We obtain a set of small cubes with a rib size of 10^{-2} cm. The surface area of each newly formed cube is $W_1 = 6 \times 10^{-4}$ cm^2, and their total number will be $N = 10^6$. The total surface area of all cubes is $W_{tot} = 6 \cdot 10^2$ cm^2, a hundred times the surface area of the original cube. The total surface of the substance increases as the degree of fragmentation increases (the number of particles formed). In the latter case, when the size of the edge of the particle is 10^{-7} cm, the surface area of each of the 10^{21} elementary cubes is $W_1 = 6 \cdot 10^{-14}$ cm^2, and their total surface will have an area of $6 \cdot 10^7$ cm^2 = $6 \cdot 10^3$ m^2 = 0.6 hectares. A small cube of solid material, when dispersed, increases its surface to thousands of square meters. In a finely divided substance, a significant fraction of the atoms are in the surface layers. In the disperse systems with particle sizes of 10^{-5}–10^{-6} cm. it is impossible to ignore surface phenomena in such systems, since it helps to explain many regularities in the behavior of particles. Therefore, surface phenomena must always be borne in mind and taken into account, and in some cases, the surface properties of the particles are the determining. It is necessary to take into account such characteristics as surface tension σ, edge angle θ, adhesion W_a, and adsorption Γ_i, at the interface. At very small particle sizes, the characteristics themselves begin to depend on the linear parameters of the micro objects [8]. On a flat surface, the resultant of the forces that pull the atom into the volume of the phase is maximal. If the particle of matter is small, then because of the considerable curvature of the surface, the part of the bonds of the surface atom with the bulk atoms disappears. The force field closes to a lesser degree and the surface tension decreases [9]:

$$\sigma_r = \sigma_0\left(1 - \frac{2\sigma}{r}\right) \tag{5.3}$$

σ_0—surface tension for a large drop; r—radius of a droplet or half the thickness of a thin film; δ is the thickness of the transition layer. Consider the interface between gas and liquid. Within the phase the density is constant. When passing through the interphase boundary, the values vary from the values characteristic of one phase to those characteristic of the other phase.

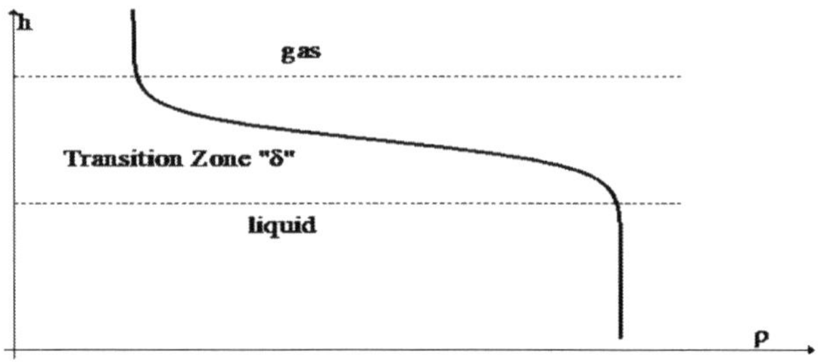

Figure 5.1: Gibbs energy change.

Change in Gibbs energy with increasing surface on Δω is usually expressed by the equation [10]:

$$\Delta G = \sigma \cdot \Delta_{\omega} \tag{5.4}$$

The surface (interfacial) tension is greater than zero: $\sigma > 0$—always. Therefore, if the surface increases by Δω, then the Gibbs energy of the system increases. And since the minimum state of Gibbs energy corresponds to the equilibrium state, the system is removed from equilibrium when the surface is increased. For P and T = const, a system having a strongly developed surface, is thermodynamically unstable. It tends to spontaneously reduce its surface. In emulsions, particle enlargement is observed. The greater the surface tension at the interface, the more unstable the disperse system. For a conventional thermodynamic system, change of Gibbs energy is found from the expression

$$\Delta G = \Delta H - T\Delta S \tag{5.5}$$

This equation describes the behavior of a significant number atoms or molecules. Thermal particles vibration of atomic size provides the possibility of a system of atoms or molecules, to realize certain thermodynamic states. By inference, the mobility of colloidal particles due to their Brownian motion makes it possible to analyze the behavior of disperse systems using thermodynamic regularities. For disperse systems with a developed surface, the analogue of the energy term ΔH is the product $\sigma \cdot \Delta_{\omega}$, describing the change in the energy of the system with a surface reduction: $\Delta G = \sigma \cdot \Delta_{\omega} - T\Delta S$. Let us consider how the components of the Gibbs energy of a disperse system behaves as the particle sizes decrease [11]:

$$\Delta G = \frac{\alpha \sigma L^2}{L^3} - \frac{\beta T}{L^3} \tag{5.6}$$

where L is the linear particle size; n is the number of particles; Δ_ω is the increased surface; $\Delta_\omega = \alpha \cdot L^2 \cdot n$; α is a coefficient that depends on the shape of the particles; ΔS is the change in the entropy of the set of particles. If the grinding is small (number n), the entropy growth is proportional to the number of particles: $\Delta S \approx \beta \cdot n$, where $n \sim 1/L^3$. If we plot the variation ΔG = f (L), then it turns out that at some temperatures we obtain a region with negative values of ΔG. Given the temperature, it is possible to determine at what values of surface tension σ this thermodynamic stability is possible ($\Delta G \leq 0$). We find the boundary value of σ by equating to zero the expression for ΔG [12, 13]:

$$\Delta G = \alpha \sigma - \beta T \frac{1}{L^2} = 0; \quad \sigma_{cr} = \frac{\beta T}{\alpha L^2} = \varphi \frac{RT}{L^2} \tag{5.7}$$

where φ is a coefficient of proportionality. Approximate calculations show that when σ is small (of the order of 10^{-1}–10^{-2} mJ/m^2), then at a temperature of 20°C and $\varphi \approx 15$, the values of ΔG are negative. Then the disperse system can be thermodynamically stable. Consequently, in a number of cases the process of self-dispersing, of a substance is thermodynamically justified. If we take silicate melts, where the temperature reaches $T \approx 1500$°C, then the critical values of σ can reach the values of 0.1–3.0 mJ/m^2. Even if the values of $\sigma > \sigma_{cr}$ and the system is not thermodynamically stable, while interphase (surface) the stresses σ are not very large, the presence of small perturbations can impede its enlargement. Therefore, in many cases even finely dispersed systems can last for a long time without self-consolidation. Small pulsations, and convective currents do not allow the system to be enlarged. If the tension value is sufficiently large ($\sigma > 1$ mJ/m^2—for aqueous solutions and $\sigma > 10$ mJ/m^2—for melts), then the system is unstable and spontaneously must undergo particle coarsening. There are two types of enlargement. The first is when particles can be connected without the interface between them (two bubbles, two drops). This coarsening is called coalescence. The second occurs after the solid particle and the bubble are joined, a drop of liquid and a bubble develops an interphase boundary. This enlargement is called coagulation. Nevertheless, even with high values of interfacial (surface) tension σ (100–1000 mJ/m^2), sometimes disperse systems exist for a very long time. In these cases, they are given stability by kinetic factors.

5.2.3 *Sensitivity of tension stresses to concentration surfactants*

Rehbinder [14] drew attention to the delay in the diffusion of adsorption processes, during the flow of liquid solution from the zone of contact of converging solids. When the particles approach each other, the liquid from the thin part is mechanically squeezed out into thick layers. The film is thinned and stretched. Adsorption equilibrium cannot be restored in time. Surface-active components will move to the point of contact of solid particles. This is equivalent to an additional flow of a substance that prevents film thinning and its stability. When the saturation of the surface layer of the surface—the active component is not influenced by the adsorption of certain substances, then the surface layer has an increased viscosity, an increased shear resistance, a rupture, and so on. Similar phenomena occur in melts.

5.3 The rate of chemical reaction

Chemical reactions can occur in systems consisting of one (homogeneous reaction) or several (heterogeneous) phases. The rate of a homogeneous chemical reaction **v** over a given substance is the change in the amount of this substance n_i (in moles) per unit time τ and in unit reaction volume V:

$$v = \pm\frac{1}{V}\frac{dn_i}{d\tau} = \pm\frac{d}{d\tau}\left(\frac{n}{V}\right) = \pm\frac{dC_i}{d\tau} \tag{5.8}$$

That is, for a homogeneous chemical process proceeding at a constant volume, the speed of the process over a certain substance is called the change in the concentration of the i-th substance per unit of time. The velocity dimension in this case is mole/sec or mol/мin.

The experimental dependence $C_i = f(\tau)$ is presented in the form of an analytical one, or as a graphical dependence and is called the kinetic curve. Often, when there are no precise methods for quantifying substances involved in the chemical process, the chemical transformation is monitored by the change in some property of the system (refractive index, electrical conductivity, and for gas reactions, volume or pressure). It is only necessary that it be one of the notional dependence $C_i = f$ (property). For the continuous analysis of substances directly in the reaction volume, various methods [15–17] of physical and chemical analysis are used: conductivity metric, potentiometric, paleography, and so on. Since the reaction rate is always positive, and the concentration participants of the reaction can both decrease and increase, in general and in the above equations, the corresponding sign is placed. A plus sign is used if the rate of reaction is judged by the increase is in the concentration of the reaction product, and the minus sign if the change (decrease) is in the concentration of the starting substances. The starting materials are consumed, and the reaction products are formed in the equivalent quantities (according to the stochiometric coefficients), therefore, in determining the reaction rate, there is no need to monitor the change in the concentration of all interacting substances. So, for example, for the synthesis of ammonia: $N_2 + 3H_2 = 2NH_3$ the following relations are correct

$$v = \frac{1}{(-1)}\frac{dC_{N_2}}{d\tau} = \frac{1}{(-3)}\frac{dC_{H_2}}{d\tau}(\frac{n}{V}) = \frac{1}{2}\frac{dC_{NH_3}}{d\tau} \tag{5.9}$$

Thus, the rate of the chemical reaction is always a positive value and has the same value for the given reaction irrespective of the change in the concentration of which reagent is expressed.

The main factors affecting the rate of chemical reactions are the temperature, the concentration of reactants, the presence of a catalyst in the system. According to the law of effective masses at T = const: the reaction rate is proportional to the product of the concentrations of the reacting substances in certain degrees, therefore, in general, the relationship of the velocity with these concentrations can be represented for the reaction a(A) + b(B) = c(C) + d(D) as follows:

$$v = K[C_A^{n_A}][C_B^{n_B}] \tag{5.10}$$

In this equation, called the kinetic equation, the proportionality coefficient K is a rate constant, and the exponents n_A and n_B partial orders of reaction for the substance A and B. The sum of the exponents at the concentrations in the kinetic equation of the reaction is called the total (general) order $n_A + n_B = n$. The rate constant of the reaction in terms of physical meaning is equal to the rate of chemical reaction at fixed unit concentrations of reacting substances. It depends on the nature of the reacting substances, temperature, and presence of catalyst in the system. The fact that, in contrast to the rate of chemical reaction, the rate constant K does not depend on concentration and time makes it convenient in technological practice. Chemical reactions from the kinetic point of view are classified according to the molecularness and order of the reaction. By the order of the chemical reaction they are subdivided into the reactions of I, II and III orders, and in addition, there are reactions of zero and fractional order.

5.4 Temperature effect on chemical reaction rate

Temperature has a strong effect on the rate of chemical reaction. It has been experimentally established that, with an increase in temperature by ten degrees, the rate of homogeneous reaction increases by a factor of 2–4 (the Van's Hoff rule). The number that shows how many times the rate of constant increases with a temperature increase of ten degrees is called the temperature coefficient of the rate constant $-\gamma$: $\gamma = \frac{\tau+10}{}$. In the general case, the following relation holds as:

$$K_{T_2} = K_{T_1}\gamma^{\frac{T_2 - T_1}{10}} \tag{5.11}$$

where K_{T1} and K_{T2}—rate constants, respectively, with temperature T_1 and T_2. The Van's Hoff rule gives an approximate dependence of the velocity reaction from temperature. A more accurate mathematical temperature dependence of the constant was established by Arrhenius, whose differential equation has the form:

$$\frac{d\ln K}{dT} = \frac{E}{RT^2} \tag{5.12}$$

where—E is the activation energy. Taking E independent of temperature and integrating, we obtain the Arrhenius equation in the form:

$$K = Ae^{-\frac{E}{RT}} \tag{5.13}$$

where A is the pre-exponential factor; E is the activation energy. Constant A and E are usually determined experimentally from the temperature dependence of the reaction rate constant. For this, Equation (13) must be written in linear form:

$$\ln K = \ln A - \frac{E}{RT} \tag{5.14}$$

In the coordinates [lnK; –1/T], this dependence is expressed by a straight line with a negative slope, from which the activation energy and lnA can be found. The factor A in the Arrhenius equation, according to the theory of active collisions, is equal to the number of collisions per unit volume per unit time. The factor exp (–E/RT) determines the fraction of active collisions. In order to eliminate the discrepancy between the experimental and calculated data, a correction p (a *steric factor*; $p < 1$ generally) is introduced into Equation (5.13), taking into account the orientation of the molecules in space.

It is assumed that the rate of the chemical reaction is directly dependent on the likelihood of rapprochement and the correct orientation of the molecules of the reacting substances relative to each other, and that the ability to reorient the molecule in space depends on its mass and branching. For large and branched molecules, the probability at the same temperature, to orient correctly to form an activated reaction complex is lower than for small molecules. In addition, the more molecules of active groups capable of forming bonds, the more likely it is that these groups will be next to each other when the molecules meet, and thus the higher the probability of the appearance of new chemical bonds. All this was taken into account by the choice of the coefficients K. It was assumed that if the components of the reaction mixture are able to react with each other in four ways, then the rate of chemical reaction between them, other things being equal, will be four times higher than for molecules that can react with one another in a unique way. The corresponding combinatorial analysis is not difficult to realize. As a result, it turned out that the chemical reaction rate coefficients should be specified using two constants k_+ and k_-, taking into account the reaction features.

Two phases appear in the liquid forms. In a homogeneous liquid, separation can occur if the total energy of the separated phases is less than the energy of the initial phase. The liquation is observed in many crystallites, in particular binary systems. The cause of segregation is the heterogeneous structure of the initial phase. Micro inhomogeneity can be caused by:

- different concentration of components;
- different coordination states of actions;
- structured heterogeneity;
- the difference in the binding energies of the components, and so on.

There is a stable and metastable phase separation. Stable segregation is observed at temperatures above the liquids temperature. When the temperature falls below the critical temperature, the initial melt decomposes into two phases (Fig. 5.2).

Line 'a–b' characterizes the composition of one phase, the line 'b–c'—the composition of the second phase. At a temperature below the line level 'ac' there is a melt of composition 'ab' and crystals of pure component B. The viscosity of melts at high temperatures is usually low and the hetero-phase system of two phases is quickly divided into two layers—stratified. A liquid with a lower density is collected from above. In systems prone to crystallization formation, the laws of liquid phase separation in liquids in a metastable state should be the same as in the supercilious region. In a subliminal region, phase separation can be observed even if there is no stable segregation in the system. At $T < T_{cr}$, the system breaks up into two phases, 'ab' is the composition of one phase; 'bc' is the composition of the second phase. *From the point of view of thermodynamics there is no difference between stable and metastable segregations. However, the kinetics of the*

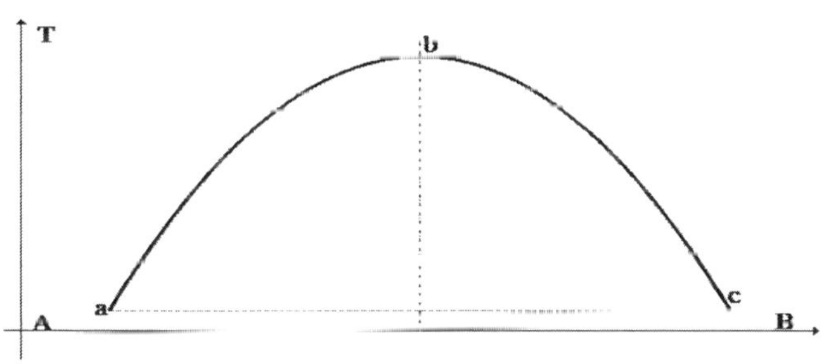

Figure 5.2: Decomposition into two phases.

phenomena is very different. The division into two phases is connected with the mutual displacement components. Because of lower temperatures, the rate of phase separation for metastable phase separation is several orders of magnitude lower than for stable segregation. Metastable phase separation does not lead to stratification, but to phase separation. In this case, the two-phase glass is a micro-inhomogeneous material consisting of two uniformly distributed different substances with different properties. The diameter of dispersed formations is several tens and hundreds, and rarely—thousands of nanometers. The shape of the particles depends on the composition and mechanism of the decay. Precisely for this reason the step-function f34, f35 and f37 had been introduced in creep constitutive integral Equation (4.16—see Chapter 4).

5.5 Phenomenological kinetics

The aim of chemical kinetics is to obtain an equation of the so called kinetic curve on the basis of the formulated laws and postulates. Such an equation contains the constants that are characteristic for a given transformation (rate constants, activation energies, and sometimes equilibrium constants) under given experimental conditions, and allows for the quantity of matter to be expressed at a given time. A peculiarity of the course of chemical transformations is that the formation of products in appreciable amounts in different reactions takes place over substantially different times. For example, the hydrolysis of sucrose requires several hours, and the recombination of methyl radicals is several microseconds. This requires the use of a variety of measurement methods.

5.5.1 Basic definitions and postulates

The present approach to chemical reaction kinetics is based on so-called Extended Irreversible Thermodynamics (EIT) [18] that is coupled with the effective medium approach (EMA) [19, 20], which allows assimilating the nanocomposites to effective homogenized media.

The main purpose is to model the heat transport associated to dispersion of nanoparticles in a bulk material, called the matrix. The description of such a heterogeneous two-component medium can be simplified by appropriately homogenizing it, as described within the effective–medium approach. Following the lines of thought of Hasselman [21] and later on by Nan et al. [22], the effective homogenized nanocomposite media is characterized by the effective modulus of elasticity of the bulk material (given by the so-called "rule of mixture") on one hand and the chemical reaction mechanism (see below) of metal solid phases composed of nano- to micro-crystalline particles on the other hand. It is examined also that the significance of various effects on the constitutive creep equation of nanocomposites, namely the particle's size, the volume fraction of particles, and the boundary matrix-particle interface resistance; the influence of nanoparticles' clusters and their progressive agglomeration with the temperature increase. For the sake of simplicity, it is assumed that the particles are spherical and mono–disperse. Many experiments have been performed on investigating the role of agglomeration [23–26], which will therefore be given a special attention in this chapter.

The proposed creep models are taking into account this nanoparticles' agglomeration and introduce an agglomerate increase effect due to a change in agglomerate's size and high temperature rate changes. These models also examine the influence of volume fraction on the creep rate degree. This point will receive a particular attention in this work and the relation between the degree of agglomeration; the particle volume fraction and the chemical transformation process in time. The results are used as one of the inputs in our creep constitutive equation model (see Chapters 4 and 7). In order to describe the chemical transformation process in time, the introduction of the concept of the rate of chemical reaction is required. A chemical reaction can take place (apart from spontaneous decay) only when the reacting particles are in direct contact (collision). If, as a result of the collision, only the intended products are immediately obtained, this reaction is called simple (using the concept elementary) reaction. In this case, only one energy barrier is usually is present. If the transformation consists of several simple reactions (reversible, parallel, sequential and their various combinations), this reaction is called complex. The actual sequence of the individual elementary reactions in a unit of the reaction space is known as reaction mechanism. The reaction mechanism of metal oxides and other stable solid phases composed of nano- to micro-crystalline particles is generally a complex process and it can be simplified by distinguishing two main contributions to the chemical reactivity of the solid: the potential rate coefficient k(E), and the conversion-dependent function f(y). Both k(E) and f(y) can be obtained experimentally and theoretical consequences of their particular forms are discussed below. One of the basic principles of nano- to micro-crystalline particles evolution is the assumption that two contributions to the reactivity of microcrystalline solids may be separated: the influence of reaction temperature and that of the increase of solid particle consumption (conversion), y [27].

$$\frac{dy}{dt} = k(T)f(y) \tag{5.15}$$

where dy/dt is a reduced reaction rate, k(T) is a potential-dependent rate coefficient, y is a reaction conversion. If during the transformation the volume of the reaction space does not change (a closed system of constant volume), then dy/dt can be replaced by dC/dt, where C is the concentration of reactants.

$$\frac{dC}{dt} = k(T)f(C) \tag{5.16}$$

All previous attempts to describe the kinetics of chemical reactions of solid particles [28] have been based on a similar separation of potential- and conversion-dependent functions. It should be noted here that utilization of Equation 5.16 is meaningful, only if the heat transfer is significant in the overall chemical reaction. The second term on the right hand side of Equation 5.16, f(y) may be chosen from a wide variety of models offered by the experimental and theoretical studies. The models should describe a possible development of formation of agglomeration of nanoparticles and reactivity during the reaction. Many particular f(y) functions can be related to a known kinetic law [27, 28], and hence a reaction mechanism can be postulated. From the variety of the models, particularly the shrinking sphere model or as frequently denoted as general *reaction order kinetics*, have been most frequently used [29–32] (with the assumption of an immediate reaction on the whole reactive surface of the solid):

$$f(y) = (1 - y)^{\gamma} \tag{5.17}$$

Equation 5.17 yields monotonously decreasing curves of dy/dt vs. time. The solution of the kinetic equation for the given initial condition makes it possible to establish a relationship between the change in concentration in time and the initial concentrations, i.e., obtain the equation of the kinetic curve. According to the Cauchy-Lipchitz-Horowitz theorem for the equation dC/dt = f (k, C) (f is an autonomous function, since t does not enter the right-hand side of the equation and given a value of C_0 at t = 0), there exists a *unique solution.* This is *the mathematical basis of phenomenological chemical kinetics*. Let's note that it is necessary to consider a limiting component, i.e., the initial amount of which is so small that, in accordance with the stochiometry of the reaction, it can react in its entirety, while others, which are larger, will remain in appreciable quantities. The reaction rate in such a case is better expressed through a limiting substance. Determining the rate of the chemical reaction can be graphically differentiated by the kinetic curve. In accordance with the form of recording, the rate of the chemical reaction has a dimension of mole/s or mole/h, and the dimension of the rate constant depends on the values of the exponents in which the reagent concentrations enter the kinetic Equation (5.16). Equation 5.16 is exactly valid only for a shape-preserving surface reaction of mono dispersed particles, and is approximately valid for general surface reaction of poly dispersed crystals. The term γ is a function of a particle shape and a size distribution. Whenever the initial reaction rate increases up to its local maximum at $0 < y < 1$ the Avrami-Erofeev equation [33, 34] is commonly used

$$(1 - y) = \exp[-(kt)^{\beta}] \tag{5.18}$$

Based on the above a thermodynamic model for transient creep in nanocomposites is proposed (see Chapter 4). The model takes into account the particle's size, the particle's volume fraction, and interface characteristics with an emphasis on the effect of agglomeration of particles on the *effective engineering creep* of the nanocomposite. The originality of the present uniaxial integral type creep model is based on extended irreversible thermodynamics, combining nano- and continuum-scales but without invoking molecular dynamics. The analysis is limited to spherical nanoparticles. The dependence of the degree of agglomeration with respect of the volume fraction of particles is also discussed and the combination of Arrhenius and the power-law relation is established through a kinetic mechanism. It is shown that the effective solid particle consumption (conversion) and agglomeration value, y, may increase or decrease with the degree of agglomeration. From thermodynamic considerations it follows that any reaction must take place in both directions the formation of products and their conversion to the starting compounds. However, in the some cases the amount of initial substances in equilibrium is negligible and often cannot be determined experimentally. In accordance with the accuracy of the analysis and the method that is used here, one can stop noticing that the initial substances in the reaction medium at different values of their initial concentrations. In such cases, it can be said that the reaction was complete, i.e., it is irreversible. Despite the conventionality of the concept of "irreversible reaction," this technique allows simplifying mathematical transformations and it is widely used in chemical kinetics.

5.6 Kinetics of simple irreversible reactions

It is considered in this section that the reactions occurring in a homogeneous medium with a constant volume of the reaction space. Let each moment of time of the concentrations of all substances be constant throughout the volume (the diffusion rate is much greater than the rate of the chemical reaction). It is also assumed that the conditions of the experiment make it possible to maintain the temperature constant over time and throughout the reactor volume. For irreversible reactions, the presence of certain quantities of products in the system at the initial instant of time does not affect the reaction rate and under the above conditions the current concentration (the value at a given time) the concentration of the resulting substances will be greater by the value of the initial concentration determined by the equation of the kinetic curve.

5.6.1 *First-order chemical reactions*

Consider a first-order reaction: $A \rightarrow \sum \nu_i P_i$, where P_i means one of the products, and the → sign indicates that the reaction proceeds in one direction. If [A] is the current concentration of the substance, then according to the basic postulate of chemical kinetics, the conversion rate A will be determined as:

$$r = -d[A]/dt = k[A] \tag{5.19}$$

Separating the variables and integrating under the initial conditions: $[A] = [A]_0$, @ t = 0, we obtain the solution

$$[A] = [A_0]\, e^{-kt} \tag{5.20}$$

Here k is the rate constant of the first-order reaction.

Sometimes other designations are used: a is the initial concentration of the substance A, x is the concentration of A, converted into products vs. time. In this case (a – x) will correspond to the concentration of substance A at time t. Then the expression for the reaction rate has the form:

$$r = -\frac{d(a-x)}{dt} = \frac{dx}{dt} k(a-x) \tag{5.21}$$

After separating the variables and integrating with the initial condition: t = 0; (a – x) = a, we obtain

$$x = a[1 - \exp(-kt)] \tag{5.22}$$

The following conclusions can be drawn from the obtained formulas:

- On the graph in coordinates $\ln\{[A]/[A]_0\}$ vs. t, there is a straight line.
- The conversion time of the starting substance by ½ does not depend on the initial concentration, $t_{1/2} = (\ln 2)/k$.
- The concentration of the initial substance with infinite time will be zero.
- The reaction rate will also be expressed by the exponential: $r = [A_0]\, e^{-kt}$.
- The quantity 1/k has the dimension of time. After substituting 1/k into the kinetic equation the initial concentration A_0 decreases by "e" times.

Recalling that the rate is determined by the tangent to the kinetic curve, another important relation is obtained. The initial transformation rate A is $[A]_0/t_0$, where t_0 is the time that the tangent intersects with the time axis. Substituting this into a kinetic equation one can see that: $[A]_0/t_0 = k\,[A]_0$, i.e., $t_0 = 1/k$. Therefore for any initial concentrations $[A]_0$, the tangents will intersect the t axis at the same point–value corresponding to the average lifetime of the molecules (Fig. 5.3).

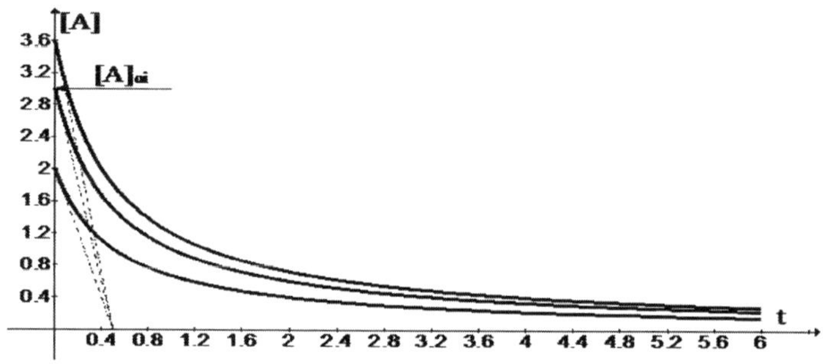

Figure 5.3: Initial transformation rate A.

Consider the basic formula for the kinetics of a first-order reaction: $-\frac{d[A]}{dt} = k[A] \cdot$

Strictly speaking, the process of transformation consists of two stages: physical—the redistribution of energy and the establishment of a statistical equilibrium distribution of energy and chemical—the actual conversion of molecules. In chemical kinetics, we assume that the particles are statistically independent; an equilibrium distribution of the molecules is established in terms of energy. Then, as the concentration increases by n times, the reaction rate also increases by the factor of n. Therefore after multiplying concentration [A] by a constant coefficient n the first-order reaction equation is as follows:

$$-\frac{dn[A]}{dt} = nk[A] \tag{5.23}$$

Since n is a constant value, the equation is converted to the original one. This is followed by two important conclusions. First, when studying the kinetics of a first-order reaction, one can measure not any concentration, but any of the parameters, which is constantly proportional to it throughout the concentration range. Secondly, the kinetic curves measured at different initial concentrations can be reduced to one if the concentrations are divided by the ratio of their ratio to one of them selected as the standard. Such a method of processing experimental data will be called an *invariant of the first kind*. This, as well as the independence of $\tau_{1/2}$ from the initial concentration, can be a criterion for the feasibility of the kinetics of the first-order reaction.

In practice, it is rather difficult to determine the time of the onset of the reaction, so calculation by the above formulas is not always correct. From the expression for the current concentration of the initial substance, it is easy to obtain the relations

$$\frac{[A_1]}{[A_2]} = \exp[-k(t_1 - t_2)] \tag{5.24}$$

and

$$\ln\left(\frac{[A_1]}{[A_2]}\right) = -k\Delta t \tag{5.25}$$

which do not include the reaction start time and the initial concentration.

Consider a first-order reaction at the initial instants of time, when x is small. Then, assuming that $kt << 1$, after expansion of the exponent in a series, one can write: $x = akt$ and $dx/dt = ak = const$. It is usually difficult to observe deviations from a straight line when the starting substance is converted to 10–15%.

5.6.2 *Second-order reactions*

The reaction of the second order in the simplest form can be written as $A + B \rightarrow \sum P_i$ and its rate in accordance with the basic postulate of chemical kinetics will be calculated by the formula:

$$r = k(a - x)(b - x) \tag{5.26}$$

where a and b are the initial concentrations of the initial substances, x—the concentration of the transformed A and B. After separating the variables, Equation (5.26) can be rewritten as:

$$\frac{dx}{(a - x)(b - x)} = kdt \tag{5.27}$$

And the result of integration is:

$$kt = \left(\frac{1}{a - b}\right)\ln\frac{b(a - x)}{a(b - x)} \tag{5.28}$$

It is obvious that the rate constant of the second-order reaction has the dimension l/mol.s. The new variable is introduced:

$$\frac{b(a - x)}{a(b - x)} = D \tag{5.29}$$

In coordinates, ln (D) vs. t, the time dependence of the concentration has the form of a straight line. It is useful to note that to study the kinetics of such a reaction it is not necessary to measure the concentrations of both substances. From the equation of material balance, one has

$$[B] = [B_0] - [A_0] + [A] \text{ and } \ln([A]/[B_0] - [A_0] + [A]) \tag{5.30}$$

and it depends linearly on time. The half-transformation time of substance A in this case:

$$t_{1/2} = \frac{\ln b(2b - a)}{k(a - b)} \tag{5.31}$$

It is clear that b must be greater than a/2. If the initial concentrations of the substances coincide, then the previous equation gives an uncertainty. For these cases, the kinetic equation can be written in a slightly different form. If the initial substances are different in nature, but their concentrations are equal, then the solution of the equation

$$r = k[A]^2 \text{ is } 1/[A] - 1/[A_0] = kt \quad (5.32)$$

If one substance A reacts and the change in its concentration is measured during the experiment, then $r = 2k[A]^2$ and the solution has the form $1/[A] - 1/[A_0] = 2kt$. For the half-transformation time of two different substances with equal initial concentrations, the expression $t_{1/2} = 1/k\ [A]_0$, i.e., in this case the half-transformation time depends on the initial concentration.

If the concentration of one of the substances is taken in a considerable excess ($[B]_0 >> [A]_0$), then in the equation

$$r = k([A]_0 - x)([B]_0 - x) \quad (5.33)$$

the term $([B]_0 - x)$ will be practically constant, and the pseudo-first order reaction is:

$$r = k_{ef}([A]_0 - x) \quad (5.34)$$

The expression for the effective rate constant takes the form: $k_{ef} = k([B]_0 - x)$.

Consider now a second-order reaction at the initial moments of time, when x is small. Expanding the exponent in the above solution in series and assuming that $a - x \approx a$, the equation is:

$$k(a - b)t = \frac{(a - b)x}{ab} \quad (5.35)$$

i.e., x = abkt = const(t). As for a first-order reaction with a small transformation, the concentration of the starting substance decreases linearly with time. An analogous expression can also be obtained from the solution of the equation under the condition of identical initial concentrations. As for a first-order reaction with a small transformation, the concentration of the starting substance decreases linearly with time. An analogous expression can also be obtained from the solution of thc cquation under the condition of identical initial concentrations

$$kt[A]_0^2 = [A]_0 \quad (5.36)$$

5.6.3 *Third-order reaction*

Reactions of the type $A + B + C \rightarrow K$ practically do not occur in the gas phase, but arc possible in solutions. The solution of the general case for a third-order reaction is rather difficult. Without a conclusion, we give general solutions.

$$A + B + C \rightarrow \sum P_i;$$

$$kt = \frac{1}{(a-b)(b-c)(c-a)} \ln\left\{ [\frac{a}{(a-x)}]^{b-c} + [\frac{b}{(b-x)}]^{a-c} + [\frac{c}{(c-x)}]^{a-b} \right\} \tag{5.37}$$

5.6.4 *Zero-order reactions*

The zero order is characterized by the independence of the rate of reaction from the concentration of the reacting substances and the rate is equal to the rate constant: $r = k$. The kinetic equation $-d[A]/dt = k$ has the solution $kt = [A]_0 - [A]$. Therefore the concentration of A decreases linearly with time. Recall the analysis of reactions of the first and second orders at small degrees of conversion and one can conclude that in all cases the concentration changes linearly at the beginning of the reaction and it is impossible to distinguish reactions of different orders from experimental data under such conditions. This leads to the conclusion that it is important for practice that the study of reaction kinetics at conversion degrees of less than 0.15 does not make sense. The half-transformation time in the zero-order reaction is calculated from the equation $t_{1/2} = [A]_0/2k$. In a number of cases, reactions of fractional orders are also possible. Thus, in solutions, if the reaction occurs between the molecule and the ion formed during dissociation, the order of 3/2 is likely.

5.7 Determination of the chemical reactions order

In solving the inverse problem, it is important to determine the order of a simple reaction. For the established order, we can write down the equation of the kinetic curve and determine the value of the rate constant. There are different methods. The first and least interesting is the substitution of measurement results in any of the known equations. The correctness of the choice of the equation is verified by the linearity of the resulting graph in the corresponding coordinates. Obviously, the method is laborious, and if the order is not integer, it is generally difficult to determine its true value. More reliable is the method for processing experimental data, based on an analysis of the above kinetic equations. Determining the orders for all substances at once is almost impossible. Usually proceed as follows, all substances, except one, are taken in excess, the value of which depends on the desired accuracy of processing the results. In this case, one can take the zero order in these substances and determine the order for one of the participants in the reaction, the amount of which is relatively small. Or, assuming a certain stochiometry of the reaction, the initial concentrations are set in the ratios of the stochiometric coefficients. In this case, the total order of the reaction is determined. However, in real experiments, stochiometry is not exactly known, therefore, the first method is used more often and the rate is calculated by the equation $r = k_{ef} [A]^n$. In addition to the actual rate k constant, the concentrations of other substances, except for A, are also included in the value of k_{ef}.

5.7.1 *Method for determining the order of chemical reaction*

Consider a method for determining the order of the reaction with respect to the conversion rates (the Van's Hoff method). From the general expression for the reaction rate of the initial substance, which changes during the reaction, we obtain: $\ln r = \ln[\frac{\Delta C}{\Delta t}] = \ln k + n \ln C$. Having determined the rates and concentrations at two different values of t, we obtain an expression for calculating the order of the reaction:

$$n = \ln\{(\Delta C'/\Delta t) / \Delta C''/\Delta t)\}/\ln(C'/C'') \tag{5.38}$$

Substituting the order values in the equation, we can determine the rate constant. The method is obviously not exact due to a graphical differentiation error. It should be noted that in the processing of results there should be no systematic deviation of the calculated value from the straight line, otherwise we can speak of an error in determining the value of the reaction order. A variant of this method is the use of initial rates, but the error in this case can be even greater, because of experimental difficulties, it is really impossible, as a rule, to determine the time at the beginning of the experiment. It is convenient to use the formulas obtained earlier for the half-transformation time. In general,

$$t_{1/2} = \frac{2^{n-1} - 1}{k(n-1)C_0^{n-1}} \tag{5.39}$$

Using L'Hospital's rule, from this expression it is possible to obtain a formula for the half-transformation time in the first-order reaction. However, since the results of the experiments have some error, the general formula can also be used for a first-order reaction. Data processing will give an order slightly different from one.

If one is determining the half-transformation time at different initial concentrations (for the first-order reaction, it does not depend on the initial concentration), then the following equation can be used:

$$\ln t_{1/2} = \ln\frac{2^{n-1} - 1}{k(n-1)} - (n-1)\ln C_0 \tag{5.40}$$

or compare the results of the calculation for two values of C_0' and C_0":

$$n - 1 + \frac{\ln\left(t'_{1/2} / t''_{1/2}\right)}{\ln\left(C''_0 / C'_0\right)} \tag{5.41}$$

Consider now the reaction of any order the kinetic equation in the form $r = k\,[A]^n$. Its solution will look like:

$$kt = \frac{1}{n-1}\left(\frac{1}{C^{n-1}} - \frac{1}{C_0^{N-1}}\right) \tag{5.42}$$

Denote now: $(C/C_0)^{n-1} = z$. Then

$$ktC_0^{n-1} = \frac{1}{n-1}\left(\frac{1}{z} - 1\right) \tag{5.43}$$

It is obvious that the transformation times up to a certain value of z are referred to as

$$t''/t' = (1/z'' - 1)/(1/z' - 1) \tag{5.44}$$

5.8 Activation energy

If the temperature is increasing, the reaction rate usually also increases and is often quite noticeable. Due to what reason is this happening? This may be due to an increase in the concentration of the reacting substances or the rate constant. The concentration in the gas mixture at constant volume does not depend on the temperature change, and in solutions it is almost independent of temperature (the volume of the solution remains almost constant). Therefore only the reaction rate constant can increase noticeably. At the very beginning of the kinetic studies it was noted that the rate (or, obviously, the rate constant) increases with increasing temperature not linearly. Van's Hoff, an empirical rule was introduced, according to which, with an increase in temperature by 10°C, the rate constant increases 2 to 4 times (see above— p. 5.4). However, this rule is feasible only for a small number of reactions and in the temperature range near room temperature. Since the chemical interaction occurs during particle collisions, the growth of the constant can be associated with an increase in the number of collisions with increasing temperature due to an increase in the velocity of particle motion. But the speed increases in proportion to $T^{1/2}$. In addition, the rate constant for particles of the same mass and similar dimensions must be practically constant, which is not observed in kinetic experiments. Experimentally, Hood first obtained a more accurate exponential dependence of the rate constant on temperature, and later on was theoretically proven and presented by Arrhenius in equation: $k = A[\exp(-E_a/RT)]$, where E_a is the so-called activation energy. In the semi-logarithmic relationship, this expression looks like this: $\ln k = \ln A - E_a/RT$. The chemical transformation itself requires the breaking of bonds in the molecule. It is clear that for this purpose the energy of the reacting molecule must be greater than the average equilibrium energy for a given temperature. The activation energy is the excess energy in comparison with the average energy of the molecules at a given temperature, which is necessary for the chemical reaction to occur. Its magnitude is determined by the properties of the reacting particles and their energy state. For further reasoning, it is assumed that the equilibrium is established in the first stage, the concentration of active particles is assumed to be small, which is quite logical: particles with an energy greater than statistically equilibrium should be small, and therefore their conversion does not affect the equilibrium, and the decay rate of the active particles, i.e., the rate constant, does not depend on temperature. Then the reaction rate can be expressed as r = k'[AB], where k' is the rate constant of the decay of active particles. The concentration [AB] can be expressed in terms of the equilibrium constant, K_c: [AB] = K_c[A][B]. Denote the value of k'K_c as k. Then r = k[A][B]. Since k' does not depend on temperature, we have the obvious equality:

$$\frac{d\ln k}{dT} = \frac{d\ln K_C}{dT} = \frac{\Delta U}{RT^2} = \frac{E}{RT^2} \tag{5.45}$$

Here, ΔU corresponds to a change in internal energy during the formation of the active particles, i.e., E has the meaning of the activation energy (sometimes it is called the heat of activation, since this is a thermodynamic quantity at a certain temperature, which differs from the activation energy in the theory of the activated complex). Obviously, integrating from T to ∞, we get a coincidence with the experiment: r = k[exp($-E_a$/RT)]. In this case, the activation energy is assumed to be temperature independent. According to the meaning of the derivation, it is obvious that the equation is valid for significant values of the activation energy, since number of active particles should be small. From statistical thermodynamics it follows that this is feasible for particles with high energy. It is obvious that in an increase in 2 to 4 times the rate constant with an increase in temperature by 10°C is possible only in a certain interval of values of the activation energy and a small change in temperature. Consider now the temperature dependence of the rate constant. The Arrhenius equation is a function with a limiting value of the constant, and the curve of the dependence of k on T has an inflection point. The second derivative is equal to zero for T = E/2R, i.e., the curve has an inflection point at this temperature. The activation energy of most reactions is tens and hundreds kJ/mol, i.e., the inflection point corresponds to temperatures when most of the molecules break up into atoms and ions. But at low activation energy, the inflection point on the curve can be observed even at a significant lower temperature.

5.8.1 *Kinetics of complex reactions*

Complex reactions are called reactions, during which several simple reactions occur simultaneously, while at least one of the starting materials or products of simple reactions is a participant in any other simple reaction. Thus, for example, the simultaneous occurrence of two processes A + B → C + D and E + F → G + H is not a complex reaction. They are not connected with each other, as a result of their flow, only some property of the medium can change (ionic strength of the solution, for example), i.e., only the physical impact of one process on another will occur. For further explanation of complex reactions the concept of a simple stage as an invertible reaction of two simple reactions is introduced here. A complex reaction can consist of a large set of simple steps and simple reactions. Thus, in simple reactions that are complex, one of the participants must necessarily be a substance or an unstable intermediate compound that participates in at least one of the other simple reactions. Minimally complex reactions are: reversible reactions: A ↔ B, parallel reactions: A → B and A → C, successive reactions: A → B → C. Various combinations of the main three types of complex reactions form more complex schemes of chemical transformations. In analyzing complex reactions, the equations of material balance and energy conservation, the statistically equilibrium energy distribution in the system and the basic postulate of chemical kinetics are used. But for the analysis of the kinetics of complex reactions, additional principles are required.

5.8.2 *The principle of independence*

The principle of independence or coexistence of various reactions means that with the simultaneous occurrence of several simple reactions, the speed of each of them can be written in accordance with the basic postulate of chemical kinetics. The rate constants of each of these reactions do not depend on the flow of any other. Using this principle, one can compose the kinetic equations of a complex reaction as a set of kinetic equations of simple ones. The justification for this position of chemical kinetics is the feasibility of a statistically equilibrium energy distribution in the reaction system. This allows us to consider the participants in a chemical reaction as statistically independent particles whose reactivity is independent of the presence of others in the system. Hence it is clear that the principle of independence is not absolute. It is obvious that when the properties of the medium change as a result of chemical interactions, especially in solutions involving ions, in nonideal systems, the rate constant of one reaction may depend on the flow of the other. For very fast reactions capable of disturbing the distribution of molecules over Maxwell-Boltzmann energies, the velocity of the slower chemical stage will also change. However, theoretical considerations and experimental data show that the principle of independence is fairly well satisfied for most chemical transformations. In thermodynamics it is postulated that in equilibrium, all flows in the opposite directions should compensate each other, and it is asserted that when equilibrium is achieved in a complex reaction mixture, it is established at each stage.

5.8.3 The concept of a rate determining stage

Obviously, for complex reactions, the most common case will be the flow of several simple stages (reactions) with different rates. The difference in rates leads to the fact that the kinetics of product production can be determined by the laws of only one of the stages. For example, for parallel reactions, the rate of the entire process depends on the rate of the fastest stage, and in the sequential ones—the slowest. Accordingly, in the analysis of kinetics in parallel reactions, with a significant difference in the constants, the rate of the slow stage can be neglected, and in successive ones—it is not necessary to determine the fast rate. In successive reactions, the slowest reaction is called the *limiting reaction*. For example, the kinetics of decomposition of hydrogen peroxide in an aqueous solution with Fe_2^+ ions at room temperature is well described by the kinetic equation of an irreversible second-order reaction.

Of course, the quantitative coincidence of the exact and approximate solutions is determined by the accuracy of the analysis, which, in turn, determines the possibility of neglecting the rate of a particular process. At close values of the rates of individual stages of a complex reaction, a complete analysis of the entire kinetic scheme is necessary. The introduction of the concept of the stage determining the rate in many cases simplifies the mathematical side of the consideration of similar systems and explains the fact that sometimes the kinetics of deliberately complex, multistage reactions is well described by simple equations, for example, of the first or second order.

5.8.4 Reversible reactions of the first order

Kinetics of this type are applicable to many simple reactions and the basic equation can be written as: $A \overset{k_1}{\underset{k_{-1}}{\leftrightarrow}} B$. The expression for the conversion rate of substance A will be written on the basis of the independence principle, denoting the current concentrations as [A] and [B]: By the equation of material balance.

$$[B] = [B]_0 + [A]_0 - [A]: \ \frac{d[A]}{dt} = -k_1[A] + k_{-1}[B] \tag{5.46}$$

the solution will have the form:

$$[A] = \frac{k_{-1}([A]_0 + [B]_0)}{k_1 + k_{-1}} + \frac{k_1[A]_0 - k_{-1}[B]_0}{k_1 + k_{-1}} e^{-(k_1+k_{-1})t} \tag{5.47}$$

Now the condition of equality of the rates of direct and reverse reactions in equilibrium can be used:

$k_1[A]_{bal} = k_{-1}[B]_{bal}$ or: $K = k_1/k_{-1} = [B]_{bal}/[A]_{bal}$.

Taking this equation into account, the solution for changing the concentration of substance A in time can be represented as:

$$[A] = \frac{([A]_0 + [B]_0)}{K+1} + \frac{K[A]_0 - [B]_0}{K+1} e^{-(k_1+k_{-1})t} \tag{5.48}$$

In order to calculate the change of [B] one can easily get the appropriate expression using the equation of material balance. It is seen that when $k_{-1} = 0$ the previous formula goes to the equation of the kinetics of the first-order of irreversible reaction. A quantitative assessment of the possibility of neglecting the reverse reaction can be carried out on the basis of the following reasoning. The rate of the reversible reaction $r = -k_1[A] + k_{-1}[B]$ (for an accuracy of 1%) the second term should be 100 times smaller. At the initial concentration B equal to zero, it can be neglected at the initial instants of time, and for comparable quantities of substances A and B, the rate constant of the reverse stage should be 100 times smaller. For independent determination of the values of the rate constants of forward and backward reactions, a coupling equation is required, given by the form of the recording of the equilibrium constant ($K = k1/k2$), or by using the value of equilibrium concentrations.

5.8.5 Reversible second-order reactions

In such transformations, at least one of the reactions, direct or inverse, is of the second order. Its velocity should be proportional to the product of the concentration of the reacting molecules. Let's take a look at a few examples: $A1 + A2 \leftrightarrow B1 + B2$. The reaction rate is: $r = k_1[A1]\,[A2] - k_{-1}[B1]\,[B2]$. At zero initial concentrations of products, the solution will look like:

$$k_1 = \frac{1}{2\beta(1-K)} \ln \frac{1 - \dfrac{x}{\alpha+\beta}}{1 - 1 - \dfrac{x}{\alpha-\beta}}; \quad K = \frac{k_{-1}}{k_1}$$

$$x = \frac{ab}{1-K}\{\alpha + \beta \mathrm{cth}[\beta t(k_1 - k_{-1})]\}^{-1} \tag{5.49}$$

$$\alpha = \frac{a+b}{2(1-K)}; \quad \beta = \frac{[(a-b)^2 + 4abK]^{1/2}}{2(1-K)}$$

$A \leftrightarrow^{k1}_{k2} B + C$

The solution of the differential equation for calculating the reaction rate is:

$$k = \frac{1}{2\alpha} + \ln\left\{\frac{a + x\left[\frac{k_2}{k_1}\alpha - \frac{1}{2}\right]}{a - x\left[\frac{k_2}{k_1}\alpha + \frac{1}{2}\right]}\right\} \tag{5.50}$$

$$\alpha = \sqrt{a\frac{k_1}{k_2} + \frac{1}{4}\left(\frac{k_1}{k_2}\right)^2}; \quad x = a\left[\frac{1}{2} + \frac{k_2}{k_1}\alpha\,\mathrm{cth}(k_2\alpha t)\right]^{-1}$$

5.8.6 Kinetics of parallel reaction with reversibility in one stage

Consider the scheme:

$$B \underset{\rightarrow k_1}{\overset{\leftarrow k_{-1}}{}} A \xrightarrow{k_2} C$$

(A is original). The system of kinetic equations is

$$\begin{cases} \frac{dC_A}{dt} = -(k_1 + k_2)[A] + k_{-1}[B] \\ \frac{dC_B}{dt} = k_1[A] - k_{-1}[A] - k_{-1}[B] \\ \frac{dC_C}{dt} = k_2[A] \end{cases} \tag{5.51}$$

As is known, the system of Equation (5.51) always has a solution and it is unique under the condition that the coefficients on the right-hand side are constant quantities and also the time t does not appear in explicit form on the right-hand side of the equation. The ultimate goal of this kind of analysis is to obtain a functional dependence of the concentration of the product of reaction C from time, which is then substituted into the general equation of the energy balance of the system in order to obtain a temperature dependence on time, which in turn is used in the analysis of creep of nanocomposites. Below are some examples that illustrate different types of chemical reactions and its applications to creep deformation analysis of nanocomposites.

Example 5.1

Find the lifetime and half life of concentration [A] of the first order chemical reaction from the data given in Fig. 5.3.

1. First determine the rate constant k. Then the lifetime is simply t = 1/k, while the half-life is $t_{1/2}$ = ln(2)/k.
2. The slope of the line is k = –7.94 s–1. Thus, the lifetime is t = 1/(7.94 s–1) = 126 ms, and the half life is $t_{1/2}$ = ln(2)/(7.94 s–1) = 87 ms.

Example 5.2

Consider the case, for which the simplest second order reaction is 2A → Product, or using the *differential rate law*

$$\frac{d[A]}{dt} = -k[A]^2 \tag{5.52}$$

Of course, a simple method for obtaining the *integrated rate law* would be to integrate from t = 0 when [A] = [A(0)] to the final time when [A] = [A(t)]. We would obtain

$$\frac{1}{[A(t)]} - \frac{1}{[A(0)]} = kt \tag{5.53}$$

Equation (5.53) suggest that a plot of 1/[A(t)] as a function of time should yield a straight line whose intercept with vertical axis is 1/[A(0)] and whose slope is the rate constant k, as shown in Fig. 5.3.

Example 5.3

For the reaction of methyl acetic ether with alkali in case of equal concentrations of ether and alkali at 298 K, the following dependence of the concentration of alkali (C) on time (t) was obtained from experimental data (see table below):

Table 5.1: The concentration of alkali (C) on time (t).

T (time, min)	0	3	5	7	10	15	25
C (mol/liter)	0.01	0.0074	0.00634	0.0055	0.00464	0.00363	0.00254

Determine the reaction order.

Solution

To determine the order of chemical reaction, we use the method of selecting the kinetic equation. Calculate the rate constants of the reaction using the kinetic equations of the reactions of zero, first and second orders, respectively (see table below).

It follows from the data given that the numerical values of the rate constants, remain constant within the error of calculation for the second-order reaction.

Table 5.2: Rate constants of the chemical reaction.

Kinetic equation	time, min					
	3	5	7	10	15	25
$k = \frac{C_0 - C}{t}$ mol/l · min	0.00087	0.00073	0,00064	0,00054	0,00043	0,0003
$k = \frac{1}{t}\ln\frac{C_0}{C}$ min^{-1}	0.100	0.091	0.085	0.077	0.068	0.055
$k = \frac{1}{t}\left(\frac{1}{C} - \frac{1}{C_0}\right)$ l/(mol · min)	11.71	11.55	11.69	11.55	11.70	11.75

Example 5.4

Substance A is mixed with B and C in equal concentrations (C_0 = 1 mol/l). After 1000 s, 50% of A remains. How much substances A will remain after 2000 s if the reaction has zero, first, second order?

Solution of problem

1. for the reaction of zero order:

$$k = \frac{C_0}{2t_{1/2}} = \frac{1}{2(1000)} = 5(10^{-4})$$

$$C = C_0 - kt = 1 - 5(10^{-4})2000 = 0$$

2. for the first-order reaction:

$$k = \frac{\ln 2}{t_{1/2}} = \frac{0.693}{(1000)} = 6.93(10^{-4})$$

$$C = C_0 \exp(-kt) = 0.25$$

3. for the second-order reaction:

$$k = \frac{1}{C_0 t_{1/2}} = \frac{1}{(1000)} = 1.0(10^{-3})$$

$$1/C - 1/C_0 + kt - 0.33[\text{mol/liter}]$$

Example 5.5

Decomposition of acetone proceeds according to the reaction: $(CH_3)_2CO \rightarrow C_2H_4 + H_2 + CO$. The dependence of the total pressure on the reaction time was obtained (see Table 5.3 below). Determine the order of the reaction with respect to acetone and the reaction rate constant.

Table 5.3: Total pressure on the reaction time.

t [min]	0	6.5	13.0	19
p [N/m^2]	41589.6	54386.4	65050.4	74914.6

Solution of the problem

Assume that the reaction has the first order in acetone. This reaction takes place in the gas phase and, since for gases at low pressures the concentration is proportional to the pressure, then for the first-order reaction the expression for the rate constant has the form:

$$k = \frac{1}{t}\ln\frac{C_0}{C} = \frac{1}{t}\ln\frac{p_0}{p}$$

where p_0 is the initial pressure of acetone. Denote at the time t:

$$p_{C_2H_4} = p_{H_2} = p_{CO} = x;\; p_{(CH_3)CO} = p_0 - x$$

The total pressure at time t: $p = p_0 - x + 3x = p_0 + 2x$. From the last equality implies that:

$$k = \frac{1}{t}\ln\frac{C_0}{C} = \frac{1}{t}\ln\frac{p_0}{3p_0 - p}$$

After substituting the numerical values of the current pressures and time, we obtain the following values of the rate constants: $k_1 = 0.0256$ [min^{-1}]; $k_2 = 0.0255$ [min^{-1}]; $k_3 = 0.0257$ [min^{-1}]. The close values of the rate constants calculated at different instants of time lead to the conclusion that this reaction has the first order in acetone. The average value of the rate constant:

$k_{av} = 0.0256$ [min^{-1}].

Example 5.6

The reaction proceeds according to the scheme: 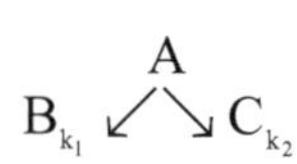

Determine k_1 and k_2 if it is known that 35% of substance B is contained in a mixture of reaction products, and the concentration of substance A is halved in 410 s.

Solution

The sum of the constants $k_1 + k_2$ is related to the half-life of substance A as follows: $k_1 + k_2 = \ln 2/t_{1/2}$; $k_1 + k_2 = 0.69/410 = 1.68(10^{-3})$ s^{-1}. The ratio of the constants is determined from the yield of substances B (35%) and C (65%): $k_1/k_2 = 35/65 = 0.537$. From the ratio of the constants and their sum, we obtain: $k_1 + 0.537\,(k_2) = 1.68(10^{-3})(1.537\,k_2) = 1.68(10^{-3})$; $k_2 = 1.09(10^{-3})\ s^{-1}$; $k_1 = 0.59(10^{-3})\ s^{-1}$.

Example 5.7

The following values of the rate constants were obtained for the reaction $2HI \underset{k_2}{\overset{k_1}{\rightleftarrows}} H_2 + I_2$ at different temperatures (see Table 5.4 below):

Table 5.4: Rate constants.

T [° K]	k_1 [$cm^3\,mol^{-1}\,s^{-1}$]	k_2 [$cm^3\,mol^{-1}\,s^{-1}$]
666.8	0.259	15.59
698.6	1.242	67.0

Find the dependence of k_1 and k_2 on the temperature and the value of the equilibrium constant of the reaction K_c at a temperature of 553 K.

Solution

According to the Arrhenius equation: $\ln(k_1) = -E_1/RT + \ln A_1$

$$\begin{cases} -1.35 = E_1/8.31(666.8) + \ln A_1 \\ 0.217 = E_1/8.31(698.6) + \ln A_1 \end{cases}$$

Solving the system of equations, obtain: $\ln A_1 = 33.08$; $E_1 = 190.87$ kJ/mol. Thus, the equation for calculating the rate constant of the direct stage will have the form: $\ln k_1 = -22957.66/T + 33.08$.

Similarly, calculate the parameters of the Arrhenius equation for the reverse stage: $\ln k_2 = -21237.67/T + 34.60$.

Compute the value of the equilibrium constant at a temperature of 553° K according to the equation: $K_c = k1/k2$ and $K_c = 9.75(10^{-3})$.

Example 5.8

In the system, a sequential reaction takes place: $A \xrightarrow{k_1} B \xrightarrow{k_2} C$. The maximum concentration of substance B is ($0.77\ C_{0A}$), where C_{0A} is the initial concentration of substance A, and is reached 170 minutes after the start of the reaction. Calculate k_1 and k_2.

Solution

Assume that at the initial instant of time the concentration of the initial substance A for the first-order reaction is a, the substances B and C are absent.

The formula for calculating the maximum concentration of the intermediate product C_{Bmax} can be obtained from the solution of corresponding general differential rate equations:

$$C_{B\max} = a\left(\frac{k_2}{k_1}\right)^{\frac{k_2/k_1}{1-k_2/k_1}}$$

It follows from this equation that the maximum concentration of the intermediate product is determined only by the ratio of the rate constants k_2/k_1 and does not depend on their values. According to the data:

$$0.77C_{0A} = C_{0A}\left(\frac{k_2}{k_1}\right)^{\frac{k_2/k_1}{1-k_2/k_1}} \Rightarrow k_2/k_1 = 0.1$$

The time to reach the maximum concentration of intermediate B can be determined as

$$t_{max} = \frac{\ln(k_2/k_1)}{k_2 - k_1}$$

Finally: $170 = \dfrac{\ln(0.1)}{k_1(0.1-1)} \Rightarrow k_1 = 2.5(10^{-4})s^{-1}; k_2 = 2.5(10^{-5})s^{-1}$

If the purpose of the practical examples given above is to introduce the methods of kinetic experiment, determine the kinetic parameters of the reactions: the rate constants, the reaction order, the temperature effect on the pre-exponential coefficient of the Arrhenius law, the activation energy, etc., then the following examples are aimed at obtaining integral kinetic curves for various first- and second-order reactions, which in turn are then used in integral creep equations for nanocomposites. It should also be noted that the integral kinetic curves (and their analytical approximate expressions) should be presented in dimensionless form (as is the case with the remaining components of the integral creep equation for nanocomposites). In the following examples, similar dimensionless variables and parameters are used that were used in the previous chapters, and therefore are considered self-explanatory.

Example 5.9

Data: A↔ B; $k_+ = 120$ and $k_- = 1$

$t = [h^2/a]\ \tau = 65.4\ \tau$; $\rho = 1000$ [kg/m^3] $\alpha = 10(10^{-6})$; $c_p = 1000$ [J/kg*K]; $k = 1$ [W/m*K]; $a = k/c_p\ \rho$ [m^2/sec]; $a = 1/1000(1000) = 1.0(10^{-6})$; $h = 2$ mm $= 2.0\ (10^{-3})$ m;

$t = (2)^2(10^{-6})/1.0(10^{-6}) = 4\tau$; initial conditions: A(0) =1.0; B(0) = 0

If the rate constants k_+ and k_- are measured from experimental data and the initial concentrations of A and B are known, then it is possible to solve the ODEs and obtain concentrations [A] and [B] versus time (integral kinetic curves).

To obtain the temperature-time dependence, which is necessary for further analysis of the creep process of the nanocomposites, it is necessary to add to the corresponding kinetic equations the conservation of energy equation of the entire system (in which the nanocomposites are at a given time), including all external heat sources. For the sake of simplicity of further discussion, such an equation of heat balance will be used (shown in the equation describing the fire dynamics and heat transfer equation that had been used in the author's earlier works [35, 36]).

Calculated values of DEQ variables

	Variable	Initial value	Minimal value	Maximal value	Final value
1	A	1.	0.0787229	1.	0.0787229
2	P	0	0	0.9212771	0.9212771
3	t	0	0	0.12	0.12
4	y	0	0	11.12286	11.12286

Differential equations

1 d(P)/d(t) = 4*(12*A – 1*P)

2 d(A)/d(t) = 4*(–12*A + 1*P)

3 d(y)/d(t) = 20*(1 – P)^0*P^1*exp(y/(1 +.1*y)) – 0.233*y^4

y(0) = 0

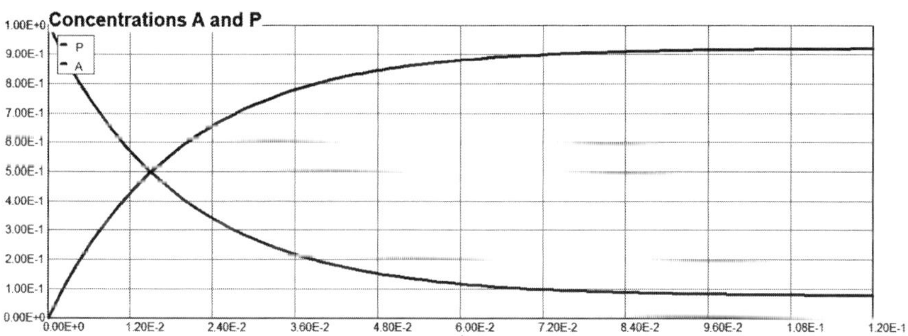

Figure 5.4. Integral kinetic curves.

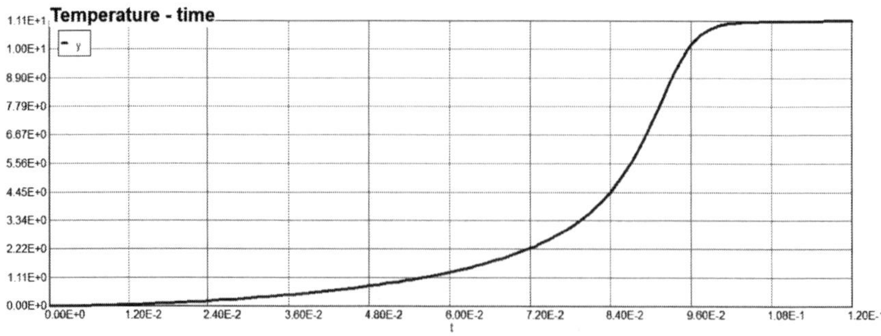

Figure 5.5: Temperature-time curve.

Model: y = a1*t + a2*t^2 + a3*t^3 + a4*t^4 + a5*t^5

Variable	Value
a1	–86.67578
a2	9542.553
a3	–2.995E+05
a4	3.707E+06
a5	–1.48E+07

$$\theta = -86.68\tau + 9542.6\tau^2 - 2.995(10^5)\tau^3 + 3.707(10^6)\tau^4 - 1.48(10^7)\tau^5 \tag{5.54}$$

Model: t = a1*y + a2*y^2 + a3*y^3 + a4*y^4 + a5*y^5

Variable	Value
a1	0.0868221
a2	–0.0385558
a3	0.0080312
a4	–0.0007564
a5	2.609E-05

$$\tau = 0.0868\theta - 0.0386\theta^2 + 0.0080\theta^3 - 0.000756\theta^4 + 2.6(10^{-5})\theta^5 \tag{5.55}$$

$$0 < \theta < 11$$

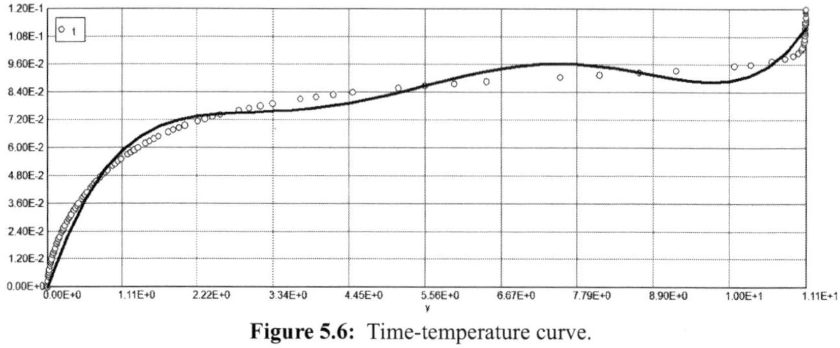

Figure 5.6: Time-temperature curve.

Example 5.10 Parallel Reaction; First Order

$$B \underset{k_{-1}}{\overset{k_1}{\leftrightarrow}} A \xrightarrow{k_2} C$$

$$\frac{d[A]}{dt} = -(k_1 + k_2)[A] + k_{-1}[B]$$

$$\frac{d[B]}{dt} = k_1[A] - k_{-1}[B]$$

$$\frac{d[C]}{dt} = k_2[A]$$

Differential equations

d(A)/d(t) = –(k1 + k2)*A + k3*B
d(B)/d(t) = k1*A – k3*B
d(C)/d(t) = k2*A
d(y)/d(t) = 20*(1 – P)^0*P^1*exp(y/(1 +.1*y)) – 0.233*y^4
y(0) = 0; A(0) = 1; B(0) = 0; C(0) = 0

Calculated values of DEQ variables

	Variable	Initial value	Minimal value	Maximal value	Final value
1	A	1.	0.0123531	1.	0.0123531
2	B	0	0	0.3027186	0.2945152
3	C	0	0	0.6931316	0.6931316
4	k1	10.	10.	10.	10.
5	k2	20.	20.	20.	20.
6	k3	1.	1.	1.	1.
7	t	0	0	0.2	0.2
8	y	0	0	3.340899	3.340899

Differential equations

1 d(A)/d(t) = –(k1 + k2)*A + k3*B
2 d(B)/d(t) = k1*A – k3*B
3 d(C)/d(t) = k2*A
4 d(y)/d(t) = 20*(1 – P)^0*P^1*exp(y/(1 +.1*y)) – 0.233*y^4

Explicit equations

1 k1 = 10
2 k2 = 20
3 k3 = 1

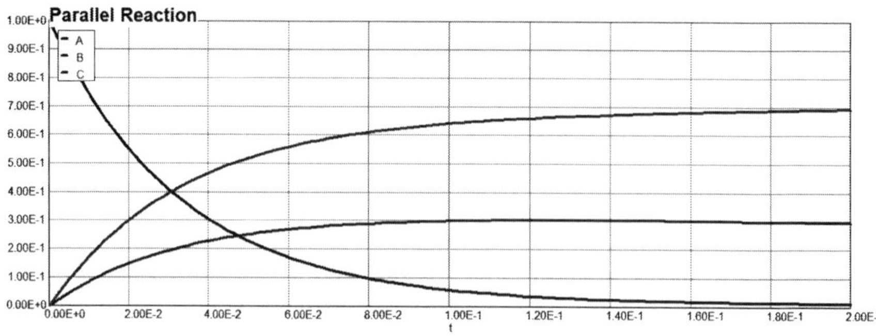

Figure 5.7: Parallel reaction.

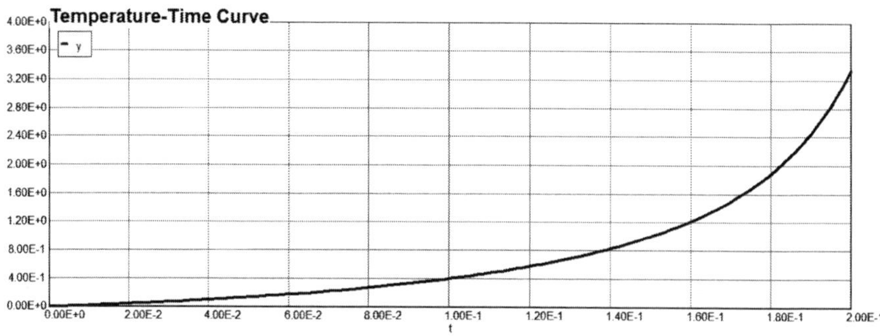

Figure 5.8: Temperature-time curve (parallel reaction).

Model: y = a1*t + a2*t^2 + a3*t^3 + a4*t^4 + a5*t^5

Variable	Value
a1	5.441994
a2	–185.6953
a3	3683.124
a4	–2.662E+04
a5	7.105E+04

$$\theta = 5.44\tau - 185.7\tau^2 + 3.68(10^3)\tau^3 - 2.66(10^4)\tau^4 + 7.1(10^4)\tau^5 \tag{5.56}$$

Model: t = a1*y + a2*y^2 + a3*y^3 + a4*y^4 + a5*y^5

Variable	Value
a1	0.3927627
a2	–0.4416955
a3	0.2651387
a4	–0.0761205
a5	0.0082143

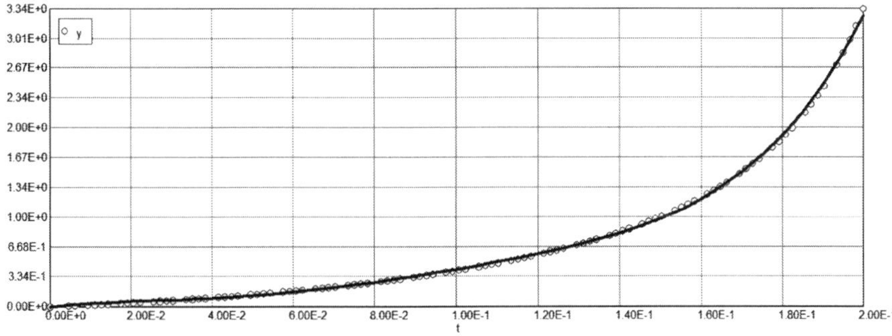

Figure 5.9: Time-temperature curve (parallel reaction).

$$\tau = 0.393\theta - 0.442\theta^2 + 0.265\theta^3 - 0.076\theta^4 + 8.2(10^{-3})\theta^5 \quad (5.57)$$

Simple Reaction Autocatalytic Process

For elementary reactions of the form **A + B = C** in which two reactants must interact to form a product, the law of mass action states that the rate of change of the product concentration [C] is proportional to the (mathematical) product of the individual reactants. Mathematically, the rate of change of [C] is equal to k [A][B] where k is some rate constant. (Schematically, one writes this as **A + B** $\rightarrow^{k}$ **C**.) The Equation (3.4) in this case has the following form:

$$\frac{d[C]}{dt} = k[A][B] \quad (5.58)$$

With the preceding equations in mind, consider now an ***autocatalytic process*** in which a species promotes its own production; schematically, $A + P \underset{k_-}{\overset{k_+}{\rightleftarrows}} P + P$

Further suppose that [A] is held constant throughout this process (the flow rate of [A] is constant); e.g., a huge abundance of A makes its depletion negligible during the reaction. Applying the law of mass action, the ODE governing [P] is:

$$\frac{d[P]}{dt} - k_+[A][P] - k_-[P]^2 \quad (5.59)$$

[A] is regarded as constant by assumption. In order for the ODE to be dimensionally consistent, both k_+ and k_- must have units of $(\text{concentration})^{-1}$ $(\text{time})^{-1}$. The ODE (5.59) in this case happens to be solvable in closed form:

$$[P] = \frac{c[A]k_+ \exp([A]k_+t)}{1 + ck_- \exp([A]k_+t)} \quad (5.60)$$

Remember that it is rarely possible or necessary to find an exact mathematical formula for the solution of an ODE, so this example is atypical in that regard.

For practical purposes, computer simulations can provide approximate solutions of ODEs, often with a level of precision that the user is able to specify.

Still, the rare fortune of having an exact formula Equation (5.60) is worth exploiting, as it suggests what type of functions might be suited for fitting experimental data. Figure 3.4 (see Chapter 3) shows a least squares regression fit of a function of the form Equation (5.60), to experimental data from an autocatalytic process similar to the one described in this example. Importantly, performing a regression fit produces estimates for the kinetic constants k_+ and k_- (the so-called *reversed problem*).

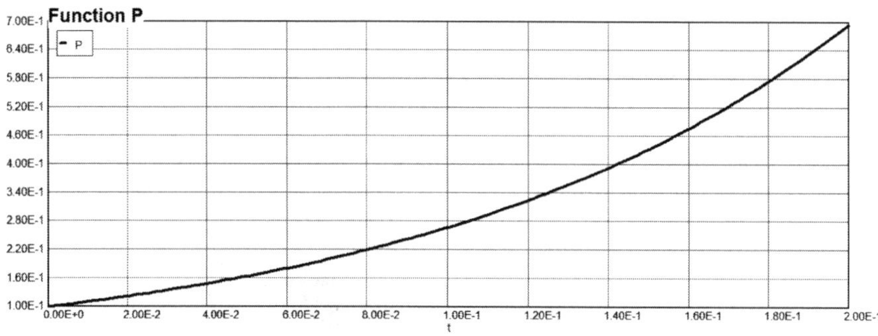

Figure 5.10: Product function P.

It is interesting to note that the closed analytical solution is obtained for only concentration [A] = constant case, but computer code us allows to solve the differential kinetic Equation (5.60) for any continuous given rate of [A(t)]. For example: [A] = exp(–t):

Example 5.10a:

same as Example 5.10, but [A] = exp(–t)

Differential equations

d(P)/d(t) = k1*A1*P – k3*P^2

P(0) = 0.1

A1 = 1*(exp(–t))

Calculated values of DEQ variables

	Variable	Initial value	Minimal value	Maximal value	Final value
1	A1	1.	0.8187308	1.	0.8187308
2	k1	10.	10.	10.	10.
3	k3	1.	1.	1.	1.
4	P	0.1	0.1	0.5789907	0.5789907

5	t	0	0	0.2	0.2
6	y	0	0	3.340899	3.340899

Differential equations

1 d(P)/d(t) = k1*A1*P – k3*P^2

Explicit equations

1 k1 = 10

2 k3 = 1

3 A1 = 1*(exp(–t))

One can see that $[P]_{max} = 0.579 < 0.70$ in this case.

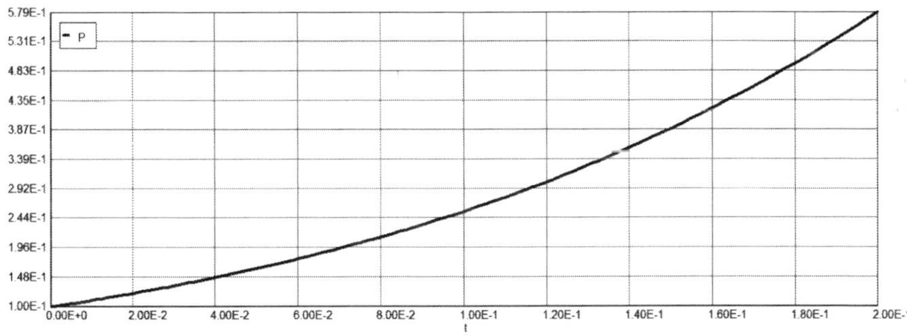

Figure 5.11: Product function P (continuous rate [A(t)]).

Example 5.11

Below is an example that is different from the two previous ones: using the law of mass we develop the model for (E)-mediated conversion of complex (S) into a product (P) through the intermediate complex (C). The reaction diagram for that process is given by $S + E \underset{k_-}{\overset{k_+}{\rightleftarrows}} C \xrightarrow{k_2} P + E$. Each of concentrations S, E, C, and P will be treated as a dependent variable and each of which will contribute an ODE to the total kinetic model. The full set of differential equations is:

$$\begin{aligned}
\frac{d[S]}{dt} &= k_-[C] \quad k_+[S][E] \\
\frac{d[E]}{dt} &= k_-[C] - k_+[S][E] + k_2[C] \\
\frac{d[C]}{dt} &= k_+[S][E] - k_-[C] - k_2[C] \\
\frac{d[P]}{dt} &= k_2[C]
\end{aligned} \tag{5.61}$$

The solution of Equation (5.61) is presented below (using POLYMATH software) via example.

Data: t = [h^2/a] τ = 65.4 τ; ρ = 4000 [kg/m^3] α = 10(10^{-6}); c_p = 1000 [J/kg*K]; k = 1 [W/m*K]; a = k/c_p ρ [m^2/sec]; a = 1/4000(1000) = 0.25(10^{-6});

h = 4.04 mm = 4.04(10^{-3}) m; t = $(4.04)^2$(10^{-6})/0.25(10^{-6}) = 65.4 τ

k_+ = 10; k_- = 1; and k_2 = 1 and initial conditions S(0) = 1.0 E(0) = 0.1, C(0) = 0.0, and P(0) = 0.0

Using computer POLYMATH software we have:

Calculated values of DEQ variables

	Variable	Initial value	Minimal value	Maximal value	Final value
1	C	0	0	0.5590785	0.5590785
2	E	1.	0.4409215	1.	0.4409215
3	P	0	0	0.8086244	0.8086244
4	S	1.	0.3600591	1.	0.3600591
5	t	0	0	0.2	0.2

Differential equations

1 d(S)/d(t) = (C–10*E*S)
2 d(E)/d(t) = (C–10*E*S+1*C)
3 d(C)/d(t) = (10*E*S–C–1*C)
4 d(P)/d(t) = 10*C

Model: P = 0.18*t + 39.82*t^2 – 150.6*t^3 + 241.1*t^4

Variable	Value
a1	0.1797004
a2	39.82461
a3	–150.6298
a4	241.129

Model: S = 0.995 – 9.143*t + 66.42*t^2 – 277.2*t^3 + 473.05*t^4

Variable	Value
a0	0.9949771
a1	–9.14307
a2	66.41824
a3	–277.2245
a4	473.053

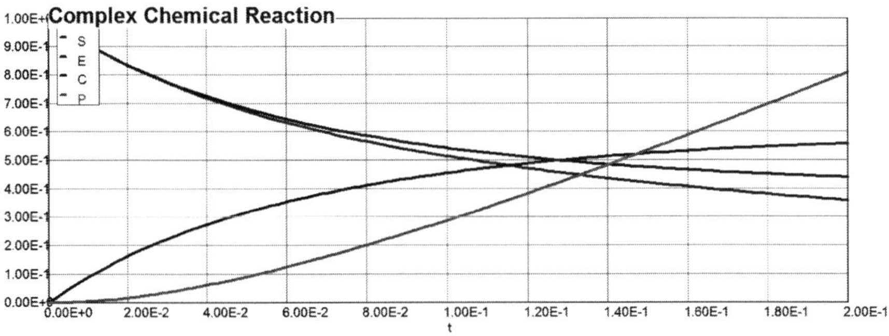

Figure 5.12: Complex chemical reactions.

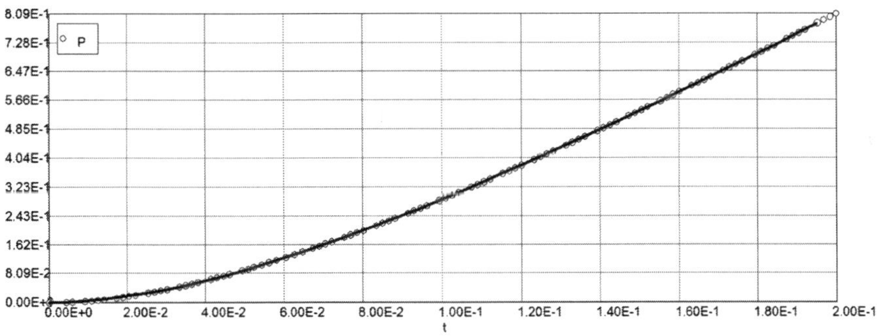

Figure 5.13: Product of chemical reaction vs. dimensionless time.

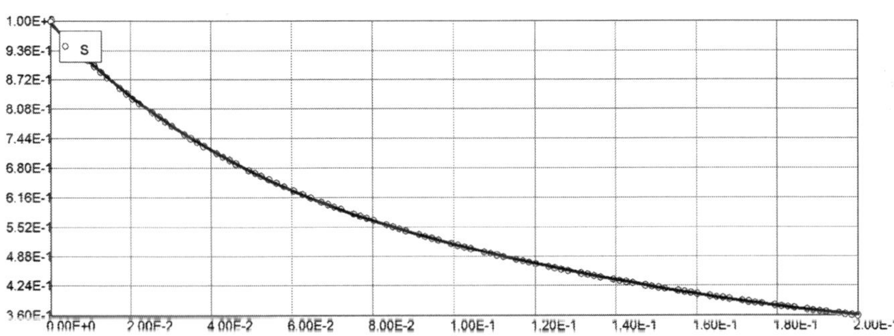

Figure 5.14: Reactant of chemical reaction vs. dimensionless time.

Readers interested in more advanced mathematical modeling techniques may wish to read about asymptotic methods, principal components analysis, sensitivity analysis, scaling and the use of non-dimensional variables and parameters, as a means for reducing the number of unknown parameters in a model. For a mathematical reference on chemical kinetics see, for example [36, 37].

5.9 Autocatalytic chemical reactions

Autocatalytic chemical reactions are those in which the rate increases as the materials react. The most important single factor in predicting effects is the temperature. The rate of chemical reaction changes with the temperature change. The fundamental equation that describes an autocatalytic process is the following:

$$\frac{d[C]}{d\tau} = \gamma\delta[C]^{p}(1-[C])^{q} Ze^{-\frac{E}{RT}} \tag{5.62}$$

[C] is the fraction reacted at any specific time, t. The derivative, d[C]/dt, is the rate of the reaction. E is the "Arrhenius activation energy," and Z is the "***Arrhenius pre-exponential.***" Each applies only to a single specific, consistent reaction being studied. The value of the "rate constant," k = d[C]/dt is different at each specific temperature: It is a constant only at one temperature, and it applies only to one specific reaction. The values of E and Z are determined from a large number of 'k' measurements at different temperatures. E, Z, and k are the most important values in a discussion of rates and associated lifetimes of materials. All of these values have fundamental meaning in the chemical reaction. The exponent's p and q allow the prediction of the position of the maximum rate in an autocatalytic process, i.e., the amount reacted at the maximum rate—at constant temperature. An example of autocatalytic rate curve is shown in the Fig. 5.15. Notice that the rate increases with time in the autocatalytic curve, at constant temperature, until it reaches a maximum reaction rate. Then the rate decreases. However, the initial rate at any temperature is much lower than the maximum rate.

The activation energy, E, is closely related to the strength of the reacting chemical bond. Strong bonds show high activation energies when they react, weaker bonds show lower activation energies. Activation energies in solids,

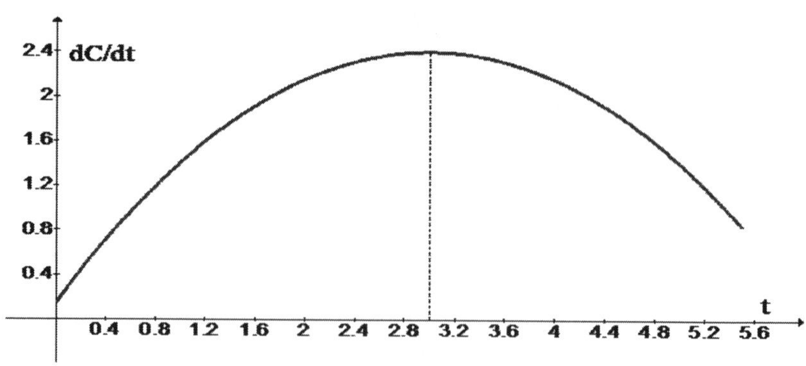

Figure 5.15: Maximum reaction rate.

especially crystalline solids, are higher than the values for the same material in a solution or melt, because a crystalline lattice is stabilized by its ordered structure. Some zones between the small crystals are amorphous, and they act more like in a melted phase. The major cause for autocatalysis in the decomposition of a composite material is the destruction of crystalline order when the material is heated to a high temperature. When crystalline order is destroyed by heating, autocatalysis involves a process like "melting with decomposition." Rates are much higher in a melt than in the solid; therefore, melting increases decomposition rates. When the material is cooled below the melting point, autocatalysis stops entirely.

Many of kinetics of nanoparticles synthesis processes can be described as a first order chemical reaction ($p = 0$), except for autocatalytic reactions: they are chemical reactions in which at least one of the products is also a reactant. The rate of these equations for autocatalytic reactions is fundamentally nonlinear and they should be presented by the conservation of mass differential equation similar to Equation (5.62). In order to simplify the heat and mass conservation equations the following assumptions are made in this book:

- That heat transfer due to conduction can be neglected due to nanometer scale of elements.
- The increase in energy flux (in addition to heat combustion release) is due to natural convection.
- The loss of energy is due to irradiation only (the conductive heat loss to the atmosphere are neglected due to much weaker dependence on temperature).

Let's consider now the spatial averaging of temperature and unsteady first-order chemical reaction rate. The Equations (5.62) are written now in dimensionless form and simplified further due to assumptions made above [36]:

$$\begin{aligned} \frac{\partial\theta}{\partial\tau} &= \delta(1-C)^k \exp\left(\frac{\theta}{1+\beta\theta}\right) - \alpha\theta^4 \\ \frac{\partial C}{\partial\tau} &= \gamma\delta(1-C)^k \exp\left(\frac{\theta}{1+\beta\theta}\right) \end{aligned} \qquad (5.63)$$

The solutions of Equation (5.63) are presenting temperature-time functions m; m1; m2 and m21 that are used in constitutive creep equations for nanomaterials. The more accurate analysis of temperature-time functions obviously can be obtained if the second equation in (5.63) will be substituted by the corresponding solution of kinetic equations (for a given chemical reaction of a particular nanomaterials), for example (see Example 5.2 above).

Example 5.12

Data: see Example 5.11 and $k_+ = 10$; $k_- = 1$; $k_2 = 1$; $\beta = 0.02$

The model for (E)-mediated conversion of a substrate **(S)** into a product **(P)** via an intermediate substrate complex (C). The reaction diagram for that process is given by:

$$S + E \underset{k_-}{\overset{k_+}{\rightleftarrows}} C \xrightarrow{k_2} P + E$$

For the first order chemical reaction the dimensionless temperature-time relationship can be described by the system of ODE's as follows:

$$\frac{\partial\theta}{\partial\tau} = \delta(1-P)\exp\left(\frac{\theta}{1+\beta\theta}\right) - \alpha\theta^4$$

$$\frac{d[S]}{dt} = k_-[C] - k_+[S][E]$$

$$\frac{d[E]}{dt} = k_-[C] - k_+[S][E] + k_2[C] \tag{5.64}$$

$$\frac{d[C]}{dt} = k_+[S][E] - k_-[C] - k_2[C]$$

$$\frac{d[P]}{dt} = k_2\gamma\delta[C]$$

The coefficient '65.4' in four Equations (5.64) reflects the scale change of real time independent variable 't' to dimensionless time variable τ that is used in fifth dimensionless ODE describing the conservation of thermal energy of the exothermal unsteady complex autocatalytic chemical reaction. It has been assumed in this example that nanocrystalline materials are grown from h = 1 nm to h = 4.04 μm and the diffusivity coefficient a = 1/4000(1000) = 0.25 (10^{-6}) [m^2/sec]. Therefore $t = [h^2/a]\ \tau = (4.04)^2(10^{-6})/0.25(10^{-6}) = 65.4\tau$

Solution of Equation (5.64) using POLYMATH software is:

Calculated values of DEQ variables

	Variable	Initial value	Minimal value	Maximal value	Final value
1	C	0	0	0.5779855	9.203E-06
2	E	1.	0.4220145	1.	0.9999908
3	P	0	0	0.9999898	0.9999898
4	S	1.	1.011E-06	1.	1.011E-06
5	t	0	0	0.2	0.2
6	y2	0	0	0.2012992	0.2012987

Differential equations

1 d(S)/d(t) = (C–10*E*S)*65.4
2 d(E)/d(t) = (C–10*E*S+1*C)*65.4
3 d(C)/d(t) = (10*E*S–C–1*C)*65.4
4 d(P)/d(t) = 1.0*C*65.4
5 d(y2)/d(t) = 20*(1–P)^1*P^1*exp(y2/(1+.02*y2))–1* 0.233*y2^4

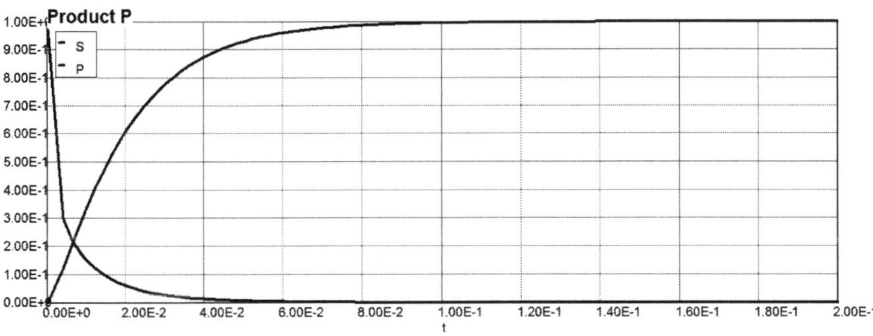

Figure 5.16: Reactant and product of chemical reaction vs. dimensionless time.

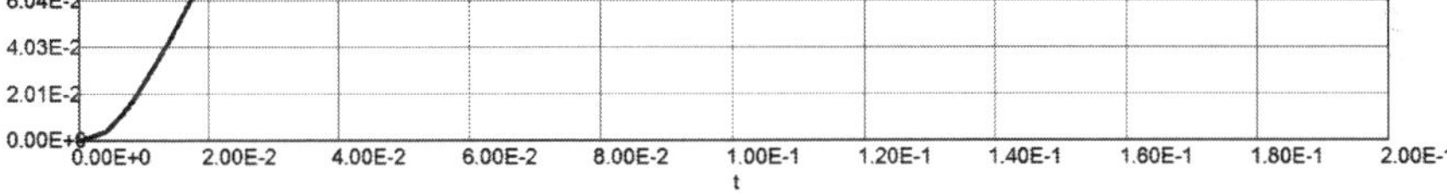

Figure 5.17: Temperature-time curve.

Model: y = a1*t + a2*t^2 + a3*t^3 + a4*t^4 + a5*t^5

Variable	Value
a1	–0.3707
a2	388.37
a3	–1.162E+04
a4	1.375E+05
a5	–5.946E+05

The time-temperature relationship is:

Model: t = a1*y + a2*y^2 + a3*y^3 + a4*y^4 + a5*y^5

Variable	Value
a1	0.747
a2	–17.61
a3	234.65
a4	–1385.4
a5	3053.6

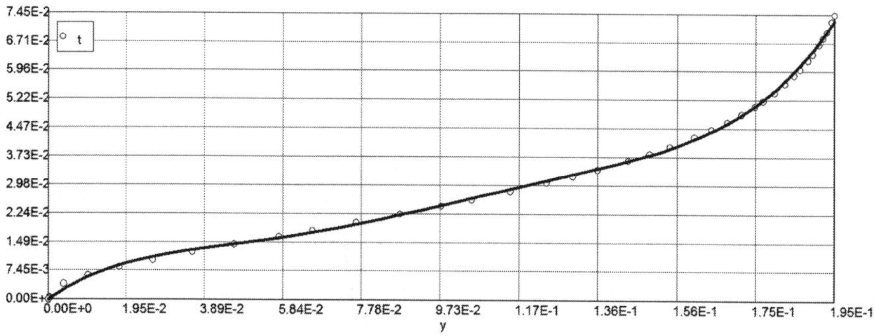

Figure 5.18: Time-temperature curve.

Analytical polynomial expression of function y2

y2 = θ = – 0.3707*τ + 388.37*τ^2 – 1.162*10^4*τ^3 + 1.375*10^5*τ^4 – 5.946*10^5*τ^5

Analytical polynomial function and first derivative

τ = m2 = 0.747*θ – 17.61*θ^2 + 234.65*θ^3 –1385.4*θ^4 +3054*θ^5 (5.65)

d(τ)/d(θ) = m21 = 0.747 – 35.22*θ + 707*θ^2 – 5542*θ^3 + 15270*θ^4 (5.66)

It should be noted that the temperature-time, θ(τ), and the reverse function, τ(θ), are different from m and m1. In case of chemical reaction (simple or autocatalytic) they are m2 and m21 accordingly. Of course, the activation energy and the base temperature are also different, so the dimensionless parameter β has different value.

Example 5.13

Data: see Example 5.2 and $k_+ = 20$; $k_- = 1$; $k_2 = 2$; [A] ≡ 1; [P_0] = 0.1; β = 0.02

$A + P \underset{k_-}{\overset{k_+}{\rightleftarrows}} P + P$

[A] = 1; [P_0] = 0.1; $k_+ = 20$; $k_- = 2$

For the *second-order chemical reaction* in accordance with Equations (3.10) and (3.13) we have:

Calculated values of DEQ variables

	Variable	Initial value	Minimal value	Maximal value	Final value
1	C	0	0	0.5037247	0.5037247
2	c	0.0555556	0.0555556	0.0555556	0.0555556
3	P3	0.1	0.1	1.	1.
4	t	0	0	0.2	0.2
5	y3	0	0	0.1479689	0.1479597

Differential equations

1 d(y3)/d(t) = 20*(1 – P3)^1*P3^1*exp(y3/(1 + .067*y3)) – 1* 0.233*y3^4

Explicit equations

1 c = 0.1/(2 – 0.1*2)

2 P3 = (2*c*exp(2*65.4*(10^ – 0)*t))/(1 + 2*c*exp(2*65.4*(10^ – 0)*t))

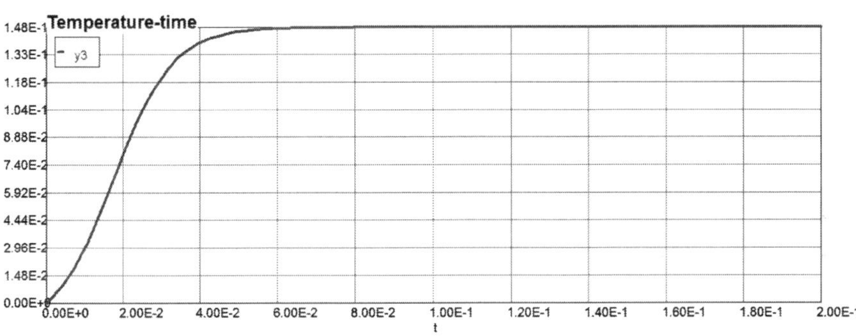

Figure 5.19: Temperature-time curve.

Model: y3 = a1*t + a2*t^2 + a3*t^3 + a4*t^4

Variable	Value
a1	5.603956
a2	–73.23093
a3	398.1486
a4	–769.8791

Analytical polynomial expression of function y2 and first derivative

y3 = θ = 5.6*τ – 73.23*τ^2 + 398*τ^3 – 769.9*τ^4

Model: t = a1*y3 + a2*y3^2 + a3*y3^3

Variable	Value
a1	2.347936
a2	–52.6876
a3	286.7798

Analytical polynomial function time-temperature and first derivative

$$\tau = m2 = 2.35*\theta - 52.69*\theta\text{^}2 + 286.78*\theta\text{^}3 \quad (5.67)$$

$$d(\tau)/d(\theta) = m21 = 2.35 - 105.38*\theta + 860.3*\theta\text{^}2 \quad (5.68)$$

Example 5.14

Simple autocatalytic reaction

Data: $\delta = 20$; $\beta_3 = 0.02$; $0 < \tau < 3.3$; $E_a = 60$ kkal/mol; $R = 2$; $T_* = 600°$ K; $\gamma\delta = 1$

$A \underset{k_-}{\overset{k_+}{\rightleftarrows}} P \quad k_+ = 40; \;\; k_- = 1$

Equation (3.4) in this case is:

$$\frac{d[A]}{dt} = k\_[P] - k_+[A] \text{ and } \frac{d[P]}{dt} = k_+[A] - k\_[P]$$

For an autocatalytic process the solution of the simplified equation in this example is:

Calculated values of DEQ variables

	Variable	Initial value	Minimal value	Maximal value	Final value
1	A	1.	0.0243902	1.	0.0243902
2	P	0	0	0.9756098	0.9756098
3	t	0	0	3.3	3.3
4	y	0	0	1.876697	1.876697

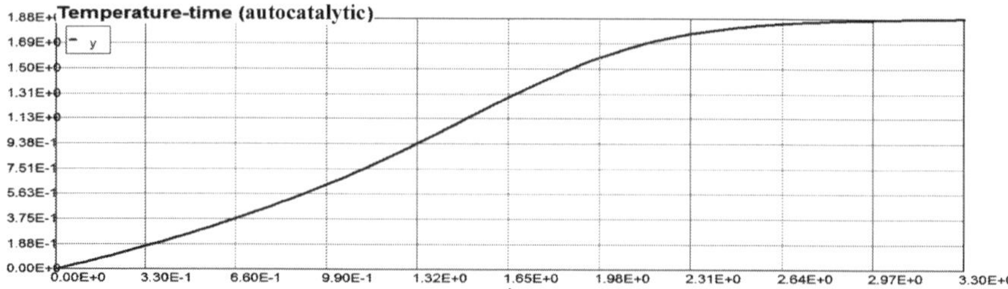

Figure 5.20: Temperature-time curve (autocatalytic reaction).

Differential equations

1 d(P)/d(t) = 1*(40*A – 1*P)*65.4
2 d(A)/d(t) = 1*(–40*A + 1*P)*65.4
3 d(y)/d(t) = 20*(1 – P)^1*P^1*exp(y/(1 + .02*y)) – 0.233*y^4

Model: y = a1*t + a2*t^2 + a3*t^3 + a4*t^4

Variable	Value
a1	0.2097711
a2	0.6790665
a3	–0.2317836
a4	0.0173965

$\theta = 0.21*\tau + 0.679*\tau\text{^}2 - 0.231*\tau\text{^}3 + 0.0174*\tau\text{^}4$

Model: t = a1*y + a2*y^2 + a3*y^3 + a4*y^4

Variable	Value
a1	1.265897
a2	2.120473
a3	–3.076895
a4	1.085109

$$\tau = m2 = 1.266\theta + 2.12\theta^2 - 3.077\theta^3 + 1.085\theta^4 \quad (5.69)$$

$$m21 = 1.266 + 4.24\,\theta - 9.231\,\theta^2 + 4.34\,\theta^3 \quad (5.70)$$

For simplicity these functions m2 and m21 will be used in the creep constitutive equation for nanocomposite materials.

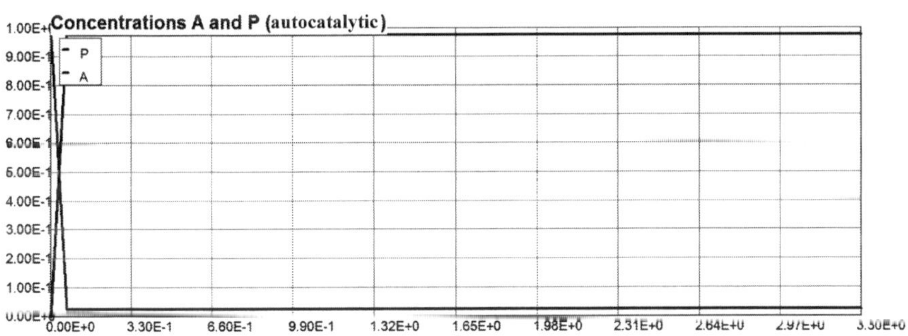

Figure 5.21: Reactant and product of chemical reaction vs. dimensionless time (autocatalytic).

Check: A + P = 0.9756098 + 0.0243902 = 1.0 → ***total mass does not change!***

References

[1] Laidler, K.J. 1987. Chemical Kinetics (3rd ed.). Harper & Row. NY., 277 p.

[2] Frank-Kamenetskii, D.A.1967. Diffuziia i teploperedacha v khimicheskoi kinetike, 2nd ed. (in Russian), Moscow.

[3] Bird, R.B., Stewart, W.E. and Lightfoot, E.N. 2002. Transport Phenomena. Wiley, NJ, USA.

[4] Boudart, M. 1968. Kinetics of Chemical Processes. Englewood Cliffs, Prentice-Hall, NJ, USA.

[5] Nawla, H.S. (ed). 2000. Handbook of Nanostructured Materials and Nanotechnology, Volume 5, Academic Press: New York, NY, USA.

[6] Jones, R.G. Wilks, E.S., Val Metanomski, W., Kahovec, J., Hess, M., Stepto, R. and Kitayama, T. (eds.). 2009. Compendium of Polymer Terminology and Nomenclature (IUPAC Recommendations 2008) (2nd ed.). RSC Publishing, Cambridge, UK.

[7] Cussler, E.L. 2009. Diffusion: Mass Transfer in Fluid Systems. Cambridge University Press, Cambridge, UK.

[8] Jackson, J.D. 1975. Classical electrodynamics. John Wiley & Sons, NJ, USA.

[9] Miller C.A. and Neogi, P. 2008. Interfacial Phenomena. CRC Press, Boca Raton. FL. USA.

[10] Woodruff, D.P. (ed.). 2002. The Chemical Physics of Solid Surfaces. Vol. 10, Elsevier, NY, USA.

[11] Zhang, Z., Jin, H., Zhang, L.H., Sui, M.L. and Lu, K. 2000. Superheating of confined thin films. Phys. Rev. Lett. 85: 1484—Published.

[12] Grb, L., Bohr, J., Andersen, H.H., Johansen, A., Johnson, E., Sarholt-Kristensen, L. and Robinson, I.K. 1992. Melting, growth, and faceting of lead precipitates in aluminum. Phys. Rev. B 45: 2628.

[13] Jeuss Kirpatric, R. 1975. Crysral growth from the melt: A review. American Mineralogist 60: 798–614.

[14] Malkin, A.I. 2012. Regularities and mechanisms of the Rehbinder's effect. Colloid Journal 74(2): 223–238.

[15] Davis, M.E. and Davis, R.J. 2003. Fundamentals of Chemical Reaction Engineering. McGraw-Hill Companies Inc., NY, USA.

[16] Cain, J.W. 2014. Chemical Reaction Kinetics: Mathematical Underpinnings. Department of Mathematics and Computer Science, University of Richmond, Richmond, VA, Molecular Life Sciences, Springer Science+Business Media, New York.

[17] Grygar, T. 1998. Phenomenological kinetics of irreversible electrochemical dissolution of metal-oxide microparticles. J. Solid State Electrochem. 2: 127 ± 136.

[18] Czirók, E., Bácskai, J., Kulesza, P.J., Inzelt, G. and Malik, M.A. 1996. Quartz crystal microbalance study of the growth of indium hexacyanoferrate films during electro deposition and coagulation. Journal of Electro Analytical Chemistry 405(1-2): 1–264.

[19] Ramteke, K.H., Dighe, P.A., Kharat, A.R. and Patil, S.V. 2014. Mathematical models of drug dissolution: A review. Scholars Academic Journal of Pharmacy (SAJP), ISSN 2320–4206 (Online), Sch. Acad. J. Pharm. 3(5): 388–396.

[20] Machrafi, H., Lebon, G. and Iorio, C.S. 2016. Effect of volume-fraction dependent agglomeration of nanoparticles on the thermal conductivity of nanocomposites: applications to epoxy resins, filled by SiO_2, AlN and MgO nanoparticles. Composites Science and Technology 130: 1–96.

[21] Hasselman, D.P.H. and Johnson, L.F. 1987. Effective thermal conductivity of composites with interfacial thermal barrier resistance. J. Compos. Mater. 21: 508–515.

[22] Nan, C.W., Birringer, R., Clarke, D.R. and Gleiter, H. 1997. Effective thermal conductivity of particulate composites with interfacial thermal resistance. J. App. Phys. 81: 6692.

[23] Chen, H., Witharana, S., Jin, Y., Kim, C. and Ding, Y. 2009. Predicting thermal conductivity of liquid suspensions of nanoparticles (nanofluids) based on Rheology Particuology 7: 151–157.
[24] Chen, H., Ding, Y. and Tan, C. 2007. Rheological behavior of nanofluids. New Journal of Physics 9.
[25] Anoop, K.B., Kabelac, S., Sundararajan, T. and Das, S.K. 2009. Rheological and flow characteristics of nanofluids: Influence of electroviscous effects and particle agglomeration. J. Appl. Phys. 106: 034909.
[26] Behrang, A., Grmela, M., Dubois, C., Turenne, S. and Lafleur, P.G. 2013. Influence of particle-matrix interface, temperature, and agglomeration on heat conduction in dispersions. J. Appl. Phys.
[27] Sestak, J. 1984. Thermophysical Properties of Solids, their Measurements and Theoretical Analysis. Elsevier, Amsterdam.
[28] Jou, D., Casas-Vazquez, J. and Lebon, G. 2010. Extended Irreversible Thermodynamics. 4th edition, Springer, N.Y.
[29] Nan, C.W., Birringer, R., Clarke, D.R. and Gleiter, H. 1997. Effective thermal conductivity of particulate composites with interfacial thermal resistance. J. Appl. Phys. 81: 6692–6699.
[30] Minnich, A. and Chen, G. 2007. Modified effective medium formulation for the thermal conductivity of nanocomposites. Appl. Phys. Lett.
[31] Tavman, I.H. and Akinci, H. 2000. Transverse thermal conductivity of fiber reinforced polymer composites. Int. Comm. Heat Mass Transfer 27: 253–261.
[32] Chen, H., Witharana, S., Jin, Y., Kim, C. and Ding, Y. 2009. Predicting thermal conductivity of liquid suspensions of nanoparticles (nano-fluids) based on Rheology 7: 151–157.
[33] Avrami, M. 1939. Kinetics of phase change. I. General theory. Journal of Chemical Physics 7(12): 1103–1112.
[34] Avrami, M. 1940. Kinetics of phase change. II. Transformation-time relations for random distribution of nuclei. Journal of Chemical Physics 8(2): 212–224.
[35] Razdolsky, L. 2012. Structural Fire Loads: Theory and Principle. McGaw-Hill, N.Y.
[36] Razdolsky, L. 2014. Probability Based Structural Fire Load. Cambridge University Press, London.
[37] Cain, J.W. 2014. Chemical Reaction Kinetics: Mathematical Underpinnings. Springer Science+Business Media, New York.

6
Phenomenological Creep Models of Fibrous Composites (Probabilistic Approach)

6.1 Introduction

6.1.1 Basic concepts and definitions of applied probability theory

Known regularities describing objects behavior in composites industry can be conditionally divided into two groups: (1) deterministic (uniquely defined); (2) under uncertainty. The boundary that separates a random event from a non-random event is very fuzzy. In a pure form, unambiguously defined processes do not seem to exist. When describing fairly complex processes, regularities always have a stochastic character. Reasons for the appearance of uncertainty: (a) the object's performance depends on a large number of factors, some of which may be unknown to the researcher; (b) when constructing a model, the selection of the factors are usually limited to the most significant variables (in the opinion of the engineer or due to objective circumstances), which leads to a coarsening of the model; (c) mathematical errors that occur when the model is linearized or when series are used in a series with a restriction on the number of terms in the series; errors in measurements, errors in the experiment, etc.

Mathematically, uncertainty can be described stochastically and statistically, from the viewpoint of the theory of fuzzy sets, and also intermittently. A stochastic description is used when the undefined parameters have a probabilistic (random) character, and it is necessary that the law of distribution of such random parameters be determined. The statistical description is, in fact, a particular case of stochastic description. This form of description is used when only selective estimates of any characteristics of a random variable are given. When describing from the positions of fuzzy sets, the indeterminate parameter is given by a certain set of possible values that characterize the belonging (using the special function) to the object. The special function can take a value from 1 (full membership) to 0 (complete non-membership). An interval description can be used when undefined parameters are specified only by ranges of possible values (upper and lower bounds), where the parameter can take any value inside the interval and it cannot assign any probability measure.

6.1.2 *The distribution function and the distribution density of a random variable*

The experiment is the realization of some set of conditions that can be reproduced many times. An event is understood as the result of experience or observation. Events can be elementary (indecomposable) and composite (decomposable). An elementary event occurs as a result of a single experience. A compound event is a collection of elementary events.

The totality of events that can be realized as a result of an infinite number of similar experiences is called a general population. A sample is a collection of randomly selected events from the general population. The aggregate volume is the number of events N of this set. A random variable is a variable that can take on different values as a result of the experiment. Random variables are usually denoted by large letters, for example X. The values of the random variable that it takes as a result of the experiment are denoted by small letters $x_1; x_2;\dots,x_n$. With a large number of tests, each of the possible values of the random variable $x_1; x_2;\dots,x_n$ can occur $m_1, m_2, \dots, m_T$ times. These numbers are called frequencies. The entire set of values of the random variable forms the general population N. The gross error values sampled from the general population N_X form a sample of the volume N. If all N_X experiments were carried out, then as a result of the sampling one gets: $\sum_{i=`}^{n} m_i = N$, and the ratio m_i/N is called the frequency or the relative frequency. The probability of an event is a measure of its "favor". Events are called equally possible if the measure of their "favor" is the same. In this case, the frequency W of the event A W (A)—is defined by the formula: W (A) = n/N.

The probability P (A) of an arbitrary event A varies from 0 to 1. In this case, the zero probability corresponds to an impossible event (which can never happen), and the unit probability to an authentic event (which will necessarily happen). For large samples, the probability of an event is equal to its frequency: P (A) ≈ W (A). For independent events, the probability of a product is equal to the product of their probabilities (the multiplication theorem):

$$P\left(\prod_{i=1}^{n} A_i\right) = \prod_{i=1}^{n} P(A_i) \tag{6.1}$$

6.1.3 The Poisson probability distribution

In probability theory and statistics, the *Poisson distribution* is a discrete probability distribution that expresses the probability of a given number of events occurring in a fixed interval of time and/or space if these events occur with a known average rate and independently of the time since the last event [1]. (The Poisson distribution can also be used for the number of events in other specified intervals such as distance, area or volume.)

The Poisson random variable satisfies the following conditions:

1. The number of successes in two disjoint time intervals is independent.
2. The probability of a success during a small time interval is proportional to the entire length of the time interval. Apart from disjoint time intervals, the Poisson random variable also applies to disjoint regions of space.
3. The probability distribution of a Poisson random variable X representing the number of successes occurring in a given time interval or a specified region of space is given by the formula:

$$P(X = m) = \frac{a^m}{m!} e^{-a} \tag{6.2}$$

where: m = 0,1,2,3... And e = 2.71828 ….; a = mean number of successes in the given time interval or region of space. The Poisson (λ) distribution is an approximation to the Binomial (n, p) distribution for the case that n is large, p is small, and $\lambda = np$. In other words, if Y is Binomial (n, p), and n is large and p is small, and X is Poisson (λ) with $\lambda = np$, then Poisson distribution can be derived by taking the appropriate limits of the binomial distribution.

$$P_{m,n} =_n \lim_{\longrightarrow\infty} C_m^n p^m (1-p)^{n-m} = \frac{a^m}{m!} e^{-a} \tag{6.3}$$

where: $n \to \infty$; $p \to 0$, but: np = a = const.

Mean and variance of Poisson distribution

If μ is the average number of successes occurring in a given time interval or region in the Poisson distribution, then the mean and the variance of the Poisson distribution are both equal to μ.

$E(X) = \mu = a$; and $D(X) = \sigma^2 = \mu = a$

Note: In a Poisson distribution, only one parameter, μ is needed to determine the probability of an event.

6.1.4 Correlation and dependence

In statistics, *dependence* refers to any statistical relationship between two random variables or two sets of data. *Correlation* refers to any of a broad class of statistical relationships involving dependence. Correlations are useful because they can indicate a predictive relationship that can be exploited in practice. For example, an electrical utility may produce less power on a mild day based on the correlation between electricity demand and weather. In this example there is a causal relationship, because extreme weather causes people to use more electricity for heating or cooling; however, statistical dependence is not sufficient to demonstrate the presence of such a causal relationship. Formally, *dependence* refers to any situation in which random variables do not satisfy a mathematical condition of probabilistic independence. In loose usage, *correlation* can refer to any departure of two or more random variables from independence, but technically it refers to any of several more specialized types of relationship between mean values. There are several *correlation coefficients*, often denoted ρ or r, measuring the degree of correlation. The most common of these is the Pearson correlation coefficient, which is sensitive only to a linear relationship between two variables (which may exist even if one is a nonlinear function of the other). Other correlation coefficients have been developed to be more robust than the Pearson correlation—that is, more sensitive to nonlinear relationships [2, 3]. The most familiar measure of dependence between two quantities is the Pearson product-moment correlation coefficient, or "Pearson's correlation." It is obtained by dividing the covariance of the two variables by the product of their standard deviations. The population correlation coefficient ρ_{XY} between two random variables *X* and *Y* with expected values μ_X and μ_Y and standard deviations σ_X and σ_Y is defined as:

$$\rho_{X,Y} = \mathrm{corr}(X,Y) = \frac{\mathrm{cov}(X,Y)}{\sigma_X \sigma_Y} = \frac{E[(X-\mu_X)(Y-\mu_Y)}{\sigma_X \sigma_Y} \tag{6.4}$$

where E is the expected value operator, *cov* means covariance, and, *corr* a widely used alternative notation for Pearson's correlation. The Pearson correlation is defined only if both of the standard deviations are finite and both of them are nonzero. The correlation cannot exceed 1 in absolute value. The correlation coefficient is symmetric: corr(*X*, *Y*) = corr(*Y*, *X*). If the variables are independent, Pearson's correlation coefficient is 0, but the converse is not true because the correlation coefficient detects only linear dependencies between two variables. For example, suppose the random variable *X* is symmetrically distributed about zero, and $Y = X^2$. Then *Y* is completely determined by *X*, so that *X* and *Y* are perfectly dependent, but their correlation is zero; they are uncorrelated. However, in the special case when *X* and *Y* are jointly normal, uncorrelativeness is equivalent to independence. If we have a series of *n* measurements of *X* and *Y* written as x_i and y_i where $i = 1, 2, ..., n$, then the *sample correlation coefficient* can be used to estimate the population Pearson correlation *r* between *X* and *Y*. The sample correlation coefficient is written

$$r_{x,y} = \frac{\sum_{i=1}^{n}(x_i - \overline{x})(y_i - \overline{y})}{(n-1)s_x s_y} = \frac{\sum_{i=1}^{n}(x_i - \overline{x})(y_i - \overline{y})}{\sqrt{\sum_{i=1}^{n}(x_i - \overline{x})^2 \sum_{i=1}^{n}(y_i - \overline{y})}} \tag{6.5}$$

where x and y are the sample means of *X* and *Y*, and s_x and s_y are the sample standard deviations of *X* and *Y*. If *x* and *y* are results of measurements that contain measurement error, the realistic limits on the correlation coefficient are not −1 to +1 but a smaller range [4].

6.2 Continuous probability distributions

6.2.1 *Normal probability distributions*

The Normal Probability Distribution is very common in the field of statistics. Normal random variables are encountered in a wide variety of problems. From the central limit theorem we know that the sum of a large number of identical independent random variables is approximately normal. Actually this theorem even holds under much weaker conditions the variables do not have to be identical and independent. It is this theorem that explains why normal random variables are so often encountered in nature. When we have an aggregate effect of a large number of small random factors, the resulting random variable is normal. A random variable X whose distribution has the shape of a *normal curve* is called a *normal random variable.*

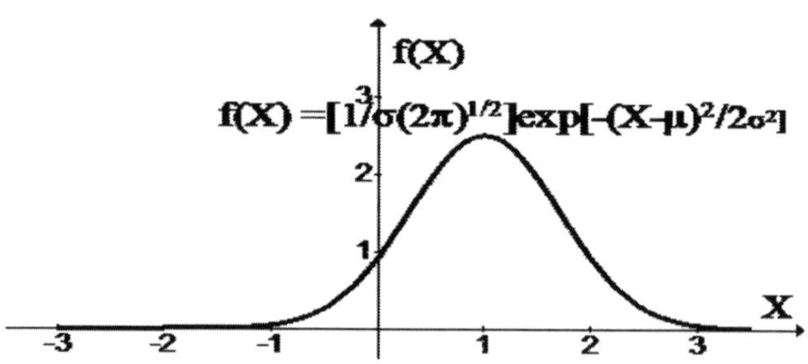

Figure 6.1: Normal distribution curve.

This random variable X is said to be normally distributed with mean μ and standard deviation σ if its probability distribution is given by

$$f(X) = \frac{1}{\sigma\sqrt{2\pi}} e^{-\frac{(x-\mu)^2}{2\sigma^2}} \tag{6.6}$$

Properties of a normal distribution

The normal curve is symmetrical about the mean μ; the mean is at the middle and divides the area into halves; the total area under the curve is equal to 1; it is completely determined by its mean and standard deviation σ (or variance σ^2)

Note: In a normal distribution, only 2 parameters are needed, namely μ and σ^2. The probability of a continuous normal variable X found in a particular interval [a, b] is the area under the curve bounded by x = a and x = b and is given by

$$P(a < X, b) = \int_a^b f(X)dx = \int_a^b \frac{1}{\sigma\sqrt{2\pi}} e^{-\frac{(x-\mu)^2}{2\sigma^2}} dx \tag{6.7}$$

The standard normal distribution

It makes life a lot easier for us if we standardize our normal curve, with a mean of zero and a standard deviation of 1 (one). If we have the standardized situation of μ = 0 and σ = 1, then:

$$f(X) = \frac{1}{\sqrt{2\pi}} e^{-\frac{x^2}{2}} \tag{6.8}$$

All the observations of any normal random variable X with mean μ and standard deviation σ can be transformed to a new set of observations of another normal random variable Z with mean 0 and standard deviation 1 using the following transformation: $z = (x - \mu)/\sigma$. The new distribution of the normal random variable Z with mean 0 and variance 1 (or standard deviation 1) is called a standard normal distribution. Standardizing the distribution like this makes it much easier to calculate probabilities. Since all the values of X falling between x_1 and x_2 have corresponding Z values between z_1 and z_2, and it means: the area under the X curve between $X = x_1$ and $X = x_2$ equals the area under the Z curve between $Z = z_1$ and $Z = z_2$. Hence, we have the following equivalent probabilities: $P(x_1 < X < x_2) = P(z_1 < Z < z_2)$.

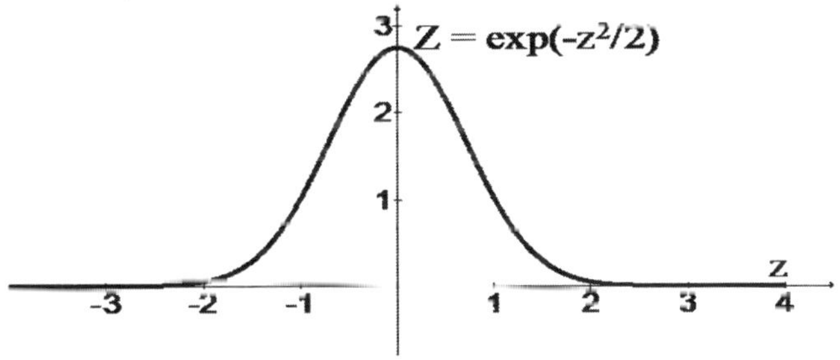

Figure 6.2: Graph of standardized (mean value is 0 and variance is 1) normal curve.

The total area from $-\infty < z < \infty$ is 1. The z-Table indicates the area to the right of the vertical centre-line of the z-curve (or standard normal curve) for different standard deviations. This table is very useful for finding probabilities when the event in question follows a normal distribution. The $\Phi^*(z)$ Table presented in [5] is more convenient, because it covers both, positive and negative z-values:

$$\Phi^*(z) = \frac{1}{\sqrt{2\pi}} \int_{-\infty}^{z} e^{-\frac{z^2}{2}} dz \tag{6.9}$$

Example 6.1

Distance measured by radar is systematically overestimated by 3 m (incorrect zeroing leading to a zero error). The σ—standard deviation is 8 m. The width of the target is 20 m. Someone takes aim at the center of the target. Find the probability of shooting down the target at first trial (experiment).

Data: $\mu = -3$; $\sigma = 8$.

The random variable X has normal distribution, therefore:

$$P(-10 < X < 10) = \Phi^*(\frac{13}{8}) - \Phi^*(\frac{-7}{8}) = \Phi^*(1.625) - \Phi^*(-0.875) \approx 0.757$$

Example 6.2

Suppose someone observed in creep test that the range of maximum temperatures (normal pdf) is between 1200°C and 1100°C, with mean value $\mu = 1000$°C. Find the standard deviation σ of random variable X that gives the maximum probability P.

$$P(1100 < X < 1200) = \Phi^*(\frac{1200-1000}{\sigma}) - \Phi^*(\frac{1100-1000}{\sigma}) = \varphi(\sigma)$$

$$\text{First derivative } \varphi'(\sigma) \text{ is: } [\Phi^*(\frac{200}{\sigma})]' - [\Phi^*(\frac{100}{\sigma})]' = \left\{ \frac{1}{\sigma^2\sqrt{2\pi}}[200e^{-\frac{200^2}{2\sigma^2}} - 100e^{-\frac{100^2}{2\sigma^2}}] = 0 \right.$$

$$\text{or}: 2e^{-\frac{200^2}{2\sigma^2}} - e^{-\frac{100^2}{2\sigma^2}} = 0; \Rightarrow \sigma = 147$$

Example 6.3

Data: same as Example 6.2. Suppose decision has been make to have 3 more tests. The results of these tests will be omitted, if the maximum temperature is less then 1100°C. Find the probabilities of following events: (a) one test will have the maximum temperature more than 1200°C; (b) all three tests will have the maximum temperature less than 1200°C; (c) at least one test result will be rejected; (d) one test result will be rejected and two other test result will be excepted.

a) $P(A) = P(X > 1200) = 1 - \Phi^*(\frac{200}{147}) = 1 - \Phi^*(1.36) \approx 0.0869$

b) $P(B) = [1 - p(A)]^3 = 0.761$

c) $p_o = P(X < 1100) = \Phi^*(\frac{100}{147}) = \Phi^*(0.68) \approx 0.7517$; and $P(C) = 1 - (1 - p_o)^3 \approx 0.9847$

d) $P(D) = C_1^3 (0.7517)(0.2483)^2 = 0.139$

6.2.2 *Weibull distribution*

The Weibull distribution is used world wide to model Life Data. The distribution can handle increasing, decreasing or constant failure-rates and can be created for data with and without suspensions (non-failures). The Weibull distribution is flexible and fits to a wide range of data, including Normal distributed data. Only Log-Normal data does not fit in the Weibull distribution and needs separate analyses. For creating the plot you need to record the time to failure that can be expressed in mileage, cycles, minutes, strength, stiffness or similar continuous parameters. The two-parameter Weibull distribution is the most widely used distribution for life data analysis. Apart from the 2-parameter Weibull distribution, the 3-parameter and the 1-parameter Weibull distribution are often used for detailed analysis.

The 2-parameter Weibull cumulative distribution function (CDF), has the explicit equation:

$$F(t) = 1 - e^{-(t/\eta)^{\beta}} \qquad (6.10)$$

F(t) = Probability of failure (often used) at time t; t = time, cycles, miles, or any appropriate parameter.

The Weibull probability density function (PDF) is [6]:

$$f(t) = \beta \frac{t^{\beta-1}}{\eta^{\beta}} e^{-(t/\eta)^{\beta}} \qquad (6.11)$$

where $\beta > 0$ is the *shape parameter* and $\eta > 0$ is the *scale parameter* of the distribution. Its complementary cumulative distribution function is a stretched exponential function. The Weibull distribution is related to a number of other probability distributions; in particular, it interpolates between the exponential distribution ($\beta = 1$) and the Rayleigh distribution ($\beta = 2$).

If the quantity t is a "time to failure", the Weibull distribution gives a distribution for which the failure rate is proportional to a power of time. The *shape* parameter, β, is that power plus one, and so this parameter can be interpreted directly as follows:

A value of $\beta < 1$ indicates that the failure rate decreases over time. This happens if there is significant number of defective items failing early and the failure rate decreasing over time as the defective items are weeded out of the population.

A value of $\beta = 1$ indicates that the failure rate is constant over time. This might suggest random external events are causing mortality, or failure.

A value of $\beta > 1$ indicates that the failure rate increases with time. This happens if there is an "aging" process, or parts that are more likely to fail as time go on.

In materials science field, the shape parameter β of strength distribution is known as the Weibull modulus. The failure rate h (or hazard rate) is given by

$$h(t;\beta,\eta) = \frac{\beta}{\eta}\left(\frac{t}{\eta}\right)^{\beta-1} \tag{6.12}$$

Moments:

In particular, the n-th raw moment of X is given by:

$$m_n = \beta^n \Gamma(1+\frac{n}{\eta}) \tag{6.13}$$

where Γ—gamma function. The mean and variance of a Weibull random variable can be expressed as:

$$E(X) = \mu_X = \beta\Gamma(1+\frac{1}{\eta}); \text{ and } Var(X) = D(X) = \sigma_X^2 = \beta^2\left[\Gamma(1+\frac{2}{\eta}) - \Gamma(1+\frac{1}{\eta})\right] \tag{6.14}$$

Uncertainty in estimating the Weibull parameters is, as in any other distribution estimation, related to data. With this we mean the sample size and number of failures. Confidence limits should always be used in order to assess potential uncertainty.

6.2.3 *Rayleigh distribution*

In probability theory and statistics, the *Rayleigh distribution* is a continuous probability distribution. A Rayleigh distribution is often observed when the overall magnitude of a vector is related to its directional components. One example where the Rayleigh distribution naturally arises is when wind speed is analyzed into its orthogonal 2-dimensional vector components. Assuming that the magnitude of each component is uncorrelated and normally distributed with equal variance, than the overall wind speed (vector magnitude) will be characterized by a Rayleigh distribution. A second example of the distribution arises in the case of random complex numbers whose real and imaginary components are i.i.d. (independently and identically distributed) Gaussian. In that case, the absolute value of the complex number is Rayleigh-distributed. The distribution is named after Lord Rayleigh.

The Rayleigh probability density function is:

$$f(x;\sigma) = \frac{x}{\sigma^2} e^{-\frac{x^2}{2\sigma^2}} \tag{6.15}$$

$$x \geq 0$$

for parameter $\sigma > 0$ and cumulative distribution function:

$$F(x) = 1 - e^{-\frac{x^2}{2\sigma^2}} \tag{6.16}$$

$$x \in [0, \infty)$$

The raw moments are given by:

$$\mu_k = \sigma^k 2^{k/2} \Gamma(1 + k/2)$$

where $\Gamma(z)$ is the Gamma function. The mean and variance of a Rayleigh random variable may be expressed as:

$$\mu(X) = \sigma\sqrt{\frac{\pi}{2}} \approx 1.253\sigma; \text{ and } \mathrm{Var}(X) - D(X) - \frac{4-\pi}{2}\sigma^2 \approx 0.429\sigma^2 \tag{6.17}$$

Given N independent and identically distributed Rayleigh random variables, the maximum likelihood estimate of σ is:

$$\bar{\sigma} \approx \sqrt{\frac{1}{2N}\sum_{i=1}^{N} x_i^2} \tag{6.18}$$

Hence, the above formula can be used to estimate the noise variance [16, 17].

6.2.4 *Chi-squared distribution*

The chi-squared distribution is used in statistics (also *chi-square* or χ^2-distribution), with *k* degrees of freedom in the distribution of a sum of the squares of *k* independent standard normal random variables. It is one of the most widely used probability distributions in inferential statistics, e.g., in hypothesis testing or in construction of confidence intervals [5]. When there is a need to contrast it with the non-central chi-squared distribution, this distribution is sometimes called the *central chi-squared distribution*.

The chi-squared distribution is used in the common chi-squared tests for goodness to fit of an observed distribution to a theoretical one; the independence of two criteria of classification of qualitative data, and in confidence interval estimation for a standard deviation of a normal distribution from a given sample standard deviation. The chi-squared distribution is a special case of the gamma distribution.

If $Z_1, \ldots, Z_k$ are independent, standard normal random variables, then the sum of their squares, $Q = \sum_{i=1}^{k} Z_i^2$ is distributed according to the *chi-squared distribution* with k degrees of freedom. This is usually denoted as $Q \sim \chi^2(k)$ or $Q \sim \chi_k^2$. The chi-squared distribution has one parameter: k—a positive integer that specifies the number of degrees of freedom (i.e., the number of Z_i's). The probability density function (pdf) of the chi-squared distribution is:

$$f(k;x) = \begin{cases} \dfrac{x^{(k/2)-1}(e^{-x/2})}{2^{k/2}(\Gamma\dfrac{k}{2})} & \text{if } x \geq 0 \\ 0 & \text{if } \quad x \leq 0 \end{cases} \tag{6.19}$$

where Γ(k/2) denotes the Gamma function, which has closed-form values for odd k.

Its cumulative distribution function is:

$$F(x;k) = \frac{\gamma(\frac{k}{2},\frac{x}{2})}{\Gamma(\frac{k}{2})} = P(\frac{k}{2},\frac{x}{2}) \tag{6.20}$$

where γ(k,z) is the lower incomplete Gamma function and P(k,z) is the regularized Gamma function. In a special case of k = 2 this function has a simple form:

$$F(x;2) = 1 - e^{-(\frac{x}{2})}$$

Tables of this cumulative distribution function are widely available and the function is included in many spreadsheets and all statistical packages. The sum of independent chi-squared variables is also chi-squared distributed. Specifically, if $\{X_i\}$ are independent chi-squared variables with $\{k_i\}$ degrees of freedom, respectively, then $Y = X_1 + \cdots + X_n$ is chi-squared distributed with $k_1 + \cdots + k_n$ degrees of freedom.

The chi-squared distribution has numerous applications in inferential statistics, for instance in chi-squared tests and in estimating variances. It enters the problem of estimating the mean and variance problems via its role in the F-distribution.

Following are some of the most common situations in which the chi-squared distribution arises from a Gaussian-distributed sample; if $X_1, \ldots, X_n$ are i.i.d. $N(\mu, \sigma^2)$ random variables, then:

$$\sum_{i=1}^{n}(X_i - \bar{X})^2 \sim \sigma^2\chi_{n-1}^2, \text{ where: } \bar{X} = \frac{1}{n}\sum_{i=1}^{n} X_i \tag{6.21}$$

The p-value is the probability of observing a test statistic *at least* as extreme in a chi-squared distribution. Accordingly, since the cumulative distribution function (CDF) for the appropriate degrees of freedom *(df)* gives the probability of having obtained a value *less extreme* than this point, subtracting the CDF value from 1 gives the p-value. The table in [7] gives a number of p-values matching to χ^2.

6.3 Joint probability distribution

Given two random variables *X* and *Y* that are defined on the same probability space, the *joint distribution* for *X* and *Y* defines the probability of events defined in terms of both *X* and *Y*. In the case of only two random variables, this is called a *vicariate distribution*, but the concept generalizes to any number of random variables, giving a *multivariate distribution*. The equation for joint probability is different for both dependent and independent events. The joint probability function of a set of variables can be used to find a variety of other probability distributions. The probability density function can be found by taking a partial derivative of the joint distribution with respect to each of variables. The cumulative distribution function for a pair of random variables is defined in terms of their joint probability distribution $F(x, y) = P(X \leq x, Y \leq y)$. The joint probability mass function of two discrete random variables is equal to:

$$P(X = x; Y = y) = P(Y = y \mid X = x)P(X = x) = P(X = x \mid Y = y)P(Y = y)$$

In general, the joint probability distribution of n discrete random variables X_1; X_2; X_3...... X_n is equal to:

$$P(X_1 = x_1; .....Xn = xn) = P(X_1 = x_1)P(X_2 = x_2 \mid X_1 = x_1)$$

$$P(X_3 = x_3 \mid X_1 = x_1; X_2 = x_2).....P(X_n = x_n \mid X_{n-1} = x_{n-1}; ....X_1 = x_1)$$

This identity is known as the chain rule of probability. Since these are probabilities, we have:

$$\sum_x \sum_y P(X = x; Y = y) = 1 \tag{6.22}$$

Generalizing for n discrete random variables X_1; X_2; X_3...... X_n

$$\sum_{x1} \sum_{x2} .... \sum_{x_n} P(X_1 = x_1; X_2 = x_2 .... X_n = x_n.) = 1$$

Similarly for continuous random variables, the *joint probability density function* can be written as $f_{X,Y}(x, y)$ and this is $f_{X,Y}(x, y) = f_{Y|X}(y|x)f_X(x) = f_{X|Y}(x|y)f_y(y)$, where $f_{Y|X}(y|x)$ and $f_{X|Y}(x|y)$ are the conditional distributions of Y given X = x and of X given Y = y respectively, and $f_X(x)$ and $f_Y(y)$ are the marginal distributions for X and Y respectively. Again, since these are probability distributions, one has:

$$\iint_{x,y} f_{X,Y}(x,y)dxfy = 1 \tag{6.23}$$

If for discrete random variables $P(X = x; Y = y) = P(X = x)P(Y = y)$ for all x and y, or for absolutely continuous random variables $f_{X,Y}(x, y) = f(x)f(y)$ for all x and y, then X and Y are said to be independent [8].

6.4 Characteristic function

In probability theory and statistics, the *characteristic function* of any real-valued random variable completely defines its probability distribution. If a random variable admits a probability density function, then the characteristic function is the Inverse Fourier transform of the probability density function. Thus it provides the basis of an alternative route to analytical results compared with working directly with probability density functions or cumulative distribution functions. There are particularly simple results for the characteristic functions of distributions defined by the weighted sums of random variables. The characteristic function always exists when treated as a function of a real-valued argument, unlike the moment-generating function. There are relations between the behavior of the characteristic function of a distribution and properties of the distribution, such as the existence of moments and the existence of a density function. The characteristic function provides an alternative way for describing a random variable. The *characteristic function* $\varphi_X(t) = E[e^{itX}]$ also completely determines behavior and properties of the probability distribution of the random variable X. The two approaches are equivalent in the sense that by knowing one of the functions it is always possible to find the other, yet they both provide different insight for understanding the features of the random variable. Note however that the characteristic function of a distribution always exists, even when the probability density function or moment-generating function does not. The characteristic function approach is particularly useful in analysis of linear combinations of independent random variables: a classical proof of the Central Limit Theorem uses characteristic functions. Another important application is to the theory of the decomposability of random variables. For a scalar random variable X the characteristic function is defined as the expected value of e^{itX}, where i is the imaginary unit, and $t \in R$ is the argument of the characteristic function. If random variable X has a probability density function f_X, then the characteristic function is its Fourier transform [21]. Extensive tables of characteristic functions are provided in [22].

The characteristic function of a real-valued random variable always exists, since it is an integral of a bounded continuous function over a space whose measure is finite.

The characteristic function of a symmetric (around the origin) random variable is real-valued and even.

If a random variable X has moments up to k-th order, then the characteristic function φ_X is k times continuously differentiable on the entire real line. In this case $E[X^k] = (-i)^k \varphi_X^{(k)}(0)$.

If $X_1, \ldots, X_n$ are independent random variables, and $a_1, \ldots, a_n$ are some constants, then the characteristic function of the linear combination of X_i's is:

$$\varphi_{a_1 X_1 + \ldots + a_n X_n}(t) = [\varphi_{X_1}(a_1 t)] \bullet \ldots \ldots \bullet [\varphi_{Xn}(a_n t)] \tag{6.24}$$

One specific case would be the sum of two independent random variables X_1 and X_2 in which case one would have $\varphi_{X1+X2}(t) = [\varphi_{X1}(t)] \bullet [\varphi_{X2}(t)]$. Since there is a one-to-one correspondence between cumulative distribution functions and characteristic functions, it is always possible to find one of these functions if we know the other one. The formula in definition of characteristic function allows us to compute φ when we know the distribution function F (or density f). If, on the other hand, we know the characteristic function φ and want to find the corresponding distribution function, then: the following inversion theorem can be used:

$$f_X(x) = F_X' = \frac{1}{2\pi} \int_R e^{-itx} \varphi_X dt \tag{6.25}$$

Characteristic functions are particularly useful for dealing with linear functions of independent random variables. For example, if $X_1, X_2, \ldots, X_n$ is a sequence of independent (and not necessarily identically distributed) random variables, and $a_i = 1/n$ are some constants, then S_n is the sample mean. In this case, the characteristic function for the mean of X_i's is:

$$\varphi_{\bar{X}}(t) = [\varphi_X(t/n)]^n$$

Characteristic functions can also be used to find moments of a random variable. Provided that the nth moment exists, characteristic function can be differentiated n times and:

$$E\left[X^n\right] = (i)^{-n} \varphi_X^{(n)}(0) = (i)^{-n} \left[\frac{d^n}{dt^n} \varphi_X(t)\right]_{t=0} \tag{6.26}$$

Characteristic functions can be used as part of procedures for fitting probability distributions to samples of data. Cases where this provides a practicable option compared to other possibilities include fitting the stable distribution, since closed form expressions for the density are not available which makes implementation of maximum likelihood estimation difficult. Estimation procedures are available that match the theoretical characteristic function to the empirical characteristic function, calculated from the data.

6.5 Functions of random variables and their distribution

Let X be a random variable with known distribution f(x). Let another random variable Y be a function of X, where Y = F(X). When the function g is strictly increasing on the support of X, then g admits an inverse defined on the support of Y, i.e., a function $g^{-1}(y)$ such that $X = g^{-1}(y)$. Furthermore $g^{-1}(y)$ is itself strictly increasing. The distribution function of a strictly increasing function of a random variable can be computed as follows. Let say X is a random variable with support R_X and distribution function $F_X(x)$. Let g be strictly increasing on the support of X. Then, the support of Y = g(X) is R_Y, and the distribution function of Y is $R_Y = \{y = g(x): x \in R_X\}$.

$$F_Y(y) = \begin{cases} 0 & \text{if } y < x \\ F_X(g^{-1}(y)) & \text{if } y \subset R_y \\ 1 & \text{if } y > x \end{cases} \tag{6.27}$$

Therefore, in the case of an increasing function, knowledge of g^{-1} and of the upper and lower bounds of the support of Y is all we need to derive the distribution function of Y from the distribution function of X.

Example 6.4

Let X be a random variable with support and distribution function: This function is assumed to be strictly increasing and it admits an inverse on the support of $R_X = [1, 2]$ and distribution function:

$$F_X(x) = \begin{cases} 0 & \text{if } x < 1 \\ \frac{1}{2}x & \text{if } 1 \le x \le 2 \\ 1 & \text{if } x > 2 \end{cases}$$

Let $Y = X^2$

The function $y = g(x) = x^2$ is strictly increasing and it admits an inverse on the support of X: $g^{-1}(y) = \sqrt{y}$. The support of Y is R_Y [1, 4]. The distribution function of Y is:

$$F_X(x) = \begin{cases} 0 & \text{if } y < x \text{, i.e., if } y < 1 \\ F_X(g^{-1}(y)) = \frac{1}{2}\sqrt{y} & \text{if } 1 \le y \le 4 \\ 1 & \text{if } y > 4 \end{cases}$$

In cases where X is either discrete or absolutely continuous there are specialized formulae for the probability mass and probability density functions, which are reported below. When X is a discrete random variable, the probability mass function of Y = g(X) can be computed as follows. Let X be a discrete random variable with support R_X and probability mass function $p_X(x)$. Let g be strictly increasing on the support of X. Then, the support of Y = g(X) is: $R_Y = \{y = g(x): x \in R_X\}$ and its probability mass function:

$$p_Y(y) = \begin{cases} p_X(g^{-1}(y)) & \text{if } y \in R_Y \\ 0 & \text{if } y \notin R_Y \end{cases}$$

Example 6.5

Let X be a discrete random variable with support $R_X = R\{1,2,3\}$ and probability mass function $p_X(x)$. Let the support of this function be strictly increasing and the probability mass function is:

$$p_X(x) = \begin{cases} x/6 & \text{if } x \in R_X \\ 0 & \text{if } x \notin R_X \end{cases}$$

Let $Y = g(X) = 3 + X^2$. The support of Y is $R_Y = [4, 7, 12]$.

The function is strictly decreasing and its inverse is: $g^{-1}(y) = \sqrt{y-3}$

The probability mass function of is:

$$p_Y(y) = \begin{cases} \frac{1}{6}\sqrt{y-3} & \text{if } y \in R_Y \\ 0 & \text{if } y \notin R_Y \end{cases}$$

When X is an absolutely continuous random variable and g is differentiable, then Y also is absolutely continuous and its probability density function can be easily computed as follows: let g be an absolutely continuous random variable with support R_X and probability density function $f_X(x)$. The function g is strictly increasing and differentiable on the support of X. Then, the support of Y = g(X) is R_Y and its probability density function of Y is:

$$f_Y(y) = \begin{cases} f_X(g^{-1}(y))\dfrac{dg^{-1}(y)}{dy} & \text{if } y \in R_Y \\ 0 & \text{if } y \notin R_Y \end{cases}$$

Example 6.6

Let X is an absolutely continuous random variable with support $R_X = (0,1]$ and probability density function:

$$f_X(x) = \begin{cases} 2x & \text{if } x \in R_X \\ 0 & \text{if } x \notin R_X \end{cases}$$

Let Y = g(X) = ln(X). The support of Y is: $R_Y = (-\infty,0]$. The function g is strictly increasing and its inverse is: $g^{-1}(y) = \exp(y)$. The probability density function of Y is:

$$f_Y(y) = \begin{cases} 2\exp(y)\exp(y) = 2\exp(2y) & \text{if } y \in R_y \\ 0 & \text{if } y \notin R_Y \end{cases}$$

The distribution function of a strictly decreasing function of a random variable can be computed similarly to a strictly increasing case.

Example 6.7

Let X be a uniform random variable on the interval [0,1], i.e., an absolutely continuous random variable with support $R\{0,1\}$. Probability density function:

$$f_X(x) = \begin{cases} 1 & \text{if } x \in R_X \\ 0 & \text{if } x \notin R_X \end{cases}$$

Let $Y = g(X) = -\frac{1}{\lambda}\ln(X)$, where λ is a constant. The support of Y is: $R_Y = (0, \infty]$. The function g is strictly decreasing and its inverse is: $g^{-1}(y) = \exp(-\lambda y)$ and with the derivative $\frac{dg^{-1}(y)}{dy} = -\lambda\exp(-\lambda y)$. The probability density function of Y is:

$$f_Y(y) = \begin{cases} \lambda\exp(-\lambda y) & \text{if } y \in R_Y \\ 0 & \text{if } y \notin R_Y \end{cases}$$

Therefore, Y has an exponential distribution with parameter λ.

6.5.1 *One-to-one functions of an absolutely continuous random variable*

When X is an absolutely continuous random variable and g is differentiable, then Y is also absolutely continuous and its probability density function is given by the following proposition (density of a one-to-one function).

Let X be an absolutely continuous random variable with support R_X and probability density function $f_X(x)$. Let g be one-to-one and differentiable on the support of X. Then, the probability density function is:

$$f_Y(y) = \begin{cases} f_X(g^{-1}(y))\left|\frac{dg^{-1}(y)}{dy}\right| & \text{if } y \in R_Y \\ 0 & \text{if } y \notin R_Y \end{cases} \tag{6.28}$$

6.5.2 Probabilistic transformation (linearization) method

Tom Caughey was a pioneer in the development of the stochastic equivalent linearization procedure for estimating the mean and variance of a non-linear system to random variables. The stochastic equivalent linearization procedure or statistical linearization procedure was almost simultaneously introduced more than fifty years ago by three independent investigators: Booton [9], Kazakov [10], and Caughey [11]. In mathematics and its applications, *linearization* refers to finding the linear approximation to a function at a given point. In the study of dynamic systems, linearization is a method for assessing the local stability of an equilibrium point of a system of nonlinear differential equations or discrete dynamical systems. This method is used in fields such as engineering, physics, economics, and ecology. Linearization is an effective method for approximating the output of a function y = f(x) at any x = a based on the value and slope of the function at x = b, given that y = f(x) is continuous on [a,b] (or [b,a]) and that a is close to b. In, short, linearization approximates the output of a function near x = a. The concept of local linearity applies to the most of points arbitrarily close to x = a, and the slope M should be, most accurately, the slope of the tangent line at x = a.

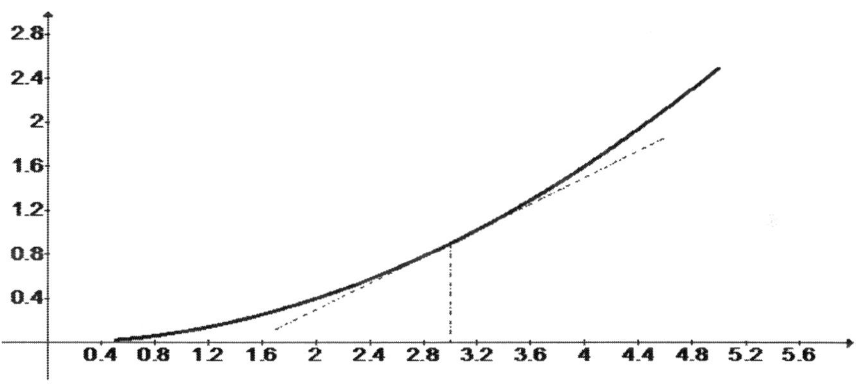

Figure 6.3: Statistical linearization.

Visually, the accompanying diagram shows the tangent line of f(x) at x. At f(x + h), where h is any small positive or negative value, f(x + h) is very nearly the value of the tangent line at the point (x + h, L(x + h)).

The final equation for the linearization of a function at x = a is: y = f(a) + f'(a)(x – a). The equation for the linearization of a function f(x,y) at a point P(a,b) is:

$$f(x,y) \approx f(a,b) + \frac{\partial f(x,y)}{\partial x}\Big|_{a,b}(x-a) + \frac{\partial f(x,y)}{\partial y}\Big|_{a,b}(y-b) \tag{6.29}$$

The general equation for the linearization of a multivariable function f(X) at a point P is:

$$f(X) \approx f(P) + \nabla f\big|_P (X - P)$$

where X is the vector of variables, and P is the linearization point of interest. Linearization makes it possible to use tools for studying linear systems to analyze the behavior of a nonlinear function near a given point. The linearization of a function is the first order term of its Taylor expansion around the point of interest. A Random Equation (RE) is an equation containing a random term. The study of RE is an exciting topic which brings together techniques from probability theory, functional analysis and the theory of equations analysis. The solution to random equations may be viewed in several manners. We can view a solution as a random field (set of random variables indexed by a multidimensional parameter). In the case where the RE is an evolution equation, the finite dimensional points of view consists in viewing the solution at a given time as a random element in a space function and thus view the RE as a stochastic evolution equation in an infinite dimensional space. In this book, a new technique is proposed in order to evaluate the probability density function of the solution, based on the combination of known probabilistic transformation methods and a developed nonlinear technique. The main disadvantage of the previous methods is that they are approximate methods and they are applicable under some conditions (e.g., the randomness should be small).

Example 6.8

Consider the non-linear function $\varphi(x) = e^{-x}$. Develop the linear approximation function in the vicinity of μ_x. To do this we need to use $\varphi(x) = \exp(-\mu_x)$ and $d\varphi(x)/dx = -\exp(-\mu_x)$. The linear approximation is then given by the following approximation:

$$\varphi(x) \approx e^{-\mu_x} - e^{-\mu_x}(x - \mu_x) = (1 + \mu_x)e^{-\mu_x} - xe^{-\mu_x}$$

Rewrite this equation in form of y = ax + b, where:

$$y = \varphi(x) \approx [\varphi(\mu_x) - \varphi'(\mu_x)\mu_x] + [\varphi'(\mu_x)]x$$

$$\text{where}: a = [\varphi(\mu_x) - \varphi'(\mu_x)\mu_x];\ b = \varphi'(\mu_x)$$

The mean value of y depends upon both “a” and “b” terms:

$$\mu_y = [\varphi(\mu_x) - \varphi'(\mu_x)\mu_x] + [\varphi'(\mu_x)]\mu_x = a + b\mu_x$$

The variance depends only upon the b term:

$$\sigma_y^2 = [\varphi'(\mu_x)]^2$$

Consider a general non-linear function of three independent random variables, $\varphi(x,y,z)$. The first order Taylor’s series is given by the following equation, where $\mu_{x,y,z}$ denotes simultaneous evaluation at all three means: μ_x; μ_y and μ_z:

$$y = \varphi(x, y, z) \approx \varphi(\mu_{xyz}) + \varphi_x^{'}(x - \mu_x) + \varphi_y^{'}(y - \mu_y) + \varphi_z^{'}(z - \mu_z)$$

The propagation of means yields an anticipated result: $\mu_\varphi = \varphi(\mu_{xyz})$ and propagation of variances yields the following equation:

$$\sigma_\varphi^2 = [\varphi_x^{'}(\mu_{xyz})]^2 + [\varphi_y^{'}(\mu_{xyz})]^2 + [\varphi_z^{'}(\mu_{xyz})]^2$$

6.6 Confidence interval

In statistics, a *confidence interval (CI)* is a type of interval estimate of a population parameter and is used to indicate the reliability of an estimate. It is an observed interval (i.e., it is calculated from the observations), in principle different from sample to sample, that frequently includes the parameter of interest if the experiment is repeated. How frequently the observed interval contains the parameter is determined by the *confidence level* or *confidence coefficient*. More specifically, the meaning of the term "confidence level" is that, if confidence intervals are constructed across many separate data analyses of repeated (and possibly different) experiments, the proportion of such intervals that contain the true value of the parameter will match the confidence level; this is guaranteed by the reasoning underlying the construction of confidence intervals [9, 10, 11].

Confidence intervals consist of a range of values (interval) that act as good estimates of the unknown population parameter. However, in infrequent cases, none of these values may cover the value of the parameter. The level of confidence of the confidence interval would indicate the probability that the confidence range captures this true population parameter given a distribution of samples. It does not describe any single sample. This value is represented by a percentage, so when we say, "we are 99% confident that the true value of the parameter is in our confidence interval", we express that 99% of the observed confidence intervals will hold the true value of the parameter. The desired level of confidence is set by the researcher (not determined by data). If a corresponding hypothesis test is performed, the confidence level corresponds with the level of significance, i.e., a 95% confidence interval reflects a significance level of 0.05, and the confidence interval contains the parameter values that, when tested, should not be rejected with the same sample. Greater levels of variance yield larger confidence intervals, and hence less precise estimates of the parameter. Confidence intervals of difference parameters not containing 0 imply that there is a statistically significant difference between the populations. Certain factors may affect the confidence interval size including size of sample, level of confidence, and population variability. A larger sample size normally will lead to a better estimate of the population parameter. Let X be a random sample from a probability distribution with statistical parameters θ, which is a quantity to be estimated, and φ, representing quantities that are not of immediate interest. A *confidence interval* for the parameter θ, with confidence level or confidence coefficient γ, is an interval with random endpoints (u(X), v(X)), determined by the pair of random variables u(X) and v(X), with the property:

$$\Pr_{\theta,\varphi}(u(X) < \theta < v(X)) = \gamma \quad \text{for all } (\theta,\varphi)$$

The quantities φ in which there is no immediate interest are called nuisance parameters, as statistical theory still needs to find some way to deal with them. The number γ, with typical values close to but not greater than 1, is sometimes given in the form 1 − α (or as a percentage 100%·(1 − α)), where α is a small non-negative number, close to 0. Here $\Pr_{\theta,\varphi}$ indicates the probability distribution of X characterized by (θ, φ). An important part of this specification is that the random interval (u(X), v(X)) covers the unknown value θ with a high probability no matter what the true value of θ actually is. Note that here $\Pr_{\theta,\varphi}$ need not refer to an explicitly given parameterized family of distributions, although it often does. Just as the random variable X notionally corresponds to other possible realizations of x from the same population or from the same version of reality, the parameters (θ, φ) indicate that we need to consider other versions of reality in which the distribution of X might have different characteristics. In a specific situation, when x is the outcome of the sample X, the interval (u(x), v(x)) is also referred to as a confidence interval for θ. Note that it is no longer possible to say that the (observed) interval (u(x), v(x)) has probability γ to contain the parameter θ. This observed interval is just one realization of all possible intervals for which the probability statement holds. In many applications, confidence intervals that have exactly the required confidence level are hard to construct. But practically useful intervals can still be found: the rule for constructing the interval may be accepted as providing a confidence interval at level γ if: $\Pr_{\theta,\varphi}(u(X) < \theta < v(X)) \approx \gamma$ for all (θ,φ) to an acceptable level of approximation. Alternatively, it is simply required that $\Pr_{\theta,\varphi}(u(X) < \theta < v(X)) \geq \gamma$ for all (θ,φ), which is useful if the probabilities are only partially identified, or imprecise.

Suppose $\{X_1, ..., X_n\}$ is an independent sample from a normally distributed population with (parameters) mean μ and variance σ^2. Let

$$\bar{X} = (X_1 + ... + X_n)/n;$$

$$S^2 = \frac{\sum_{i=1}^{n}(X_i - \bar{X})^2}{S/\sqrt{n}} \tag{6.30}$$

where $\overline{X}$ the sample mean, and S2 is the sample variance. After observing the sample we find values x for X and s for S, from which we compute the confidence interval:

$$\left[\overline{x}-\frac{cS}{\sqrt{n}};\overline{x}+\frac{cS}{\sqrt{n}}\right] \tag{6.31}$$

The quintile function of a distribution is the inverse of the cumulative distribution function. The quintile function of the standard normal distribution is called the probity function, and can be expressed in terms of the inverse error function $\Phi^{-1}(p)=\sqrt{2}\mathrm{erf}^{-1}(2p-1);\ p\in(0,1)$. For a normal random variable with mean μ and variance σ^2, the quintile function is: $F^{-1}(p)=\mu+\sigma\Phi^{-1}=\mu+\sigma\sqrt{2}\mathrm{erf}^{-1}.(2p-1)\ p\in(0,1)$.

The quintile $\Phi^{-1}(p)$ of the standard normal distribution is commonly denoted as z_p. These values are used in hypothesis testing and construction of confidence intervals. A normal random variable X will exceed $\mu + \sigma z_p$ with probability $1 - p$; and will lie outside the interval $\mu \pm \sigma z_p$ with probability $2(1 - p)$. In particular, the quantile $z_{0.975}$ is 1.96; therefore a normal random variable will lie outside the interval $\mu \pm 1.96\sigma$ in only 5% of cases. These values are useful to determine tolerance interval for sample averages and other statistical estimators with normal (or asymptotically normal) distribution [12].

6.6.1 Confidence interval (poisson distribution)

Given a sample of n measured values k_i we wish to estimate the value of the parameter λ of the Poisson population from which the sample was drawn. The maximum likelihood estimate is [11] $\overline{\lambda}_{MLE}=\frac{1}{n}\sum_{i=1}^{n}k_i$. Since each observation has expectation λ, so does this sample's mean value; therefore the maximum likelihood estimate is an unbiased estimator of λ. It is also an efficient estimator, i.e., its estimation variance achieves the lower bound. Also it can be proved that the sample mean is a complete and sufficient statistic for λ. The confidence interval for a Poisson mean is calculated using the relationship between the Poisson and Chi square distributions, and can be written as:

$$\frac{1}{2}\chi^2(\alpha/2;2k)\le\mu\le\frac{1}{2}\chi^2(1-\alpha/2;2k+2)$$

where k is the number of event occurrences in a given interval and $\chi^2(p; n)$ is the chi-square deviate with lower tail area p and degrees of freedom n. This interval is 'exact' in the sense that its coverage probability is never less than the nominal $1 - \alpha$.

6.6.2 *Confidence interval (binomial proportion)*

In statistics, a *binomial proportion confidence interval* is a confidence interval for a proportion in a statistical population. It uses the proportion estimated in a statistical sample and allows for sampling error. There are several formulas for a binomial confidence interval, but all of them rely on the assumption of a binomial distribution. In general, a binomial distribution applies when an experiment is repeated a fixed number of times, each trial of the experiment have two possible outcomes (labeled arbitrarily success and failure), the probability of success is the same for each trial, and the trials are statistically independent. The simplest and most commonly used formula for a binomial confidence interval relies on approximating the binomial distribution with a normal distribution. This approximation is justified by the central limit theorem. The formula is:

$$\overline{p} \pm z_{1-\alpha/2}\sqrt{\frac{\overline{p}(1-\overline{p})}{n}}$$

where $\overline{p}$ is the proportion of successes in a Bernoulli trial process estimated from the statistical sample, $z_{1-\alpha/2}$ is the $1-\alpha/2$ percentile of a standard normal distribution, α is the error percentile and n is the sample size. For example, for a 95% confidence level the error (α) is 5%, so:

$1 - \alpha/2 = 0.975$ and $z_{1-\alpha/2} = 1.96$.

The central limit theorem applies well to a binomial distribution, even with a sample size less than 30, as long as the proportion is not too close to 0 or 1. For very extreme probabilities, though, a sample size of 30 or more may still be inadequate. A frequently cited rule of thumb is that the normal approximation works well as long as $np > 5$ and $n(1 - p) > 5$.

Example 6.9

In the factory, the technical part can be rejected because of various elements of the technological process: the low quality of the casting mold (sand sinks, collapses, squeezes, etc.); due to the violation of the technological process of melting and out-of-furnace treatment of metal (non-metallic inclusions, gas shells, porosity, etc.); due to a violation of the mold casting mode (slag inclusions, spay, etc.). Each of these elements of the process, independently of the other, may be the cause of the final rejection in the casting. Let the probability of obtaining a qualitative casting without defects "through fault" of the form p (f) = 0.98; by the fault of the metal p (m) = 0.93; due to the fault of pouring p (s) = 0.99. It is necessary to assess the reliability of the technological process as a whole, i.e., determine the probability of obtaining a defect-free casting p. By the formula (6.1), we find: $p = p(f) \cdot p(m) \cdot p(s)$

= 0.98 · 0.93 · 0.99 = 0.90. For incompatible events (they cannot occur at the same time), the following theorem is true for the addition of probabilities:

$$P(A_1 + A_2 + ... + A_n) = P(A_1) + P(A_2) + ... + P(A_n) \quad (6.32)$$

This theorem has two consequences: (1) To complete the group of incompatible events the sum of their probabilities is equal to one; (2) The sum of the probabilities of opposite events is equal to one. The law of distribution of a random variable is any rule (table, function) that allows you to find the probabilities of all possible events. Random variables can be discrete or continuous. Discrete random variables are those that can take a finite and countable set of possible values. Continuous random variables are those which in any interval can take any value. Any continuous random variable can be specified as a discrete one if all its possible values are broken up into intervals and the probability of occurrence of these intervals (because of the limitations of the measuring means, all measurements of continuous quantities are given in a discrete form). Random variables are characterized by probability distribution functions. The integral distribution function $F(x_i)$ of a random variable X is the probability that a random variable will take values not exceeding x_i, i.e., will fall into the interval $(-\infty, x_i)$: $F(x_i) = p(X < x_i)$. $F(x_i)$ determines the law of distribution of the random variable X. In most practical cases, the distribution of random variables can be specified by introducing the probability density function f(x) (differential distribution function). Here, x is a vector whose components are the quantities x_i.

A characteristic feature of a random variable is that it is not known in advance which value it will take. The possibility of accepting a random variable X from the interval (x1, x2) is quantitatively estimated by the probability:

$$P(x_1 < X < x_2) = f(x)dx \quad (6.33)$$

where $P(x_1 < X \le x_2)$ is the probability of the event $(x_1 < X \le x_2)$; f(x) is the distribution density of a random variable; $x_2 = x_1 + dx$. The probability density satisfies two conditions: it is nonnegative and the integral of it in the complete limits of the variation of the argument x is equal to one. The distribution function F(x) is expressed in terms of the density f(x):

$$F(x) = \int_{-\infty}^{x} f(x)dx \quad (6.34)$$

On the other hand, if the density f(x) is continuous at the point x, then its value at this point is equal to the derivative of the function F(x): F'(x) = f(x). The distribution function F(x) is primitive for the density f(x), therefore:

$$F(x) = F(x_2) - F(x_1) = \int_{x_1}^{x_2} f(x)dx = P(x_1 < x < x_2) \quad (6.35)$$

Properties of the distribution function: it is non-negative, increasing and equal to 0 and 1 for the value of the argument $-\infty$ and ∞: $F(x) \geq 0$; $F(x1) < F(x2)$ for $x1 < x2$; $F(-\infty) = 0$; $F(\infty) = 1$. The graph of the distribution density f (x) is called the distribution curve of a random variable. Proceeding from the geometric interpretation of the integral as the area of the corresponding curvilinear trapezium, we conclude that for an arbitrary $-\infty < x_0 < +\infty$ the number $F(x_0)$ is equal to the area under the distribution curve lying to the left of the line $X = x_0$. Similarly, the probability $p(x_1 < x \leq x_2)$ is interpreted.

A random variable x for which a distribution density f(x) exists, is said to be continuous. If by the random variable x is meant the duration of non-failure operation of the object, then the product f(x)dx is the probability of object failure in the time interval (x1, x2). The value of the distribution function F(x) is equal to the probability of failure of the object up to the instant x. In reliability theory, the notion of probability of non-failure operation p(x) is often used, which is an additional concept to the distribution function F(x). The probability of failure-free operation at the point x is equal to the probability that the random variable X will exceed x, i.e. the product will work reliably for a time x: $P(x) = 1 - F(x) = p\{X > x\}$. The function F(x) is also called a reliability function. Exemplary graphs of the distribution function F(x) and the reliability function P(x) are shown in Fig. 6.4.

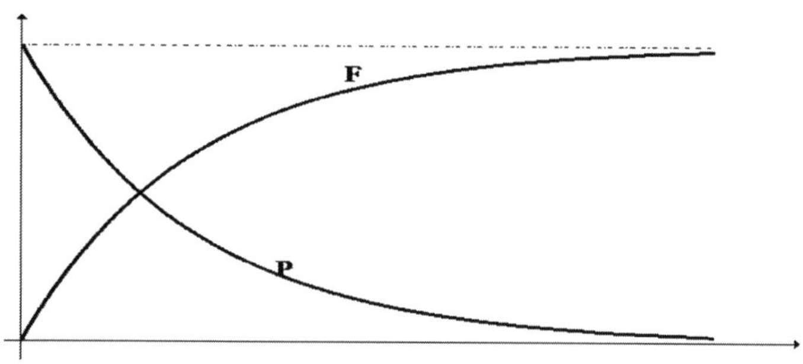

Figure 6.4: Distribution function F(x) and the reliability function P(x).

The mathematical expectation of a discrete random variable is the sum of the products of all possible values of a random variable by the probabilities of these values:

$$\mu_x = \sum_{i=1}^{n} x_i P_i \tag{6.36}$$

The mathematical expectation of a random variable X having a distribution density f(x) is calculated by the formula:

$$\mu_x = \int_{-\infty}^{+\infty} x f(x) dx \tag{6.37}$$

The statistical estimate of the mathematical expectation is the arithmetic mean of the random variable:

$$\overline{x} = \frac{1}{n}\sum_{i=1}^{n} x_i m_i \tag{6.38}$$

where n is the number of values of x; mi is the frequency of occurrence of the result xi.

The mathematical expectation (the arithmetic average) of a random variable is often called the scattering center or the center of the grouping of a random variable. The expectation is an estimate of the true value of the measured quantity.

Example 6.10

Find the mathematical expectation and mode of a random variable given by a Table 6.1 of values:

Table 6.1: Statistical data.

X	2	3	5
P	0.3	0.1	0.6

Solution:

The median of a random variable (M_e) is its value x, for which the probability of the appearance of a random variable smaller than the median, or greater than the median, is the same: $p(x < M_e) \approx p(x > M_e)$. The geometric median is the abscissa of a point at which the area bounded by the distribution curve is divided in half.

6.7 Probability distributions and concept of random success (failure)

The term "statistical experiment" is used to describe any process by which several chance observations are obtained. All possible outcomes of an experiment comprise a set that is called the sample space. We are interested in some numerical description of the outcome. A random variable is a variable whose value is determined by the outcome of a random experiment. A discrete random variable is one whose set of assumed values is countable (arises from counting). A continuous random variable is one whose set of assumed values is uncountable (arises from measurement). For discrete random variable X and real number x, the probability distribution function (pdf) (defined for both discrete and continuous random variables) or probability mass function (pmf) (defined only for discrete random variables) is:

$$p(x) = P(X = x) = P(\text{all events } s \subset S \text{ s.t. } X(s) = x)$$

The cumulative distribution function (cdf) F(x) for a discrete rv X with pmf p(x) is defined as:

$$F(x) = P(X \le x) = \sum_{y \le x} p(y) \tag{6.39}$$

Expected (Mean) Values of Discrete Random Variables: Let X be a discrete R.V. with pmf p(x1), p(x2), then the expected value of X is:

$$\mu_X = E[X] = \sum_{\substack{x=x_1;\\ x=x_2;\dots}} xp(x) \tag{6.40}$$

The mean is the "average value" of a random variable. It is one measure of the "center" of a probability distribution. We shall use Capitals X for the random variable and Lower case x_1, x_2, x_3, ... for the values of the random variable in an experiment. This $\boldsymbol{x_i}$ then represents an event that is a subset of the sample space. The probabilities of the events are given by: $P(x_1)$, $P(x_2)$, $P(x_3)$…., we also use the notation *P(X)*.

6.7.1 The binomial probability distribution

A *binomial* experiment is one that possesses the following properties: the experiment consists of n repeated trials; each trial results in an outcome that may be classified as a *success* or a *failure* (hence the name, *binomial*); the probability of a success, denoted by p, remains constant from trial to trial and repeated trials are independent. The number of successes X in n trials of a binomial experiment is called a *binomial random variable*. In probability theory and statistics, the Bernoulli distribution, named after the Swiss scientist Jacob Bernoulli, is a discrete probability distribution, which takes value 1 with success probability p and value 0 with failure probability q = 1 – p. So if X is a random variable with this distribution, we have:

$$\Pr(X = 1) = 1 - \Pr(X = 0) = 1 - q = p$$

The probability mass function f of this distribution is:

$$f(k;p) = \begin{cases} p & \text{if} \quad k = 1 \\ 1-p & \text{if} \quad k = 0 \end{cases} \tag{6.41}$$

The expected (mean) value of a Bernoulli random variable X is E(X) = p, and its variance is Var(X) = D(X) = p(1–p). The above can be derived from the Bernoulli distribution as a special case of the Binomial distribution [9]. The Bernoulli distribution is a member of the exponential family.

NOTE: The number of k-combinations from a given set S of n elements is often denoted in elementary combinatory texts by C(n, k). The same number however occurs in many other mathematical contexts, where it is denoted by C_k^n (often read as "n choose k"); notably it occurs as coefficient in the binomial formula, hence its name binomial coefficient. One can define C_k^n for all natural numbers k at once by the relation:

$$(1+X)^n = \sum_{k=0}^{n} C_k^n X^k \tag{6.42}$$

One can first consider a collection of n distinct variables X_s labeled by the elements s of S, and expand the product over all elements of S. Now setting the entire X_s equal to the unlabeled variable X, so that the product becomes $(1 + X)^n$, the term for each k-combination from S becomes X^k, so that the coefficient of that power in the result equals the number of such k-combinations.

For determining an individual binomial coefficient, it is more practical to use the formula:

$$C_k^n = \frac{n(n-1)(n-2)(n-3)......(n-k+1)}{n!} \tag{6.43}$$

The numerator gives the number of k-permutations of n, i.e., of sequences of k distinct elements of S, while the denominator gives the number of such k-permutations that give the same k-combination when the order is ignored. When k exceeds n/2, the above formula contains factors common to the numerator and the denominator, and canceling them out gives the relation:

$$C_k^n = C_{n-k}^n ;...\text{for}...0 \le k \le n$$

This expresses a symmetry that is evident from the binomial formula, and can also be understood in terms of k-combinations by taking the complement of such a combination, which is an (n − k)-combination. Finally there is a formula which exhibits this symmetry directly, and has the merit of being easy to remember:

$$C_k^n = \frac{n!}{k!(n-k)!} \tag{6.44}$$

Binomial coefficient is the # Sum of coefficients row. The number of k-combinations for all k

$$\sum_{k=0}^{n} C_k^n - 2^n$$

The probability distribution of the random variable X is called a binomial distribution, and is given by the formula:

$$P(X = m) = C_m^n p^m q^{n-m} \tag{6.45}$$

where: n = the number of trials; m = 0, 1, 2… n; p = the probability of success in a single trial; q = the probability of failure in a single trial (i.e., q − 1 − p). P(X) gives the probability of successes in n binomial trials. If p is the probability of success and q is the probability of failure in a binomial trial, then the expected number of successes in n trials (i.e., the mean value of the binomial distribution) is E(X) = μ = np. The variance of the binomial distribution is D(X) = σ^2 = npq. Note: In a binomial distribution, only 2 parameters, namely n and p are needed to determine the probability. For the special case where r is an integer, the binomial distribution is known as the Pascal distribution. It is the probability distribution of a certain number of failures and successes in a series of independent and identically distributed Bernoulli trials. For k + r Bernoulli trials with success probability p, the binomial distribution gives the probability of k successes and r failures, with a failure on the last trial. In other words, this type of binomial distribution is the probability distribution of the number of successes before the rth failure in a Bernoulli process, with probability p of successes on each trial. A Bernoulli process is a discrete time process, and so the number of trials, failures, and successes are integers. Consider the following example.

Example 6.11

Suppose we are looking for the event "A" (exceeding the maximum temperature T = 1000°C) that can occur with constant probability "p" over a composite's life span. Therefore the event "B"—the reliability of the structure is q = 1 – p. Let's say also that the occurrence of such high temperature creates substantial structural damage (partial or progressive collapse) and the operations must be stopped. Calculate the life span of the composite element (x_i = 1,2,3.....- hours number—random variable). Random series distribution in this case is as follows (see Table 6.2):

Table 6.2: Data for Example 6.11.

x_i	1	2	3	....	I	....
p_i	p	qp	pq^2	....	q^{i-1}	....

Let's compute now the mean value:

$$\mu_X = 1p + 2qp + 3q^2p + ... + iq^{i-1}p + ... = p\sum_{i=1}^{\infty} iq^{i-1}$$

The series $\sum_{i=1}^{\infty} iq^{i-1}$ represents the derivative of a geometric progression: $\sum_{i=1}^{\infty} q^i = \frac{q}{1-q}$, therefore finally:

$$\sum_{i=1}^{\infty} iq^{i-1} = \frac{d}{dq}(\sum_{i=1}^{\infty} q^i) = \frac{d}{dq}(\frac{q}{1-q}) = \frac{1}{(1-q)^2} = \frac{1}{p^2}$$

and $\mu_X = \frac{1}{p}$

Let's compute now the variance value (similar to the mean value):

$$\alpha_2(X) = \sum_{i=1}^{\infty} x_i^2 p_i = (\sum_{i=1}^{\infty} i^2 q^{i-1} p) = (p \sum_{i=1}^{\infty} i^2 q^{i-1}) = p \frac{d}{dq}[\frac{q}{(1-q)^2}] = \frac{p(1+q)}{(1-q)^2} = \frac{q}{p^2}$$

$$\text{and } D_x = \alpha_2(X) - m_x^2 = \frac{1+q}{(1-q)^2} - \frac{1}{(1-q)^2} = \frac{q}{p^2} \text{ or: } \sigma_X = \frac{\sqrt{q}}{p}$$

Now, if $p = 0.02$ and $q = 0.98$ then:

$$\mu_X = \frac{1}{0.02} = 50 \text{ hours; and } \sigma_X = \frac{\sqrt{0.98}}{0.02} = 49.5 \text{ hours}$$

Total maximum number of hours (with confidence probability $P_\alpha = 0.95$):
$N = 50 + 1.96(49.5) = 147.0$ hours.

Example 6.12

Data: The "Structural failure probability" (exceeding the maximum temperature $T = 1000^oC$) is p (event "A" see Example 6.3). The life span of the structure is given: N = 100 hours. The agreed confidence probability P_α = 0.95. Find the maximum probability value $p = p_{max}$ (the upper limit value). From formulae (6.1) we have:

$P(B) = (1 - p)^N$

Probability of failure (at least one time over period of N hours) is small: $1 - P_\alpha$. Therefore the maximum value of probability $p = p_{max}$ can be calculated as follows:

$$1 - P_\alpha = (1 - p_{max})^N \text{ or: } p_{max} = 1 - \sqrt[N]{1-\alpha}$$

$$N = \frac{\lg(1 - P_\alpha)}{\lg(1 - p_{max})} \tag{6.46}$$

From formulae (6.46) we have:

$$p_{max} - 1 - \sqrt[100]{1 - 0.95} = 0.03$$

If the agreed confidence probability $P_\alpha = 0.90$, then:

$$p_{max} = 1 - \sqrt[100]{1 - 0.9} = 0.023$$

Example 6.13

Data: If the given probability $p = 0.02$ is small enough and the number of hours $N = 100$ is large enough, then the fire event "A" probability of failure (at least ones over period of time N years) can be calculated using Poisson formulae, where mean value a = Np. In this case:

$$P(\bar{A}) \approx e^{-Np} \text{ and } P(A) = 1 - \alpha \text{ then} : p_{max} \approx \frac{-\ln(1-\alpha)}{N} \tag{6.47}$$

For the agreed confidence probability $\alpha = 0.95$ we have now:

$$p_{max} \approx \frac{-\ln(1-\alpha)}{N} = \frac{-\ln(1-0.95)}{100} = 0.03$$

These simple approximate methods have two major assumptions: probability "p" is small and number "N" is large. However this type of problems can be solved analytically by computing the cumulative discrete function without any restrictions and consequently can be applied to many other structural engineering problems.

Example 6.14

Data: The group of fiber composite elements have 10 elements with high strength and brittle fibers (e.g., SiC), ceramic oxide and 50 elements with graphite fibers that typically exhibit large variability in strength due to flaws of varying severity that are randomly distributed along their lengths. The Material Engineer has decided to test only 2 composite elements and he will be satisfied if in this case he is able to check 2 elements with high strength and brittle fibers. Find the probability of checking two of them.

Solution:

a) The total number of "events"—**(complete sample space)**: $n = C_2^{50+10} = 1770$;
b) The total number of "successful events" is: $m = C_2^{10} = 45$;
c) Therefore the probability P is: $P = \frac{45}{1770} = 0.0254$

Example 6.15

Same as Example 6.14, but the Material Engineer had decided to test 20 composite elements.

Solution:

a) The total number of "events"—**(complete sample space)**: $n = C_{20}^{50+10} = 4.2(10)^{15}$;
b) The total number of "successful events" is: $m = C_2^{10}\,C_{20-2}^{60-10} = 8.12(10)$ ^ 14;
c) Therefore the probability P is: $P = \frac{8.12(10)^{14}}{4.2(10)^{15}} = 0.193$

Example 6.16

Data: Same as Example 6.14, but the Material Engineer had decided to check 4 composite elements. Find the probability of testing: (1) 2 high strength and 2 graphite fibers composite elements; (2) 2 high strength or 2 graphite fibers composite elements.

Solution 1:

a) The total number of "events"—**(complete sample space)**: $n = C_4^{50+10} = 487635$;

b) The total number of "successful events" is: $m = C_2^{10}\, C_2^{50} = 55125$;

c) Therefore the probability P is: $P = \dfrac{55125}{487635} = 0.113$

Solution 2:

a) The total number of "events"—**(complete sample space)**: $n = C_4^{50+10} = 487635$;

b) The total number of "successful events" is: $m = C_4^{10} + C_4^{50} = 230510$;

c) Therefore the probability P is: $P = \dfrac{230510}{487635} = 0.473$

Example 6.17

Data: Suppose we have a set of "n" events (A—composite element failure due to stresses borne by broken fibers) with the probability of success "p" per each hour. Random value "R" is the frequency of such events (see Table 6.3) over the period of time ("n" hours). Compute CDF and find the mean and variance.

Table 6.3: Data for Example 6.17.

X_i	0	1/n	2/n	....	m/n	....	1
p_i	q^n	$C_n^1 pq^{n-1}$	$C_n^2 p^2 q^{n-2}$	....	$C_n^m p^m q^{n-m}$	....	p^n

Solution (CDF):

Where: $q = 1 - p$; $m_x = p$; *Variance* $D = pq/n$; $\sigma_x = (pq/n)^{1/2}$

If: $p = 10^{-5}$; $D = pq/n$; $\sigma_x = (pq/n)^{1/2}$ failures/hour and $n = 10000$ flight hours; then: $m_x = p = 10^{-5}$; $D = pq/n$; $\sigma_x = (pq/n)^{1/2} = 0.316(10^{-4})$. Therefore one might say that with 95% confidence level the probability (the maximum number of flight—failures per hour) will be: $p_n = [1 + 1.96(3.16)]\,(10^{-5}) = 7.2\,(10^{-5})$ failure/hour.

Failure of fiber-reinforced composites (probabilistic approach)

Below is the classification of fiber-reinforced composites based on the reinforcement aspect ratio that is used in common engineering practice and followed by some of examples.

1. Short Fiber Thermoplastics l/d < 100
2. Long Fiber Thermoplastics 200 < l/d < 10000
3. Continuous Fiber Thermoplastics l/d > 10000

In probabilistic analytical approaches, it has been proven and a simple modeling of composite failure evolution has been used without resorting to major simplifications, particularly in redistributing stress from broken fibers onto intact fibers. One such simplification is to assume that stresses borne by broken fibers are transferred according to a rule, such as the equal load sharing (ELS) or local load sharing (LLS) rule. In ELS, all the intact fibers share the applied stress equally and thus equally carry the loads lost from broken fibers, but in LLS, only the immediate unbroken neighbors carry these lost loads, thus causing more severe overloads than those that occur by ELS (considering a dry bundle of fibers with no matrix participation in redistribution of stresses). Under LLS or similar localized stress transfer, a common approach has been to proceed with certain assumptions on the fiber fracture sequences and their probabilities for example, at some point in plan layout the first fiber breaks, then the neighboring fiber to the right (or left) and so on until all 'n' fibers are broken (the total number broken fibers is based on test results or on a chosen "limit stress" theory). The failure of fiber reinforced composite materials can be adequately described via examples presented below.

Example 6.18

Data: This example introduces the different probabilistic approach that has been used to represent the strength of fiber reinforced composite materials and to assess the reliability of the whole composite structure.

Suppose we have a set of "N = 100" fibers, and if four (n = 4) of them break in sequential order then the whole composite structure fails too.

Compute the probability of failure—event A.

Solution:

Of the total number of N fibers one can choose n neighboring to each other fibers in different N – n + 1 different ways (groups). The probability that all n fibers will be located in one of the (N – n + 1) groups is equal: $p = [1/N]^n n!$ (since each failed fiber can be permutated—change place with each other). The probability of event A is calculated now:

$$P(A) = [\frac{1}{N}]^n n!(N - n + 1) \tag{6.48}$$

In our case: N = 100 and n = 4 → $P(A) = 10^{-8}(4!)97 = 0.233(10^{-4})$

Sterling's approximation is an approximation for *factorials*. It is a good-quality approximation, leading to accurate results even for small values of *n*. The formula as typically used in applications is:

$$n! \approx \sqrt{2\pi n}\left(\frac{n}{e}\right)^{n} \tag{6.49}$$

After substituting Equation (6.49) into Equation (6.48) we have:

$$P(A) = \sqrt{2\pi n}\left(\frac{n}{e}\right)^{n}\left[\frac{1}{N}\right]^{n}(N - n + 1) \tag{6.50}$$

If n = N = 1 then from Equation (6.48) then P(A) = 1 and from approximate Equation (6.50) P(A) = 0.92. Obviously, for small n (broken fibers) and large N (total number of fibers) the probability of the whole composite structure should be small. If N = n then Equation (6.48) can be approximated as:

$$P(A) = \sqrt{2\pi n}\,\frac{1}{1 + n + 0.5n^{2}} \tag{6.51}$$

Equation (6.51) represents the probabilities of failure of composite element as function of number of neighboring broken fibers n (see Fig. 6.5).

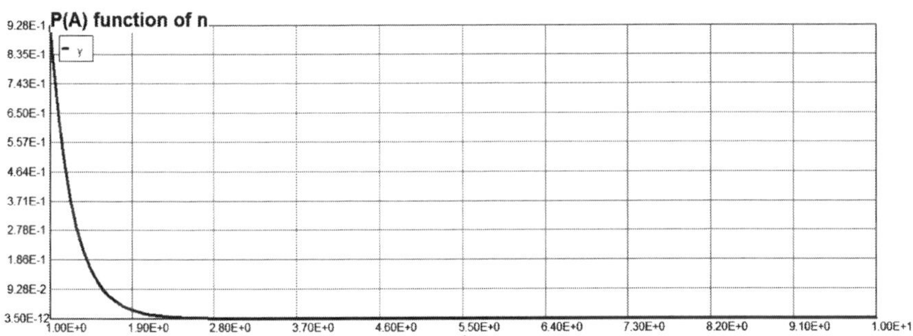

Figure 6.5: Probability function as a function of number of neighboring broken fibers n.

Approximation P(n) is given below:

Model: y = b*exp(–(a*x))

Variable	Initial guess	Value
b	51	51.18894
a	4.	4.0118

$$P(A|N = 100 \text{ and } 1 \le n \le 10) = 4.0[\exp(-51.2n)] \tag{6.52}$$

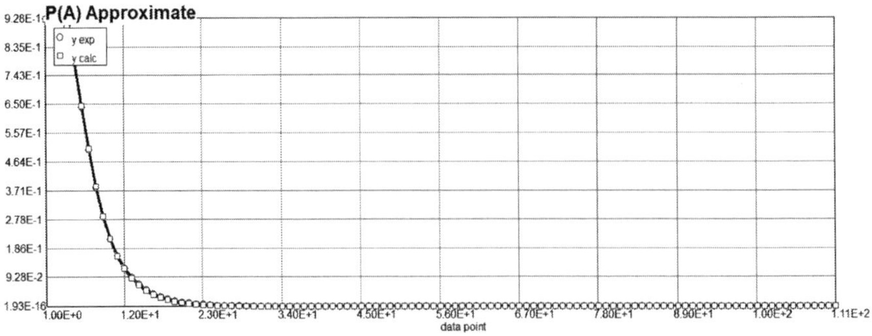

Figure 6.6: Probability function: approximation of P(n).

Example 6.19

Data: Novel properties of nanocomposites can be derived from the successful combination of the individual characteristics of parent constituents into a single material. During the assembly of the nano-sized building blocks, one should be able to control the extent of the actual distribution of nanoparticles in the building blocks as well as the risk computation of not complying with the requirements imposed by a material design engineer. Let's assume that nanoparticles are dispersed uniformly in the building block based on the Poisson distribution density with the parameter λ. Conditionally it can be assumed that nanoparticles are presented as a group of points on the plane. It is necessary to find the distribution law and the numerical characteristics μ_r and D_r of the distance r from any given point to the nearest neighboring point. Draw a circle around the given point of radius r. In order for the distance R from this point to the nearest neighbor point to be less than r, it is necessary that at least one point (except for a given point) falls into the circle. According to the properties of the Poisson law, the probability of this event does not depend on whether there is already a point in the center of the circle or not. Therefore:

$$F(r) = 1 - \exp(-\pi r^2 \lambda)$$

$$f(r) = \begin{cases} 2\pi r\lambda \exp(-\pi r^2 \lambda) & \text{if } r > 0 \\ 0 & \text{if } r < 0 \end{cases}$$

Such a distribution law is called the Rayleigh law.

$$\mu_r = \int_0^\infty r 2\pi r\lambda \exp(-\pi r^2 \lambda)\,dr = \frac{1}{2\sqrt{\lambda}}$$

$$D_r = \int_0^\infty r^2 2\pi r\lambda \exp(-\pi r^2 \lambda)\,dr - \mu_r^2 = \frac{1}{\pi\lambda} - \frac{1}{4\lambda} = \frac{4-\pi}{4\pi\lambda}$$

The distribution function of the distance R_2 from this point to the next nearest point was less than r it is calculated as:

$$F_2(r) = 1 - \exp(-\pi r^2 \lambda) - \pi r^2 \lambda \exp(-\pi r^2 \lambda) \tag{6.53}$$

By using similar arguments in the general case we have:

$$F_n(r) = P(R_n < r) = 1 - \sum_{0}^{n-1} \frac{a^k}{k!} \exp(-\pi r^2 \lambda)$$

$$a = \pi r^2 \lambda$$

The probability density is obtained by differentiating F_n with respect to r:

$$f_n(r) = \frac{dF_n}{da}\frac{da}{dr} = \frac{a^{n-1}}{(n-1)!} e^{-a} 2\pi r \lambda \tag{6.54}$$

Probability density changes (qualitatively), depending on the variations of the parameter λ—density of Poisson probability distribution law and the numbers of nanoparticles (points) in the selected area (circle with the radius r) are presented above (see Figs. 6.4 and 6.5):

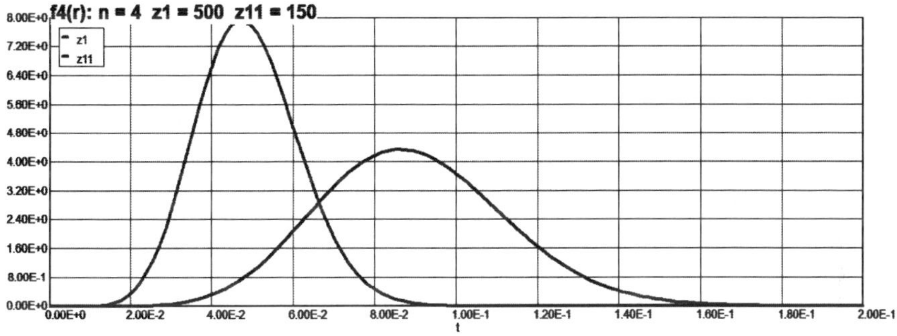

Figure 6.7: Probability density changes depending on the variations of the parameter λ and n = 1.

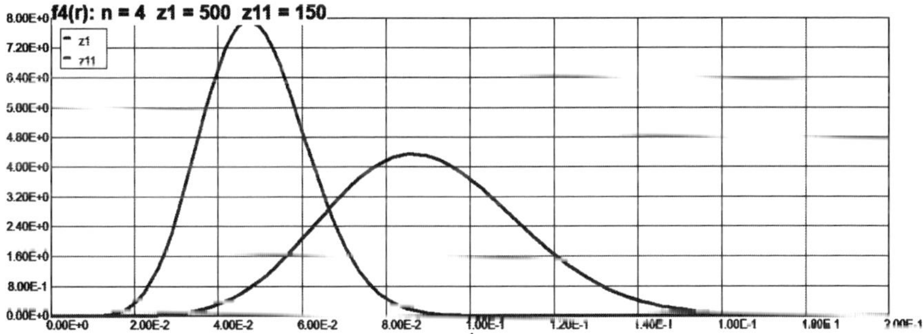

Figure 6.8: Probability density changes depending on the variations of the parameter λ and n = 4.

One can see from Figs. 6.7 and 6.8 that probability is increasing when parameter λ is bigger and number of nanoparticles in the selected area (circle with the radius r) n is smaller.

Example 6.20

Data: If each point in the previous Example 6.19 is a cross section of a continuous fiber and the pattern of their failures formation is sequential (i.e., the load from the broken fiber is redistributed only to the neighboring fiber, then on to the next neighboring fiber, and so on), then the failure process of the composite structure is described by the breaks in a predetermined number of fibers. At the same time, the probability of breaking each fiber (depending on a given load) can be established from the corresponding experimental tests results. Assuming that these probabilities $p_1, p_2, ..., p_n$ are known, the probability of the appearance of the A-fiber break event k times can be obtained using the so-called generating function of probability.

$$\varphi_n(z) = (p_1z + q_1)(p_2z + q_2)......(p_nz + q_n) \tag{6.55}$$

Probabilities $p_1, p_2, ..., p_n$ should be computed using Equation (6.54). For example, if $r = 0.05$ mm and $\lambda = 150$ [1/mm²] than: $p_1 = F_1(r) = 1 - \exp(-\pi 25(10^{-4})150) = 0.308$; and $p_2 = F_2(r) = 1 - \exp(-\pi 25(10^{-4})150) - \pi 25(10^{-4})150 \exp(-\pi(10^{-2})150) = 1 - 0.308 - 0.362 = 0.329$

The generating function of probability is:

$$\varphi_n(z) = (p_1z + q_1)(p_2z + q_2) = 0.101z^2 + 0.434z + 0.464$$

Check: $0.101 + 0.434 + 0.464 = 1.0$

Here: $P_1 = 0.101$—Probability of two fibers break; $P_2 = 0.434$—Probability of one fiber breaks; $P_3 = 0.464$—Probability of none out two fibers breaks.

Example 6.21

If the given probability of a composite element to fail is p and it is small number comparable to the life span of years N = 100, then the event "A"—probability of failure (at least ones over period of time N years) can be calculated using the Poisson formulae, where mean value a = Np. In this case:

$$P(\bar{A}) \approx e^{-Np} \text{ and } P(A) = 1 - \alpha \text{ then}: p_{max} \approx \frac{-\ln(1-\alpha)}{N} \tag{6.56}$$

For the agreed confidence probability α we have now:

$$p_{max} \approx \frac{-\ln(1-\alpha)}{N} \tag{6.57}$$

For example, if $\alpha = 0.95$ and $N = 100$, then $p_{max} \approx 0.03$.

Example 6.22

The structural system (simple frame) subjected to thermal load has been designed with almost equal capacities of beam and beam to column connection. The structural engineer had performed the standard tests of the beam and connection to the column separately and concluded by observation that the connection failure probability (statistical data) is P_1 and the beam failure probability is $P_2 = 1 - P_1$ and ($P_1 > P_2$).

Probability-based computations have shown that this structural system (simple frame) can fail with probability "p" (the maximum temperature exceeds the predicted level "a", which has been used in separate standard tests). The structural engineer is planning now to perform "n" new full simple frame tests to prove his previous assessment (connection failure). (1) How many tests are needed ($n_1 < n$) in this case? (2) What is the connection failure probability in real fire scenario?

Solution:

H_c—Hypothesis I: The structure will fail because of inadequate connection capacity.

H_b—Hypothesis II: The structure will fail because of inadequate beam capacity.

$P(H_c) = P_1$; $P(H_b) = 1 - P_1$;

Event "A"—The structure will fail because of inadequate connection capacity.

$$P(A) = P_1[1-(1-p)^{n_1}] + (1-P_1)[1-(1-p)^{n-n_1}]$$

Let's consider now P(A) as a function of continuous independent variable "n1". The first derivative is:

$$\frac{dP(A)}{dn_1} = [-P_1(1-p)^{n_1} + (1-P_1)(1-p)^{n-n_1}]\ln(1-p)$$

It can be proven in this case: (a) the second derivative is less than 0 for all n_1; (b) the first derivative is zero when:

$$n_1 = \frac{n}{2} + \frac{\ln\dfrac{1-P_1}{P_1}}{2\ln(1-p)} \tag{6.58}$$

where: $n_1 > n/2$; and $P_1 > 0.5$

Example 6.23

Data: similar to Example 6.22, but: (1) On the reliability (probability not to fail)—for connection—P_1; for beam—P_2. (2) The structural engineer had established that the structure will fail over a given period of time T.

Solution:

H_o—Hypothesis I: Connection and beam will not fail over a given period of time.

H_1—Hypothesis I: Connection did fail, but beam did not fail over a given period of time.

H_2—Hypothesis I: Connection did not fail, but beam did fail over a given period of time.

H_3—Hypothesis I: Connection and beam had failed over a given period of time.

$P(H_o) = P_1P_2$; $P(H_1) = (1 - P_1)P_2$; $P(H_2) = (1 - P_2)P_1$; $P(H_3) = (1 - P_1)(1 - P_2)$.

Event A: Structure failed: $P(A|H_o) = 0$; $P(A|H_1) = P(A|H_2) = P(A|H_3) = 1$.

Based on Byes' theorem we have now:

$$P(H_1 \mid A) = \frac{(1-P_1)P_2}{(1-P_1)P_2 + (1-P_2)P_1 + (1-P_1)(1-P_2)} = \frac{(1-P_1)P_2}{(1-P_1P_2)}$$

If: $P_1 = P_2 = 0.996$; $n = 30$ and $P = 0.1$, then: $P\,(H_1|A) = 0.5$

Example 6.24—The buffoon problem

The parallel straight lines spaced from each other at a distance 2a. On this plane throw at random a needle (a fiber with length 2L and L < a). Find the probability that the needle (fiber) crosses some straight line.

Solution:

We introduce the following notation: x is the distance from the middle of the needle to the nearest parallel; φ is the angle formed by the inclined needle with the horizontal (parallel) line. The position of the needle is completely determined by specifying certain changes in x and φ, where x takes values from 0 to a, and the possible values of φ vary from 0 to π. In other words, the middle of the needle can fall into any of the points of the rectangle with sides a and π (Fig. 6.9).

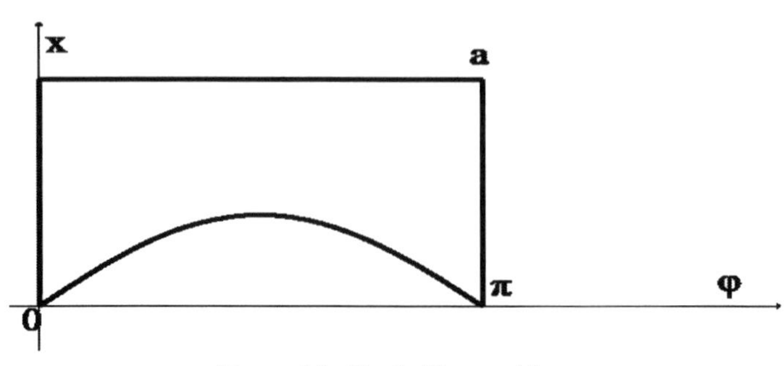

Figure 6.9: The buffoon problem.

Thus, this rectangle can be considered as figure G, the points of which represent all the possible positions of the middle of the needle. Obviously, the area of the figure G is equal to πa. Let us now find the figure g—each point of which favors the event of interest to us, that is, each point of this figure can serve as the middle of the needle that crosses the one closest to it parallel. As can be seen from Fig. 6.5, a, the needle crosses the one closest to it parallel under the condition $x \leq L(\sin\varphi)$, that is, if the middle of the needle belongs to any of the points of the curved figure in Fig. 6.5. Let's find the area of this figure G:

$$g = \int_0^{\pi} L \sin\varphi d\varphi = 2L$$

The probability that the needle will intersect the straight line is: $P = 2L/\pi a$

Example 6.25

Data: see Example 6.24

Long-fiber-reinforced thermoplastic (LFRTs) is a type of easily moldable thermoplastic used to create a variety of composites. With NY66, HDPE, and PP base polymers (polypropylene (PP) and high density polyethylene (HDPE)) that have the initial fiber length L of 12.7 mm. The rods were extruded and compression molded, and the specimens with dimensions 1270.0 mm × 128 mm × 6.0 mm (H × b × d) were sectioned and end-milled. The rods were dispersed in four strips (2a = 32 mm each). Calculate the probability that the rods are located at each strip correctly.

Solution:

$P = [2(12.7)/\pi(16)]^4 = 0.505^4 = 0.065$

6.8 Probabilistic creep models of composites

The aim of this section is to develop probabilistic mathematical creep models of structural elements; complex analysis and systematization of research results on the influence of temperature action modes and the corresponding temperature induced creep stresses of composite structural systems and their elements in the probabilistic formulation. To achieve this goal, the following objectives have been set:

1) Carry out a complex analysis of stochastic fields of rheological strains and stresses;
2) Obtaining the necessary probabilistic creep characteristics, such as mathematical expectation and variance of the maximum allowable creep stress;

3) Analysis of the creep process as a random process within the framework of the correlation theory of probability and the solution of the corresponding problems of exceedance of a given level of strains and stresses;
4) The construction of phenomenological stochastic creep equations under conditions of uniaxial stress state for the composites and their experimental verification;
5) Development of a methodology for identifying estimates of random parameters and variables of stochastic creep constitutive equations on the basis of an analysis of a corresponding deterministic creep constitutive equation with a single relaxation time value, which makes it possible to reduce the volume of computations and simplifying the comparison with the experimental creep strain data;
6) Development of a numerical solution method for the stochastic boundary value problem of creep deformation under uniaxial tension conditions.

The process of curing (or degrading) the composite at each point of the medium is characterized by a monotonically increasing (decreasing) function 'f3' (see Chapter 3) called the degree of crystallization at a given time. Functional dependence f3(θ) is a solution of a differential equation of the form:

$$\frac{d(f3)}{d\theta} = f(f3, \theta) \tag{6.59}$$

This equation is called the crystallization kinetics equation. In it, θ denotes the dimensionless temperature at the current point of space at the current time, and t denotes the time from the beginning of the curing process. The relation (6.59) in case of high thermal creep deformation can be chosen as the Arrhenius equation:

$$\frac{d(f3)}{d\theta} = Ae^{-\frac{E}{RT}} f(f3) \tag{6.60}$$

The distribution of temperatures in a continuous medium is found from the solution of the heat equation:

$$\rho\frac{\partial(cT)}{\partial t} = -\mathrm{div}\vec{q} + q \tag{6.61}$$

where ρ is the density; c is the specific heat of the material, and q is the specific heat release power. The heat flux vector q_i is related to the temperature gradient by the Fourier law:

$$q_i = -k_{ij}\frac{\partial T}{\partial x_j} \tag{6.62}$$

in the right-hand side of the expression; for which the components of the Equation (6.62) k_{ij} is the thermal conductivity tensor. Due to the chemical reaction during the curing process, a heat generation is produced, which is proportional to the rate of the polymerization reaction. For a pure binder, the specific heat release power is given by:

$$q = \rho H_{tot} \frac{\partial f3}{\partial \theta} \tag{6.63}$$

where ρ is the density of the binder, and H_{tot} is the specific heat released during complete polymerization. For averaged material, it is customary to assume:

$$q = \mu_m \rho H_{tot} \frac{\partial f3}{\partial \theta} \tag{6.64}$$

where μ_m is the volume fraction of the binder. Mechanical deformations of ε_{ij} are equal to the difference of total deformations $\varepsilon_{tot,ij}$ and deformations of non-mechanical nature ξ_{ij}.

The deformations of non-mechanical nature are composed of deformations from thermal expansion ξ_{ij}^T and chemical shrinkage ξ_{ij}^C:

$$\xi_{ij} = \xi_{ij}^T + \xi_{ij}^C \tag{6.65}$$

The need to take into account the history of the polymerization process explains the integral form of the relationships for their calculation:

$$\begin{aligned} \xi_{ij}^T &= \int_{t_{gel}}^{t} \alpha_{ij} K(t,t') \frac{\partial \sigma_{ij}}{\partial t'} dt' \\ \xi_{ij}^C &= \int_{t_{gel}}^{t} \beta_{ij} f3(t') \frac{\partial f3}{\partial t'} dt' \end{aligned} \tag{6.66}$$

A distinctive feature of the proposed model is the consideration of non-mechanical deformations from the moment of gelation of the binder (matrix), which corresponds to the vacuum forming technology in the production of the composite. In this case, the binder is considered to be liquid and capable of flowing between the fibers prior to the beginning of gelation, and its shrinkage is not accompanied by the appearance of stresses in the composite. The parameters for finding non-mechanical deformations include the coefficients of thermal expansion of the material α_{ij} and the coefficients of chemical shrinkage of the material β_{ij}, depending on the temperature and degree of crystallization. The values of f3 characterize the degree of polymerization at which the binder jellifies, that is, a significant increase in its viscosity. The dependence of the glass transition temperature T_g on the degree of polymerization can be specified either explicitly or approximated in any suitable manner. In particular, for these purposes, the model with piecewise linear modules, also called CHILE (short for Cure Hardening Instantaneous Linear Elastic), suggests that in the vicinity of the glass transition temperature the elastic modulus vary linearly, while in a completely rubber-like and vitreous state they are constant. Material parameters were found with the help of averaging approaches [13–15]. For the binder used in the work, the polymerization kinetics equation was taken as the n-th order Prout-Tompkins equation with autocatalysis [16]:

$$\frac{dC}{dt} = Ae^{-\frac{E}{RT}}(1-C)^q C^p \tag{6.67}$$

where A, E, q, p—are parameters, R is the universal gas constant.

6.8.1 *Deterministic formulation of stochastic problems: numerical modeling*

The heat Equation (6.61) at hand is separable. This means that the mechanical deformation of the structure does not affect the heat transfer processes occurring in it. Consequently, the temperature field in the computational domain of creep can be found separately, after which the mechanical portion of the creep deformation problem can be solved. The dimensionless form of uniaxial integral type constitutive creep equation is (see Chapter 3):

$$\begin{gathered}E(\theta)\alpha(\theta)\theta = \sigma(\theta) + \delta_1\int_0^\theta e^{-\frac{[E]}{RT}}K_1(\theta,\tau)\sigma(\tau)\,m\mathbf{1}d\tau + \delta_2\int_0^\theta e^{-\frac{[E-A\sigma]}{RT}}K_2(\theta,\tau)\sigma(\tau)d\tau = \\ = \sigma(\theta) + \delta_1\int_0^\theta e^{\frac{\theta}{1+\beta\theta}}K_1(\theta,\tau)\sigma(\tau)\,m\mathbf{1}d\tau + \delta_2\int_0^\tau A\sigma^s e^{\frac{\theta}{1+\beta\theta}}K_2(\theta,\tau)\sigma(\tau)\,m\mathbf{1}d\tau \\ K_2(\theta,\tau) = \varphi(\theta)f(\tau) = m\mathbf{1}(\tau)\sum_{i=1}^{N}\exp(-\alpha_i\, m(\theta-\tau))\end{gathered} \tag{6.68}$$

The integral curves of creep theory can be interpreted as random functions with the corresponding properties. The deterministic and stochastic statements of the problem can be linked using the ergodic theorem, which makes it possible to estimate the statistical stability of solutions of boundary value problems. The ergodic theorem of the random process has the formulation of it as follows [17]: if the continuous process ξ (t) (t is the time or temperature) has a finite expectation, then with probability equal to unity, there exists a limit. Consider now the task of developing a simple methodology of a parametric identification with an allowable error.

Factor analysis refers to the methodology of complex and systematic study and measurement of the interaction of factors on the magnitude of the performance indicators. In factor analysis, the models are divided into: (1) Deterministic (with unambiguously determined results); (2) Stochastic (with different, probabilistic results). Deterministic models are constructed in this way, as if the values of the parameters included in them are known with absolute accuracy. Even when it is known that there is some uncertainty inherent in the environment, it may still be useful to adopt a deterministic approach. The best estimate value of the parameter can be taken, and then the decision sensitivity to errors in the values of this parameter is checked. Sensitivity analysis is a very useful tool and should always be used for important decisions when there are doubts about the correctness of the values used in the model. In some cases, it can be established that a sufficiently large change in the values of a certain, doubt-provoking variable has very little effect on the final result. A deterministic problem can be considered as the limiting case of a probabilistic problem.

It is important to note that the parameters of the model (see Equation 6.68) are determined by the structured properties of material and they are invariants to the input data of the experiment. We will also consider them constant, which is true for structurally stable materials. The total creep deformation of materials consists of four components [18]. This division of the total deformations is to a certain extent conditional. On the other hand, taking into account all four components of deformation, unless the experiment indicates an obvious lack of any of them (for example, viscoplastic and instantly plastic components), gives the equation of states greater flexibility, which allows satisfactorily describing practically any relaxation processes. Simple uniaxial creep experiments are usually used to determine and approximate the viscoelastic parameters of the creep constitutive equation. The time-temperature dependence of the viscoelastic component of the total deformation for the case of uniaxial tension is represented as a solution of Equation (6.68).

6.8.2 Statistical data: composites and stress effect

Adding a probability factor to deterministic problems sometimes transforms the situation and gives a significant gain [19]. The introduced concepts allow formulating deterministic problems of mathematical programming whose solutions coincide with solutions of the corresponding stochastic problems with probability constraints [20]. Rational methods for the approximate solution of deterministic problems are usually insensitive to small errors in the definition of temperature-time gradients, unless these errors are much smaller than the permissible error of the solution. The reader will see that numerical (analytical) methods for solving most probabilistic problems are a direct generalization of those approaches. Accordingly, the study of these methods for solving deterministic problems turns out to be very useful for preparation of probabilistic models [21]. In order to create the required statistical data based on solutions of a corresponding deterministic creep constitutive equation, consider, for example, the parameter α_i (dimensionless inverse relaxation time) in Equation (6.68) as a random parameter from the interval $[0.001 < \alpha_i < 1000]$. To solve these problems under conditions of certainty, various methods that have been described in Chapter 3 are used; therefore the examples below don't require any additional clarification.

Example 6.26

Data: $\alpha = 1000$

Calculated values of DEQ variables

	Variable	Initial value	Minimal value	Maximal value	Final value
1	A2	1.	1.	1.	1.
2	E	1.9	0.475	1.9	0.475
3	fi	0.1	0.1	0.1	0.1
4	k	0.1	0.1	0.1	0.1
5	m	0	0	0.0705907	0.0705907
6	m1	0.0405	0.0003976	0.0405	0.0003976
7	n	1.	1.	1.	1.
8	s	1.	1.	1.	1.
9	t	0	0	8.	8.
10	Y1	0	0	5.595404	3.8
11	z	0	0	5.441443	3.334382
12	z1	0	0	2.114E+30	2.114E+30
13	z2	0	0	1.328445	1.328445

Differential equations

1 d(z1)/d(t) = (exp(t/(1 + 0.067*t)))*((exp(1000*m)))*m1*z^n

2 d(z2)/d(t) = k*(exp(t/(1 + 0.067*t)))*(A2*(z^s))*((exp(0.01*m)))*m1*z^n

Explicit equations

1 m = (0.0405*t – 0.01126*t^2 + 0.001462*t^3 – 0.00006868*t^4)

2 m1 = (0.0405 – 0.02252*t^1 + 0.004386*t^2 – 0.0002747*t^3)

3 k = 0.1

4 fi = 0.1

5 E = (0.625 – 0.375*(tanh(5*(t – 3))))*(1 – fi) + (1/k)*
(0.625 – 0.375*(tanh(5*(t – 5))))*(fi)

6 n = 1.0
7 s = 1.0
8 z = t*E–(z1 + z2)*((exp(–1000*m)))
9 Y1 = t*E
10 A2 = 1

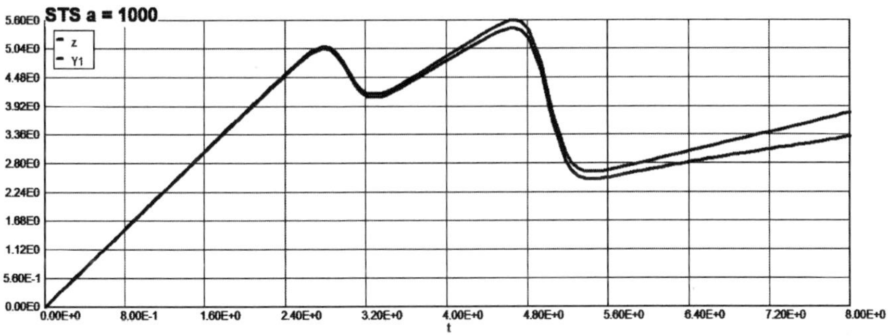

Figure 6.10: Stress-temperature-strain diagram (α = 1000).

Model: z = a1*t + a2*t^2 + a3*t^3 + a4*t^4 + a5*t^5

Variable	Value
a1	1.175453
a2	1.107412
a3	–0.4823872
a4	0.0611024
a5	–0.0024391

$$\sigma = 1.175\theta + 1.107\theta^2 - 0.482\theta^3 + 0.0611\theta^4 - 0.00244\theta^5 \quad (6.69)$$

Example 6.27

Data: α = 100

Calculated values of DEQ variables

	Variable	Initial value	Minimal value	Maximal value	Final value
1	A2	1.	1.	1.	1.
2	E	1.9	0.475	1.9	0.475
3	fi	0.1	0.1	0.1	0.1
4	k	0.1	0.1	0.1	0.1
5	m	0	0	0.0705907	0.0705907

6	m1	0.0405	0.0003976	0.0405	0.0003976
7	n	1.	1.	1.	1.
8	s	1.	1.	1.	1.
9	t	0	0	8.	8.
10	Y1	0	0	5.596112	3.8
11	z	0	0	5.018757	2.221446
12	z1	0	0	1835.54	1835.54
13	z2	0	0	0.8948635	0.8948635

Differential equations

1 d(z1)/d(t) = (exp(t/(1 + 0.067*t)))*((exp(100*m)))*m1*z^n

2 d(z2)/d(t) = k*(exp(t/(1 + 0.067*t)))*(A2*(z^s))*((exp(0.01*m)))*m1*z^n

Explicit equations

1 m = (0.0405*t – 0.01126*t^2 + 0.001462*t^3 – 0.00006868*t^4)

2 m1 = (0.0405 – 0.02252*t^1 + 0.004386*t^2 – 0.0002747*t^3)

3 k = 0.1

4 fi = 0.1

5 E = (0.625 – 0.375*(tanh(5*(t – 3))))*(1 – fi) + (1/k)*(0.625 – 0.375*(tanh(5*(t – 5))))*(fi)

6 n = 1.0

7 s = 1.0

8 z = t*E–(z1 + z2)*((exp(–100*m)))

9 Y1 = t*E

10 A2 = 1

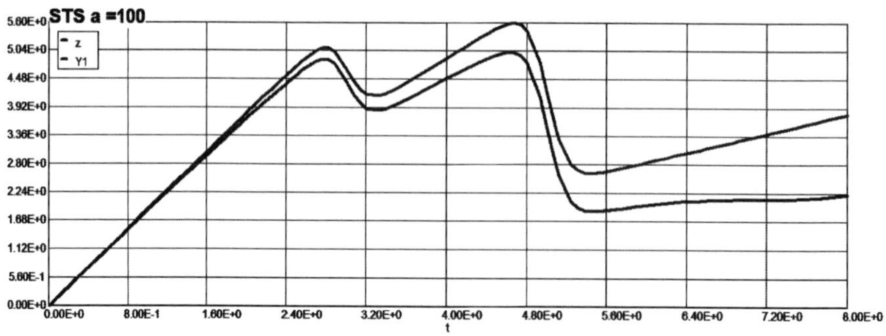

Figure 6.11: Stress-temperature-strain diagram ($\alpha = 100$).

Model: z = a1*t + a2*t^2 + a3*t^3 + a4*t^4 + a5*t^5

Variable	Value
a1	1.183698
a2	1.075306
a3	–0.4790882
a4	0.060838
a5	–0.0024319

$$\sigma = 1.184\theta + 1.1075\theta^2 - 0.479\theta^3 + 0.0608\theta^4 - 0.00243\theta^5 \qquad (6.70)$$

Example 6.28

Data: $\alpha = 10$

Calculated values of DEQ variables

	Variable	Initial value	Minimal value	Maximal value	Final value
1	A2	1.	1.	1.	1.
2	E	1.9	0.475	1.9	0.475
3	fi	0.1	0.1	0.1	0.1
4	k	0.1	0.1	0.1	0.1
5	m	0	0	0.0705907	0.0705907
6	m1	0.0405	0.0003976	0.0405	0.0003976
7	n	1.	1.	1.	1.
8	s	1.	1.	1.	1.
9	t	0	0	8.	8.
10	Y1	0	0	5.596159	3.8
11	z	0	0	4.631596	1.406865
12	z1	0	0	4.252375	4.252375
13	z2	0	0	0.59536	0.59536

Differential equations

1 d(z1)/d(t) = (exp(t/(1 + 0.067*t)))*((exp(10*m)))*m1*z^n

2 d(z2)/d(t) = k*(exp(t/(1 + 0.067*t)))*(A2*(z^s))*((exp(0.01*m)))*m1*z^n

Explicit equations

1 m = (0.0405*t – 0.01126*t^2 + 0.001462*t^3 – 0.00006868*t^4)

2 m1 = (0.0405 – 0.02252*t^1 + 0.004386*t^2 – 0.0002747*t^3)

3 k = 0.1

4 fi = 0.1

5 E = (0.625 – 0.375*(tanh(5*(t – 3))))*(1 – fi) + (1/k)*(0.625 – 0.375*(tanh(5*(t – 5))))*(fi)

6 n = 1.0

7 s = 1.0

8 z = t*E–(z1 + z2)*((exp(–10*m)))

9 Y1 = t*E

10 A2 = 1

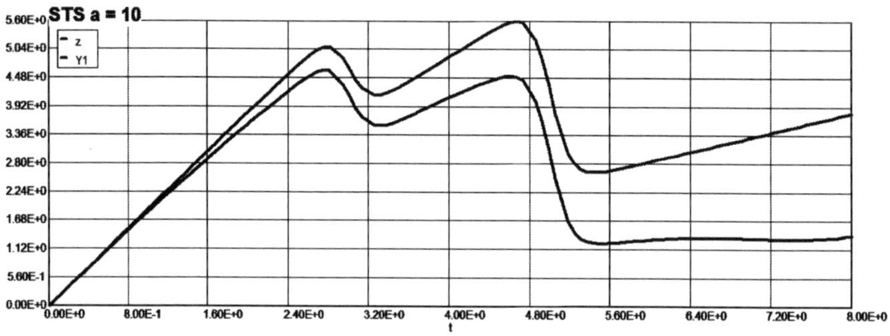

Figure 6.12: Stress-temperature-strain diagram ($\alpha = 10$).

Model: z = a1*t + a2*t^2 + a3*t^3 + a4*t^4 + a5*t^5

Variable	Value
a1	1.209451
a2	1.015406
a3	–0.4696686
a4	0.0600754
a5	–0.0023979

$$\sigma = 1.209\theta + 1.015\theta^2 - 0.470\theta^3 + 0.060\theta^4 - 0.0024\theta^5 \quad (6.71)$$

Example 6.29

Data: $\alpha = 1$

Calculated values of DEQ variables

	Variable	Initial value	Minimal value	Maximal value	Final value
1	A2	1.	1.	1.	1.
2	E	1.9	0.475	1.9	0.475
3	fi	0.1	0.1	0.1	0.1

4	k	0.1	0.1	0.1	0.1
5	m	0	0	0.0705907	0.0705907
6	m1	0.0405	0.0003976	0.0405	0.0003976
7	n	1.	1.	1.	1.
8	s	1.	1.	1.	1.
9	t	0	0	8.	8.
10	Y1	0	0	5.596159	3.8
11	z	0	0	4.557144	1.212833
12	z1	0	0	2.248973	2.248973
13	z2	0	0	0.5274249	0.5274249

Differential equations

1 d(z1)/d(t) = (exp(t/(1 + 0.067*t)))*((exp(1*m)))*m1*z^n

2 d(z2)/d(t) = k*(exp(t/(1 + 0.067*t)))*(A2*(z^s))*((exp(0.01*m)))*m1*z^n

Explicit equations

1 m = (0.0405*t – 0.01126*t^2 + 0.001462*t^3 – 0.00006868*t^4)

2 m1 = (0.0405 – 0.02252*t^1 + 0.004386*t^2 – 0.0002747*t^3)

3 k = 0.1

4 fi = 0.1

5 E = (0.625 – 0.375*(tanh(5*(t – 3))))*(1 – fi) + (1/k)*
(0.625 – 0.375*(tanh(5*(t – 5))))*(fi)

6 n = 1.0

7 s = 1.0

8 z = t*E–(z1 + z2)*((exp(–1*m)))

9 Y1 = t*E

10 A2 = 1

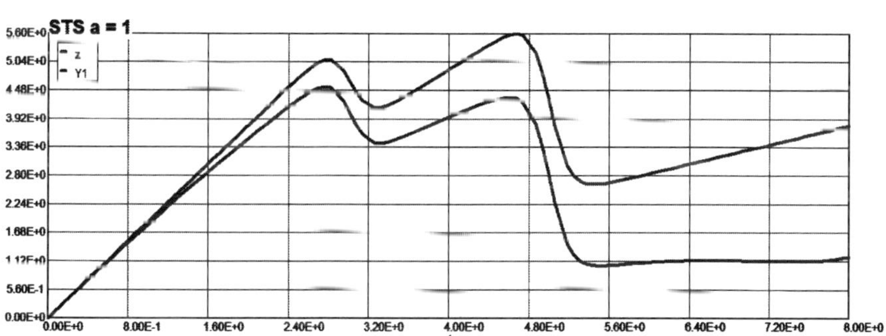

Figure 6.13: Stress-temperature-strain diagram (α = 1).

Model: z = a1*t + a2*t^2 + a3*t^3 + a4*t^4 + a5*t^5

Variable	Value
a1	1.208683
a2	1.015612
a3	–0.4762895
a4	0.0614982
a5	–0.0024785

$$\sigma = 1.209\theta + 1.015\theta^2 - 0.476\theta^3 + 0.0615\theta^4 - 0.00248\theta^5 \quad (6.72)$$

Example 6.30

Data: $\alpha = 0.1$

Calculated values of DEQ variables

	Variable	Initial value	Minimal value	Maximal value	Final value
1	A2	1.	1.	1.	1.
2	E	1.9	0.475	1.9	0.475
3	fi	0.1	0.1	0.1	0.1
4	k	0.1	0.1	0.1	0.1
5	m	0	0	0.0705907	0.0705907
6	m1	0.0405	0.0003976	0.0405	0.0003976
7	n	1.	1.	1.	1.
8	s	1.	1.	1.	1.
9	t	0	0	8.	8.
10	Y1	0	0	5.596152	3.8
11	z	0	0	4.548299	1.190925
12	z1	0	0	2.107588	2.107588
13	z2	0	0	0.5199693	0.5199693

Differential equations

1 d(z1)/d(t) = (exp(t/(1 + 0.067*t)))*((exp(0.1*m)))*m1*z^n

2 d(z2)/d(t) = k*(exp(t/(1 + 0.067*t)))*(A2*(z^s))*((exp(0.01*m)))*m1*z^n

Explicit equations

1 m = (0.0405*t – 0.01126*t^2 + 0.001462*t^3 – 0.00006868*t^4)

2 m1 = (0.0405 – 0.02252*t^1 + 0.004386*t^2 – 0.0002747*t^3)

3 k = 0.1

4 fi = 0.1
5 E = (0.625 – 0.375*(tanh(5*(t – 3))))*(1 – fi) + (1/k)* (0.625 – 0.375*(tanh(5*(t – 5))))*(fi)
6 n = 1.0
7 s = 1.0
8 z = t*E–(z1 + z2)*((exp(–0.1*m)))
9 Y1 = t*E
10 A2 = 1

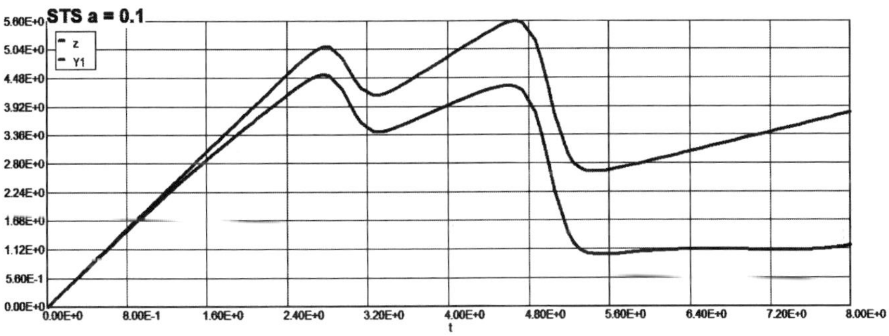

Figure 6.14: Stress-temperature-strain diagram (α = 0.1).

Model: z = a1*t + a2*t^2 + a3*t^3 + a4*t^4 + a5*t^5

Variable	Value
a1	1.208326
a2	1.016119
a3	–0.4773091
a4	0.0617072
a5	–0.0024903

$$\sigma = 1.208\theta + 1.016\theta^2 - 0.477\theta^3 + 0.0617\theta^4 - 0.00249\theta^5 \quad (6.73)$$

Example 6.31

Data: α – 0.01

Calculated values of DEQ variables

	Variable	Initial value	Minimal value	Maximal value	Final value
1	A2	1.	1.	1.	1.
2	E	1.9	0.475	1.9	0.475
3	fi	0.1	0.1	0.1	0.1

4	k	0.1	0.1	0.1	0.1
5	m	0	0	0.0705907	0.0705907
6	m1	0.0405	0.0003976	0.0405	0.0003976
7	n	1.	1.	1.	1.
8	s	1.	1.	1.	1.
9	t	0	0	8.	8.
10	Y1	0	0	5.59615	3.8
11	z	0	0	4.547406	1.188708
12	z1	0	0	2.093918	2.093918
13	z2	0	0	0.5192175	0.5192175

Differential equations

1 d(z1)/d(t) = (exp(t/(1 + 0.067*t)))*((exp(0.01*m)))*m1*z^n

2 d(z2)/d(t) = k*(exp(t/(1 + 0.067*t)))*(A2*(z^s))*((exp(0.01*m)))*m1*z^n

Explicit equations

1 m = (0.0405*t – 0.01126*t^2 + 0.001462*t^3 – 0.00006868*t^4)

2 m1 = (0.0405 – 0.02252*t^1 + 0.004386*t^2 – 0.0002747*t^3)

3 k = 0.1

4 fi = 0.1

5 E = (0.625 – 0.375*(tanh(5*(t – 3))))*(1 – fi) + (1/k)*
(0.625 – 0.375*(tanh(5*(t – 5))))*(fi)

6 n = 1.0

7 s = 1.0

8 z = t*E–(z1 + z2)*((exp(–0.01*m)))

9 Y1 = t*E

10 A2 = 1

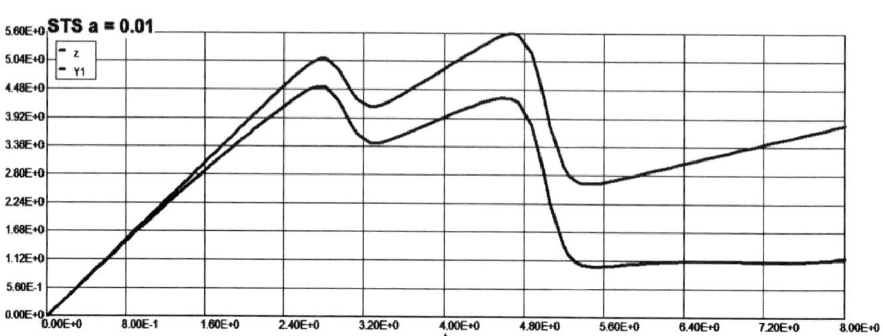

Figure 6.15: Stress-temperature-strain diagram (α = 0.01).

Model: z = a1*t + a2*t^2 + a3*t^3 + a4*t^4 + a5*t^5

Variable	Value
a1	1.208262
a2	1.016209
a3	–0.4774293
a4	0.0617313
a5	–0.0024916

$$\sigma = 1.208\theta + 1.016\theta^2 - 0.477\theta^3 + 0.0617\theta^4 - 0.00249\theta^5 \tag{6.74}$$

Consider now the "average" value of $\alpha = 50$ for computing the single realization required for obtaining mean value and autocorrelation function (ergodic process).

Example 6.32

Data: $\alpha = 50$

Calculated values of DEQ variables

	Variable	Initial value	Minimal value	Maximal value	Final value
1	A2	1.	1.	1.	1.
2	E	1.9	0.475	1.9	0.475
3	fi	0.1	0.1	0.1	0.1
4	k	0.1	0.1	0.1	0.1
5	m	0	0	0.0705907	0.0705907
6	m1	0.0405	0.0003976	0.0405	0.0003976
7	n	1.	1.	1.	1.
8	s	1.	1.	1.	1.
9	t	0	0	8.	8.
10	Y1	0	0	5.595002	3.8
11	z	0	0	4.866628	1.904939
12	z1	0	0	63.85882	63.85882
13	z2	0	0	0.7781802	0.7781802

Differential equations

1 d(z1)/d(t) = (exp(t/(1 + 0.067*t)))*((exp(50*m)))*m1*z^n

2 d(z2)/d(t) = k*(exp(t/(1 + 0.067*t)))*(A2*(z^s))*((exp(0.01*m)))*m1*
z^n

Explicit equations

1 m = (0.0405*t – 0.01126*t^2 + 0.001462*t^3 – 0.00006868*t^4)
2 m1 = (0.0405 – 0.02252*t^1 + 0.004386*t^2 – 0.0002747*t^3)
3 k = 0.1
4 fi = 0.1
5 E = (0.625 – 0.375*(tanh(5*(t – 3))))*(1 – fi) + (1/k)*
(0.625 – 0.375*(tanh(5*(t – 5))))*(fi)
6 n = 1.0
7 s = 1.0
8 z = t*E–(z1 + z2)*((exp(–50*m)))
9 Y1 = t*E
10 A2 = 1

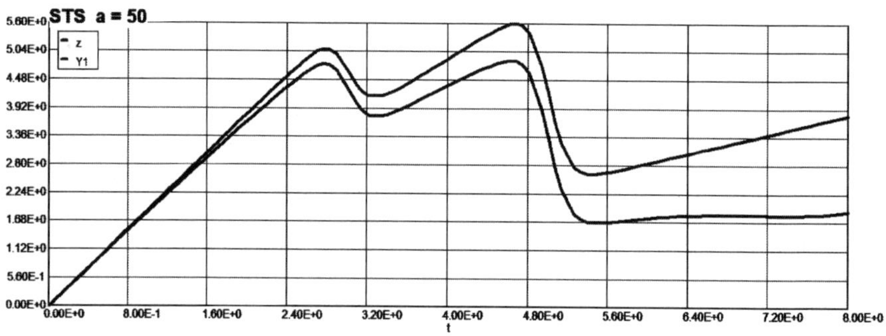

Figure 6.16: Stress-temperature-strain diagram ($\alpha = 50$).

Model: z = a1*t + a2*t^2 + a3*t^3 + a4*t^4 + a5*t^5

Variable	Value
a1	1.19148
a2	1.052269
a3	–0.4733418
a4	0.0600556
a5	–0.0023907

$$\sigma = 1.192\theta + 1.052\theta^2 - 0.473\theta^3 + 0.0607\theta^4 - 0.0024\theta^5 \tag{6.75}$$

6.9 Structural composites failures in time

The structural composites system subjected to random temperature-time loading may fail when the maximum creep stress in a critical member reaches a sufficiently high level. This type of failure is generally occurs due to overstress or by excessive permanent creep deformation rendering the structural element or system inoperative. If the structural high temperature load has a finite probability of exceeding the high level, then failure due to creep is possible, and an important problem is to find the probability that the system can operate without failure for some given time (or duration of high temperature load event). More precisely, the following problem is considered. Given a continuous and differentiable dimensionless temperature random function θ(t), one wishes to find the probability that the value σ = a will not be exceeded in the time (or temperature) interval (0, θ_{max}). This problem is called the "first-occurrence time problem" and the probability density P(a, θ) is the first-occurrence density [22]. The probability of failure in (0, θ) is unity if σ (0) > a, and the probability of failure in (0, θ_{max}) is 0 < P(σ) < 1 if σ < a:

$$P[\sigma(\theta_j) > a] = \int_a^{\infty} f(\sigma \mid \theta_j) d\sigma \tag{6.76}$$

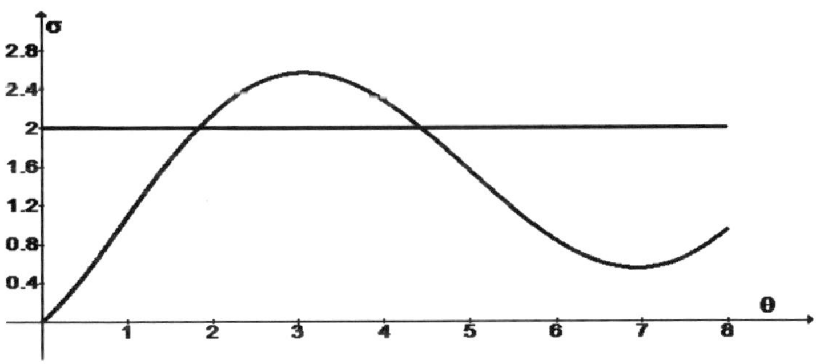

Figure 6.17: First-occurrence time curve.

In the 1970s, the analysis of the response variability of the stochastic structural systems received a lot of attention [23], consequently a new field; "Stochastic Finite Elements" was coined to stochastic mechanics of isotropic homogeneous material. Although there have been papers on computationally expensive Monte Carlo solutions and reliability considerations, most of the studies done in stochastic finite elements have been on the second moment analysis of the response of deterministic structural systems under *stochastic loading* with *deterministic* ultimate strength parameters. This chapter considers the approximate solution methods for the *probabilistic structural resistance (PSR)* of composites arising in the stochastic nature of creep process simulation problems.

For the sake of simplicity and without loss of generality, the statistical data that is required for further probabilistic analysis is based on deterministic approach of creep dynamic simulation; and it has been used in this chapter for computing the autocorrelation function parameters; and prove that the random creep deformation process can be considered as a stationary process. All random processes considered in this book are ergodic, zero-mean processes. The application of the autocorrelation theory and spectral analysis to other categories of composites described above is very similar and do not require any elaborate comments.

The step-by-step procedure of the second moment analysis of the stochastic PSR is as follows:

1. *Create statistical data*

Mainstream timing analysis uses a deterministic approach to handle the compellation of required statistical data. To start with, let's say that the stochastic dimensionless stress-temperature-time process $\sigma(\tau)$ has n realizations ($\sigma_1(\tau)$, $\sigma_2(\tau)$, ... $\sigma_n(\tau)$). In our case n = 8 (for each α_i—see computations above). Each case accounts for affects of variation of energy supply and loss during creep deformation process. Using applied probabilistic approach in obtaining the maximum stress and reliability index β, the second moment analysis quite accurately determines the worst outcome ($\sigma = a$ will be exceeded in the temperature interval [0,θ]). This makes for a better prediction of how variability will affect composites design specifically with less pessimism by targeting a slightly lower probability. Each realization function $\sigma_i(\tau)$, obviously, is just a real function (not random!) and it has a real value for any given independent value of α_k: $\sigma_i(\tau_k)$. The statistical data are compiled and represented below.

From the entire explanation given above one can see that the random function has an ergodic character, therefore only *one* random realization function is sufficient enough to obtain the correlation function in this case. Let's choose the random realization function that corresponds to $\alpha = 50$ (approximately the middle point of the interval of $0.001 < \alpha_\iota < 1000$).

$$\sigma = 1.192\theta + 1.052\theta^2 - 0.473\theta^3 + 0.0607\theta^4 - 0.0024\theta^5 \tag{6.77}$$

2. *The mean value can be calculated now is as follows:*

$$\mu_\sigma = \frac{1}{\theta_{max}}\int_0^\theta \sigma(\tau)d\tau = \frac{1}{8}\int_0^8 (\mathbf{1.192\theta + 1.052\theta^2 - 0.473\theta^3 + 0.0607\theta^4 - 0.0024\theta^5})d\theta \tag{6.78}$$

$\mu_x = 26.27925/8 = 3.285$; $\sigma = [10.75426/8]^{1/2} = 1.159$

3. *Calculate the standard deviation*

In order to calculate the standard deviation in this case the chosen function (6.85) has to be centered. Again, after using the POLYMATH software we have:

$$\sigma_\sigma^2 = \frac{1}{\theta_{max}}\int_0^\theta \sigma(\tau)d\tau = \frac{1}{8}\int_0^8 (\mathbf{1.192\theta + 1.052\theta^2 - 0.473\theta^3 + 0.0607\theta^4 - 0.0024\theta^5})d\theta$$

$\sigma = [10.75426/8]^{1/2} = 1.159$

4. *Compute the correlation function*

The correlation function of a steady time process can now be computed as follows [24]:

$$K_{\sigma}(\theta) = \frac{1}{\sigma - \tau}\int_{0}^{\theta-\tau}[\hat{\sigma}(\tau)\hat{\sigma}(\theta + \tau)]dt$$

where : $\tau = 0;1;2;................,8$

$$\hat{\sigma}(\tau) = \mathbf{1.192\tau + 1.052\tau^2 - 0.473\tau^3 + 0.0607\tau^4 - 0.0024\tau^5 - 3.285} \quad (6.79)$$

The correlation function of an ergodic process can be computed now as follows:

$$K_{\sigma}(\theta) = \frac{1}{\theta - \tau}\int_{0}^{\theta-\tau}[\hat{\sigma}(\theta)\hat{\theta}(\theta + \tau)]d\tau =$$

$$= \frac{1}{\theta - \tau}\int_{0}^{\theta-\tau}[(\mathbf{1.192\tau + 1.052\tau^2 - 0.473\tau^3 + 0.0607\tau^4 - 0.0024\tau^5 - 3.285})(1.192(\tau + x) + \mathbf{1.052(\tau + x)^2 - 0.473(\tau + x)^3 + 0.0607(\tau + x)^4 - 0.0024(\tau + x)^5 - 3.285})]dx \quad (6.80)$$

Now calculate $K_{\sigma}(\tau)$ for each $\tau = 0;1;2;.....................,8$. The results are presented in Table 6.4 below:

Table 6.4: Autocorrelation function.

τ	0	1	2	3	4	5	6	7	8
$K_{\sigma}(\theta)$	1.344	0.454	–0.472	–0.841	–0.456	0.317	0.562	–0.853	–0.045

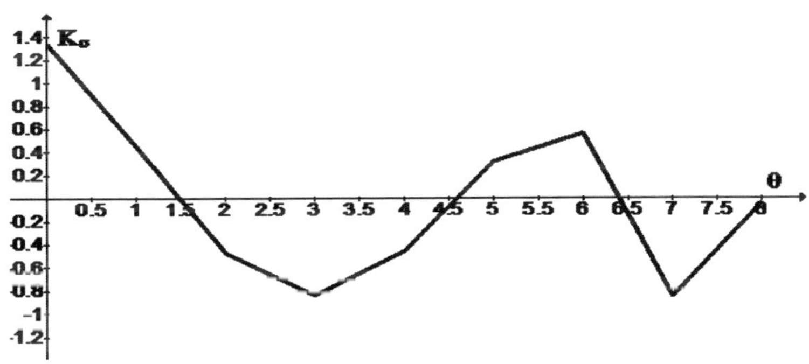

Figure 6.18: Autocorrelation function.

The normalized correlation function is (see Table 6.5):

Table 6.5: Normalized correlation function.

Temp. θ	0	1	2	3	4	5	6	7	8
$\rho_{\sigma}(\tau)$	1	0.338	–0.351	–0.626	–0.338	0.236	0.418	–0.635	0

Let's approximate the data from Table 6.5 by the formulae:

Model: y = (exp(–A*t))*(cos(B*t))

Variable	Initial guess	Value
A	0.15	0.186609
B	1.	1.152823

$$\rho_{\sigma}(\theta) = (\exp(-0.187*\theta))*(\cos(1.153*\theta) \qquad (6.81)$$

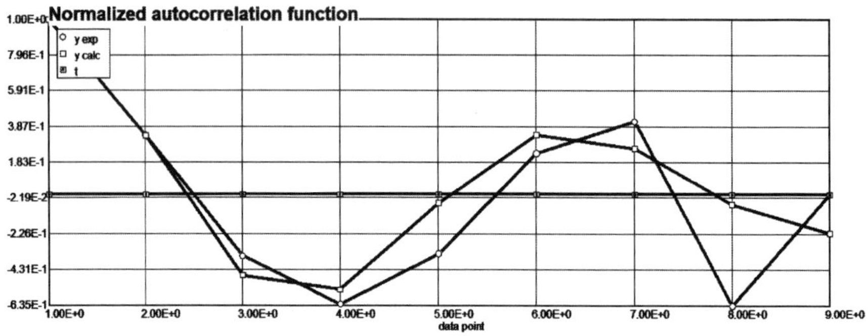

Figure 6.19: Normalized correlation function $\rho_{\theta}(\tau)$.

The ergodic function method is much simpler then the "classical" method of obtaining the correlation functions and it can be used in analysis of other creep deformation scenarios.

The corresponding spectral function is:

$$S(\omega) = \frac{1}{2}\frac{\sigma_{\sigma}^{2}}{\pi}\left[\frac{a}{(\omega - b)^{2} + a^{2}} + \frac{a}{(\omega + b)^{2} + a^{2}}\right] = \qquad (6.82)$$

$$= 0.214\left[\frac{1}{(\omega - 1.153)^{2} + 0.187^{2}} + \frac{1}{(\omega + 1.153)^{2} + 0.187^{2}}\right]$$

The graph of spectral density is (see Fig. 6.20)

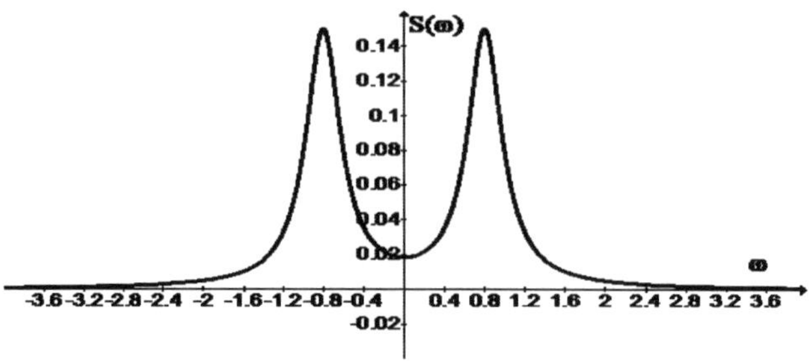

Figure 6.20: Spectral density function.

5. *The first-occurrence time problem and the probability density P (a, t)*

The average first-occurrence duration of time "t" above a given level "a" for stationary processes is defined as follows [25]:

$$\bar{t}_a = T\int_a^{\infty} f(x)dx \tag{6.83}$$

where: f(x)—probability density of the maximum temperature ordinates, which is not dependent on time for a stationary processes.

The average number of the occurrences above a given level "a" for stationary processes during the same period of time "T" is defined as follows:

$$\bar{n}_a = T\int_0^{\infty} vf(a,v)dv \tag{6.84}$$

The average duration of time for all occurrences above a given level "a" for stationary processes during the same period of time "T" is defined as follows:

$$\bar{\tau} = \frac{\int_a^{\infty} f(x)dx}{\int_0^{\infty} vf(a,v)dv} \tag{6.85}$$

The average area between the stationary random curve and the horizontal line y = a for the average first-occurrence duration of time "t" above a given level "a" is given as:

$$\overline{s\mathbf{v}}_a = \bar{x} - \int_{-\infty}^{a} xf(x)dx - a\bar{\tau}\bar{\mathbf{v}}_a$$

$$\text{where:} \tag{6.86}$$

$$\bar{\mathbf{v}}_a = \frac{\bar{n}_a}{T}$$

6. *Formulas (6.90); (6.91) and (6.92) are reduced as follows in our case of normal stationary process:*

$$\bar{t}_a = T\int_a^{\infty} f(x)dx = \frac{T}{\sigma_\theta\sqrt{2\pi}}\int_a^{\infty} \exp(-\frac{(x-\bar{x})^2}{2\sigma_0^2})dx \tag{6.87}$$

$$\bar{n}_a = T\int_0^{\infty} vf(a,v)dv = \frac{T\sigma_v}{2\pi\sigma_\theta}\exp(-\frac{(a-\bar{x})^2}{2\sigma_\theta^2})$$

$$\text{where:} \tag{6.88}$$

$$\sigma_v^2 = -\frac{d^2}{d\tau^2}K_\theta(\tau)\big|_{\tau=0}$$

$$\overline{\tau} = \frac{\int_a^{\infty} f(x)dx}{\int_0^{\infty} vf(a,v)dv} = \pi\frac{\sigma_\theta}{\sigma_v}\exp(+\frac{(a-\overline{x})^2}{2\sigma_\theta^2})[1-\Phi^*(\frac{a-\overline{x}}{\sigma_\theta})] \tag{6.89}$$

$$\overline{s} = \frac{\sigma_\theta^2\sqrt{2\pi}}{\sigma_v} + \frac{\pi(\overline{x}-a)\sigma_\theta}{\sigma_v}[1-\Phi^*(\frac{a-\overline{x}}{\sigma_\theta})][\exp(\frac{(a-\overline{x})^2}{2\sigma_\theta^2})] \tag{6.90}$$

7. *For correlation function given by Equation (6.89):*

$$\sigma_{v\sigma}^2 = K_{\dot{\sigma}}(0) = -\frac{d^2}{d\theta^2}K_{\dot{\sigma}}(\theta)\,|_{\theta=0} = \sigma_{\sigma}^2(b^2-a^2)$$

$$\overline{n}_a = \theta_{max}\frac{\sqrt{b^2-a^2}}{2\pi}\exp(-\frac{a_0^2}{2\sigma_\sigma^2}) = 8\frac{\sqrt{1.153^2-0.187^2}}{2\pi}\exp(-\frac{4.0^2}{2(\mathbf{1.159})^2}) = 0.00375$$

$$\overline{v}_a = \frac{\overline{n}_a}{\theta_{max}} = \frac{0.00375}{8} = 0.000469$$

Finally based on Poisson formulae, compute the probability of not having the minimum allowable stress ordinates crossing downwards the level "a = 4.0" is:

$$P_{rel} = P_o = \exp(-\overline{n}_a) = \exp(-0.00375) = 0.996.$$

References

[1] Haight, F.A. 1967. Handbook of the Poisson Distribution. John Wiley & Sons, NY, USA.

[2] Dietrich, C.F. 1991. Uncertainty, Calibration and Probability: The Statistics of Scientific and Industrial Measurement 2nd Edition, A. Hogler. ISBN 9780750300605.

[3] Aitken, A.C. 1957. Statistical Mathematics 8th Edition, Oliver & Boyd, Edinburgh, Scotland ISBN 9780050013007.

[4] Johnson, N.L., Kotz, S. and Kemp, A.W. 1993. Univariate Discrete Distributions (2nd Edition). Wiley, NJ, USA, ISBN 0-471-54897-9.

[5] Abramowitz, M. and Stegun, I.A. (eds.). 1965. Chapter 26. Handbook of Mathematical Functions with Formulas, Graphs, and Mathematical Tables, New York: Dover, 940 p. ISBN 978-0486612720, MR 0167642.

[6] Nelson, W. 2004. Applied Life Data Analysis. Wiley-Blackwell: NJ, USA, ISBN 0-471-64462-5.

[7] NIST. 2006. National Institute of Standards and Technology (NIST) (U.S. Department of Commerce) 100 Bureau Drive Gaithersburg, MD 20899.

[8] Oberhettinger, F. 1973. Fourier Transforms of Distributions and their Inverses: A Collection of Tables. Academic Press, Cambridge, MA.

[9] Booton, R.C. 1953. The analysis of non-linear control systems with random inputs. pp. 369–391. *In*: J. Fox (ed.). Proceedings of the Symposium on Nonlinear Circuit Analysis. Sponsored by the Polytechnic Institute of Brooklyn, Edwards Brothers, Inc., Ann Arbor, Mich.

[10] Kazakov, I.E. 1954. An approximate method for the statistical investigation of nonlinear systems. Proc. Air Force Engineering Academy named after N.E. Zhukovsky, Moscow (in Russian).
[11] Caughey, T.K. 1963. Equivalent linearization techniques. Journal of the Acoustical Society of America 35: 1706–1711. (Reference is made to presentations of the procedure in lectures delivered in 1953 at the California Institute of Technology.)
[12] Rees, D.G. 2001. Essential Statistics 4th Edition. Chapman and Hall/CRC Press, London, UK.
[13] Papoulis, A. 2002. Probability, Random Variables and Stochastic Process. McGraw-Hill Publishing Co., NY, USA.
[14] Siddall, J.N. 1982. Optimal Engineering Design. CRC Press, London/NY.
[15] Houston, A.I. and McNamara, J.M. 1999. Models of Adaptive Behavior: An Approach Based on State. Cambridge University Press, Cambridge, UK.
[16] Prout, E.G. and Tompkins, F.C. 1944. The thermal decomposition of potassium permanganate. Trans. Faraday Soc. 40: 488–498.
[17] Razdolsky, L. 2014. Probability-Based Structural Fire Load. Cambridge University Press, Cambridge, UK.
[18] Razdolsky, L. 2017. Phenomenological high temperature creep models of composites and nanomaterials. Proceedings of the AIAA SPACE 2017, Orlando, FL, USA.
[19] Hasofer, A.M. and Lind, N.C. 1974. Exact and invariant second-moment code format. Journal of the Engineering Mechanics Division, ASCE, Vol. 100, No. EM1.
[20] Nakayasu, H., Murotsu, Y., Mori, K. and Kase, S. 1978. Effect of distribution and sample size in material testing on probabilistic design of structure. ICQC, TS. D1–20, Tokyo.
[21] Mahadevan, S. 1994. Modern Structural Reliability Methods. NASA Marshall Space Flight Center, AL, USA.
[22] Rice, J.R. and Beer, F.P. 1966. First-occurrence time of high-level crossings in a continuous random process. Journal of the Acoustical Society of America 39(2).
[23] Ellingwood, B., MacGregor, J.G., Galambos, T.V. and Allin Cornell, C. 1982. Probability-based load criteria: assessment of current design practice. J. Struct. Div. ASCE 108(5): 959–977.
[24] Ovcharov, W.L. 1986. Applied Problems in Probability Theory. Translated from Russian by Irene Aleksandrova Mir Publishers, Moscow.
[25] Sveshnokov, A.A. 1978. Problems in Probability Theory, Mathematical Statistics and Theory of Random Functions. Dover Publications, New York, USA.

7

Phenomenological Creep Models of Nanocomposites (Probabilistic Approach)

7.1 Construction of a stochastic creep model of nanocomposites

Applied interest in nanomaterials is due to the possibility of significant modification or even the fundamental change in the properties of known materials, as well as the new opportunities that nanotechnology opens up in the creation of materials and products from nanomaterials-sized structural elements. Management of the fundamental properties of solids (semiconductors, metals, polymers, etc.), based on synthesizing in their volume nano-sized inclusions, crystallites, defect structures or the formation of nanoscale films and structures on the surface, is currently one of the main problems of leading scientific centers of the world working in the field of nanotechnology. The solution of these problems requires fundamentally new approaches both in the field of materials science and in the field of synthesis and formation technology. The transition to the submicron and nanometer range of element sizes requires consideration of scaling factors that reflect the influence of geometric dimensions on the properties of the material. Dispersed particles inhibit the movement of dislocations in the metal, increasing its strength at normal and elevated temperatures. *The advantage of such materials, in contrast to fibrous materials, is the isotropy of properties.* High strength is achieved at a particle size of the reinforcing agent 0.01 ... 0.1 μm. The volumetric content of particles depends on the reinforcement scheme, but usually does not exceed 5–10 volumetric percentages. Unique physicochemical, structural and technological properties of such materials allow using them in various areas of vital activity. The modern production of structural elements from nanocomposite materials is largely oriented towards the prepreg technology of manufacturing products. Prepreg are composite materials—semi-finished products—fabrics and fibers pre-impregnated with pre-catalyzed resin at high temperature and pressure. The resin in the prepreg is in a semi-solid state. Its full curing takes place during molding. Impregnation is carried out in such a way as to maximize the physicochemical properties of the reinforcing material, to provide the specified electrical, mechanical and other parameters.

As follows from above (see Chapter 4), the mechanical properties of nanocomposite structural elements depend on a large number of uncertain factors, and therefore the application of methods of applied probability theory is the most natural method for studying and analyzing the creep of this type of structural elements exposed to high temperature loads. In constructing the probabilistic creep model, the approach proposed by the author [1] is used, according to which the stochastic version of rheological relations is based on the generalization of the corresponding deterministic model with using some of the deterministic parameters as a random variables.

The procedure for selecting random parameters (inverse numbers to the relaxation times of the nanocomposites) and their intervals is formalized in [1]. Thus, the first stage is the construction of deterministic solutions of the creep equation—in our case the dimensionless Volterra integral equation of the second kind. From an analysis of the graphs of these solutions (see below), it follows, first, that the creep curves have all three stages, including the beginning of the stage of failure. The second feature of the deterministic solutions of the creep equation is the absence of qualitatively different creep strain curves, and therefore they can be considered as the realization of some ergodic random process, which in turn greatly simplifies the computational process of solving the creep problem of the nanocomposite within the framework of the correlation theory of random functions. Of course, the corresponding decisions must ultimately be confirmed and compared with a limited and selective analysis of the experimental data.

7.1.1 Selection of filler material

After selecting the polymer binder, it is necessary to determine the filler that serves to increase the mechanical properties of nanocomposite. The physical-mechanical characteristics of some nanoparticles are presented in Table 7.1 [2].

Table 7.1: Physical and mechanical characteristics of nanoparticles characteristic.

Material	Al_2O_3	BeO	Organic-adhesive	Nano-adhesive
Strength limit at tension, [MPa]	300	–	89.6	101
Strength limit at bending, [MPa]	400	–	–	145
Modulus of elasticity, [GPa]	370	317	4.83	4.60
Density, [g/cm^3]	3.96	2.30	1.15	1.90
SAP Components	6...8%	σ_b, [MPa] = 300	$\sigma_{0.2}$ [MPa] = 220	Aver. = 7%

7.1.2 Remarkable properties of nanomaterials

In the last decade, the work on synthesis and research of polymer of inorganic nanomaterials has become almost the most dynamically developing area of polymeric materials science. Works on modification of polymeric materials by the introduction of nanoparticles in them cover a wide range of polymer systems—from high-temperature films and blocks of heterocyclic polymers to polymer hydrogels. The attention of researchers is attracted to questions of the influence of nanoparticles on the nature of the mechanical behavior of various polymer materials. And among the polymer systems of various classes that are objects of such studies, invariable interest is caused by compositions based on heat-resistant aromatic polyamides. These polymeric materials are characterized by a truly unique complex of extremely high thermal and heat resistance, excellent mechanical and electrical characteristics in combination with high chemical and radiation resistance, which made them indispensable in such fields of technology as aerospace and aviation, electronics and instrument engineering, electrical engineering and others. In a number of cases a high level of mechanical properties it is an indispensable condition for the successful use of such materials. That is why the possibilities of further modification of these properties due to the introduction of various nanoparticles into polyimide matrices are currently the focus of attention of researchers.

The purpose of this chapter is to obtain the statistical information on the mechanical properties of nanocomposite materials under tension (compression) creep conditions. Nanotechnologies are based on fundamental principles and methods of research in physics, chemistry, biology, mathematics and a number of technical and human sciences, which indicates a pronounced interdisciplinary of nanotechnology. It is important to note that nanotechnology is designed to create materials and finished products with fundamentally new operational properties that cannot be achieved with the help of traditional technologies.

7.1.3 Promising nanomaterials

The development of nanotechnology is largely due to the discovery, study and already practical use of three carbon nanostructures: fullerenes, carbon nanotubes and grapheme.

Fullerenes

Fullerenes are spherical molecules consisting of 20 to 960 carbon atoms. The most studied is fullerene C60, which was discovered in 1985 in experiments on laser evaporation of a graphite target. The surface of the C60 molecule is a polyhedron consisting of 20 hexagonal and 12 pentagonal faces. The diameter of the C60 molecule is about 1 nm. Fullerenes are already used in the production of batteries; the possibility of creating on their basis optical closures, devices for recording information and elements of solar batteries is being studied. C60 molecules are also suitable for use as additives in the creation of various nanomaterials and additives for rocket fuel.

Carbon nanotubes

The carbon nanotube is actually rolled into a cylinder 1–5 nm in diameter by a mono-atomic graphite layer, which is called grapheme. Nanotubes were first discovered in 1991 when high-temperature destruction of graphite electrodes ignited between them by an electric arc. Then, the formation of nanotubes was also observed in experiments on laser evaporation of graphite, similar to those in which fullerenes were discovered, and fullerenes, in turn, were obtained by the electric arc method.

In addition to single-walled nanotubes, there are multilayer tubes, which are several nano-tubes stacked one into another. The diameter of multilayer nanotubes reaches 20–25 nm, and the distance between layers is 0.34 nm, which corresponds to the distance between carbon atoms in graphite. The length of nanotubes obtained by the electric arc method and the laser evaporation of graphite usually does not exceed 10–100 μm. The later developed methods for obtaining nanotubes by chemical deposition from hydrocarbon vapors make it possible to obtain much longer nanotubes—up to 2–3 cm. Nanotubes have very good mechanical characteristics. The strength of single walled nanotubes by different estimates ranges from 50 to 150 GPa, which is ten times higher than the strength of steel. Since the density of nanotubes is sufficiently low, the specific strength of a material made of nanotubes reaches record values. The relative elongation of nanotubes before failure is 10–15%, i.e., they have a rather high plasticity. Nanotubes also have unique electrical and thermal properties associated with the characteristics of their structure.

7.1.4 Creation of new construction materials

Nanocomposite materials

Today, for aviation, the need for new composite materials have acquired special urgency. In the newest Boeing-787 airliner, about 60% of the entire structure is made of composite materials. They provide monolithic fuselage and wings of the aircraft, while saving up to 1,500 aluminum sheets and 50 thousand fasteners for each aircraft. As a result, the mass of the aircraft significantly decreases, which allows reducing fuel consumption by 20% compared to the previous models, and the carrying capacity increases by 45%. Therefore, the aircraft industry currently is developing in directions associated with the extensive use of composite structures. In particular, studies have shown that high strength properties of composites allow large wing extensions (with the solution of a number of problems) to be realized at the same weight costs and, in the end, provide a higher level of aerodynamic characteristics of the aircraft that are not achievable on an airplane with a metal wing. In addition, the ability to control the stiffness of the composite wing and its deformations under the action of aerodynamic loads that arise in flight also makes it possible to solve the problem of optimal adaptation of the wing geometry to flight regimes, achieving maximum efficiency of the aircraft on the whole range of flight regimes. Analysis of the data presented shows that the transition to a lighter composite wing in combination with the use of composite materials in other elements leads to a reduction in the mass of the equipment making it possible to switch to an extension of the wing and reduce fuel consumption by 55.5%. This, in turn, has a positive effect on a number of other important indicators, such as an increase in the lift factor, a decrease in aircraft resistance, the level of harmful emissions into the atmosphere, and the cost of transportation.

The mechanical, electrical, thermal, optical and other characteristics of *nanocomposites* differ markedly with the properties of *ordinary composite* materials made from the same basic substances or elements. Depending on the type of the basic matrix, which occupies most of the volume of the nanocomposite material, nanocomposites are generally classified into three categories.

- Nanocomposites based on a ceramic matrix improve the optical and electrical properties of the original material (a ceramic compound consisting of a mixture of oxides, nitrides, silicates, etc.).
- In nanocomposites based on a metal matrix, the so-called reinforcing materials (nano-component) is often carbon nanotubes, which increase the strength and electrical conductivity.
- Finally, polymer nanocomposites contain a polymer matrix with nanoparticles or nano-fillers distributed over it, which may have a spherical, planar or fibrous structure. Polypropylene, polystyrene, polyamide or

nylon is used as a matrix in polymer nanocomposites [3, 4], while nano-components are aluminum or titanium oxide particles, or carbon, as well as silicon nanotubes and fibers [5–7]. Nanocomposites based on polymers differ from conventional polymeric composite materials with less weight and at the same time greater impact resistance and wear resistance, as well as good resistance to chemical influences, which allow them to be used in military and aerospace applications. Nano graphene occupies a special place in the development of nanocomposite materials [8]. It was found that the addition of graphene to epoxy composites leads to an increase in the rigidity and strength of the material compared to composites containing carbon nanotubes. Graphene is better combined with an epoxy polymer, more effectively penetrating the structure of the composite. Nanocomposites based on graphene can be used in the manufacture of aircraft components, which should remain both light and resistant to physical effects.

The introduction of carbon nanoparticles, such as fullerenes, nanotubes, etc., in the amount of more than 0.5%, in the case of carbon plastics, leads to a complex increase in mechanical and operational properties: compressive and shear strengths by 20%, shock resistance by 45%, residual strength by 1.5 times, water and fuel resistance by 1.5–2 times. At the same time, the material acquires special properties, such as electrical and thermal conductivity, X-ray and sound transparency, and lightning [9, 10].

Nanoceramics

In addition to traditional composite materials, nano-ceramics are used in aircraft structure, such as: propeller shafts, telescopic periscopes, etc. Nano-ceramics are used wherever water impermeability and corrosion protection are required. The new material is much stiffer than conventional ceramics and not so lame. With the help of nanostructures, scientists have managed to increase the rigidity of materials created on the basis of conventional silicon carbide by three times. There is already a coating for transparent polymer surfaces, consisting of nanoparticles in solution, which increase the strength of plastic several times. On the plastic surface, they form a super-hard film that protects against not only biological and chemical agents, but also from bullets.

For successful implementation of the programs connected with the development of nanotechnologies and the introduction of new nanomaterials, a number of complex fundamental and applied problems are still to be solved in related to the creation of new unique nanomaterials with the properties necessary for their application in aviation technology. To solve these problems, it is necessary to create physico-mathematical models that adequately describe the structural features of nanomaterials and the mechanisms of the effect of airspace factors on nanostructures, to develop new generation technological and testing facilities for the production of nanomaterials and to study the changes in their properties in the environment, to prepare and conduct experiments on testing nanomaterials and products made from them.

7.2 Creep models of nanocomposites: probabilistic approach

The fundamental difference between the phenomenological model of creep of nanocomposites from the model of "traditional" composites is the need to take into account four additional factors, namely: (1) allowance for free chemical energy (in the Gibbs equations of thermodynamics), which in turn leads to an additional integral in the creep Equation (4.16—see Chapter 4); (2) the presence of nanoparticles (dimensions in [nm]) or their conglomerates leads to the need to take into account the dependence of the creep stress on the size of these particles and their evolution *(the kinetics of a phase transformation in a material)* and its dependence on temperature (time); (3) the mechanical parameters of the nanocomposites structural element (for example, the effective modulus of elasticity: instantaneous modulus in Equation 4.16; and the hereditary properties of the creep function in Equation 4.16) vary with the temperature field and therefore this should also be reflected in the integral complement to Equation 4.16; (4) the mechanical parameters of the nanocomposites material also depend on its chemical composition and the technological method of its production, and hence on the physical kinetics of the chemical reaction itself (at least in a simplified final form), which determines ultimately the temperature-time dependence (different from the temperature-time dependence of the external thermal load) affecting the creep process of the nanocomposites. The first factor is taken into account in Equation 4.16 by introducing the activation energy E_a for the nanocomposites and different base temperature T_*. The second factor is taken into account by introducing the function f3 (analogy with the well-known HP principle [11, 12] for the dependence of stresses on the size of nanoparticles and the so-called degree of crystallization functions *that are encountered in innumerable publications*. The best description of kinetics of phase transformation is given by the so-called Kolmogorov-Johnson-Mehl-Avrami (KJMA) theory [13–17].

The knowledge of the kinetics of crystallization of nanomaterials is a key point in constructing the probability based creep model that depends on the microstructure. However, from the basic research point of view, it helps to validate the proposed models for phase transformations.

The third factor is taken into account by introducing the function E—an effective dimensionless instantaneous modulus of elasticity (using the rule of the mixture). Finally, the fourth factor is taken into account by introducing an additional new time-temperature dependence of the chemical reaction based on the kinetics for a given nanomaterials.

The process of thermal curing (or degrading) the nanocomposites at each point of the medium is characterized by a monotonically increasing (decreasing) function f3. Dependence f3(θ) is a solution of a differential equation of the form:

$$\frac{d(f3)}{d\theta} = f(f3, \theta) \tag{7.1}$$

In general the analysis of this differential equations is based on the two following main assumptions: (1) The transformation rate at time t during the crystallization process, df3/dt, can be expressed as a product of two separable functions, one depending solely on the temperature, θ, and the other depending solely on the fraction transformed f3:(2) The temperature dependent function, called the rate constant, follows an Arrhenius type dependency.
The relation (7.1) can be chosen as the Arrhenius equation:

$$\frac{d(f3)}{d\theta} = Ae^{-\frac{E}{RT}} f(f3) \tag{7.2}$$

The Equation (7.2) is called the crystallization kinetics equation. In it, θ denotes the dimensionless temperature at the current point of space at the current time. The distribution of temperatures in a continuous medium (as it was done in earlier author's work [18]) is found from the solution of the heat equation:

$$\rho \frac{\partial (cT)}{\partial t} = -\mathrm{div}\vec{q} + q \tag{7.3}$$

where ρ is the density, c is the specific heat of the material, and q is the specific heat release power. The heat flux vector q is related to the temperature gradient by the Fourier law:

$$q = -k_{ij} \frac{\partial T}{\partial x_j} \tag{7.4}$$

In the right-hand side of the Equation (7.4) k_{ij} is the thermal conductivity tensor k_{ij}. Due to the chemical reaction during the curing process, a heat generation is produced, which is proportional to the rate of the *crystallization of nanomaterials*. For the nanocomposites analysis used in this work, the *crystallization* kinetics equation was taken as the n-th order Prout-Tompkins equation [19] with autocatalysis:

$$\frac{dC}{dt} = Ae^{-\frac{E}{RT}} (1 - C)^q C^p \tag{7.5}$$

where A, E, q, p—are parameters, R is the universal gas constant.

The heat Equation (7.3) at hand is also separable. This means that the mechanical deformation of the structure does not affect the heat transfer processes occurring in it. Consequently, the temperature field in the computational domain can be found separately, after which the mechanical problem can be solved.

The idea of constructing the probabilistic creep model of nanocomposites is similar to the creep model of conventional composites (see Chapter 6) with the two major differences of *crystallization* kinetics function f3 involvement in the third integral in Equation (4.16) and the new temperature-time relationship based on solutions of chemical kinetics equations (see Chapter 5) coupled with the conservation of energy and mass equation [20]. Due to the fact that the probabilistic creep model of nanocomposites is based on the use of the FORM method [21–23], which in turn is based on the so-called most probable point (near the mathematical expectation), in this chapter we also consider a few more "extreme" types of functions f3 (see Examples 7.1 and 7.4 below). Analysis of the effect of these types of functions f3 on the maximum allowable creep stress makes it possible to better estimate the "average" most probable maximum creep stress from the point of view of the crystallization (or recrystallization) process of nanocomposites exposed to high (or low) temperatures. However, it must be stressed once again that this approach is approximate, in the absence of specific data on the parameters characterizing the function, f3, and in real conditions, these data on the parameters should be based on the corresponding experimental data results. A similar reasoning also applies to the new temperature-time dependence that should be based on the analysis of the physicochemical kinetics (see Chapter 5), at least in a very simplified form of the nanocomposite development. Thus, the integral type of creep constitutive equation in this case has the form:

$$
\begin{aligned}
&E(\theta)[\theta] = \sigma(\theta) + \int_{0}^{\theta} e^{\frac{\tau}{1+\beta\tau}} K_1(\theta,\tau)\sigma^n(\tau)\,m\mathbf{1}d\tau + \int_{0}^{\theta} A_2 f_2[\sigma(\tau)]K_2(\theta,\tau)\sigma^n(\tau)\,m\mathbf{1}d\tau + \\
&+\int_{0}^{\theta} \frac{E_0}{E_1} A_3 f_3[d(\tau)]K_3(\theta,\tau)\sigma(\tau)\,m\mathbf{21}d\tau \\
&K_1(\theta,\tau) = \varphi_1(\theta)f_1(\tau) = m\mathbf{1}(\tau)\sum_{i=1}^{N}\exp(-\alpha_i\, m(\theta))\exp(\alpha_i\, m(\tau)) \\
&K_2(\theta,\tau) = e^{\frac{\tau}{1+\beta\tau}} m\mathbf{1}(\tau)\sum_{i=1}^{N}\exp(-\beta_{i2}\, m(\theta))\exp(\beta_{i2}\, m(\tau)) \\
&K_3(\theta,\tau) = e^{\frac{\tau}{1+\beta\tau}} m\mathbf{21}(\tau)\sum_{i=1}^{N}\exp(-\beta_{i3}\, m\mathbf{2}(\theta))\exp(\beta_{i3}\, m\mathbf{2}(\tau)) \\
&f_2(\sigma) = \sigma^s;\;\; s = \mathbf{1,2,3}\ldots, M;\;\; f_3(\theta) = A_3[(d^{-0.5}(1+d^{-0.5})]; \\
&d = \varphi(\theta);\;\; k = E_0/E_1;\;\; A_2 = \text{const.};\;\; A_3(\theta) = A_3 f(\theta)
\end{aligned}
\tag{7.6}
$$

Numerical analysis and results, as usual, are given below in the form of examples, graphs and tables that do not require significant additional explanations.

7.2.1 *General computer code and effect of different types of function f3*

The general computer code is as follows:

Differential equations

```
1 d(z1)/d(t) = (exp(t/(1 + 0.067*t)))*((exp(0.001*m)))*m1*z^n
2 d(z2)/d(t) = (exp(t/(1 + 0.067*t)))*(A2*(z^s))*((exp(0.001*m)))*m1*z^n
3 d(z3)/d(t) = k*(exp(t/(1 + 0.067*t)))*A3*(f3*(0.1*t^p/(1 + 0.1*t^p)))*
  ((exp(0.001*m2)))*m21*z^n
```

Explicit equations

```
1 m = (0.0405*t – 0.01126*t^2 + 0.001462*t^3 – 0.00006868*t^4)
2 m1 = (0.0405 – 0.02252*t^1 + 0.004386*t^2 – 0.0002747*t^3)
3 m2 = 0.0868*t – 0.0386*t^2 + 0.0080*t^3 – 0.000756*t^4 + 2.6*
  (10^ – 5)*t^5
4 fi = 0.1; k = 0.1; n = s = 1.0; p = –0.5; A2 = 1; A3 = 1
5 E = (0.625 – 0.375*(tanh(5*(t – 3))))*(1 – fi) + (1/k)*
  (0.625 – 0.375*(tanh(5*(t – 5))))*(fi)
6 Y1 = t*E; f3 = if (t < 0) then (0) else (1); f31 = (1/16)*t*(8 – t); f32 =
  (exp(–0.1*t))
7 f33 = (sin(3.14*t/4))^2; f34 = if (t < 7.99) then (0) else (1); f35 =
  (1/8)*t^1
8 f36 = if (t < 4) then (1) else (0); f37 = if (t < 4) then (0) else (1)
9 z = t*E–(z1 + z2)*((exp(–0.001*m)))–z3*(exp(–0.001*m2))
10 m21 = 0.0868 – 0.0772*t + 0.024*t^2 – 0.003*t^3 + 1.3*(10 – 4)* t^4
```

Example 7.1 f3 = if (t < 0) then (0) else (1):

Function f3 in this case indicates that chemical free energy is thermally activated from the very beginning and lasts up to the very end of the creep process. *Therefore the maximum allowable creep stress is decreasing.* The effective instantaneous modulus of elasticity is the function of dimensionless temperature only and the dimensionless transitional temperatures parameters θ_{gm} and θ_{gf} of matrix and nanomaterials respectfully are different from each other. The ratio $k = E_m/E_f$ is given and constant. The solution of the single integral Equation (7.6) of creep process for nanocomposite is as follows:

Calculated values of DEQ variables

	Variable	Initial value	Minimal value	Maximal value	Final value
1	A2	1.	1.	1.	1.
2	A3	1.	1.	1.	1.
3	E	1.9	0.475	1.9	0.475
4	f3	1.	1.	1.	1.
5	f31	5.0E-05	0	0.9999973	0
6	f32	0.99999	0.449329	0.99999	0.449329
7	f33	6.162E-09	6.162E-09	0.9999977	1.015E-05
8	f34	0	0	1.	1.
9	f35	1.25E-05	1.25E-05	1.	1.
10	f36	1.	0	1.	0
11	f37	0	0	1.	1.
12	fi	0.1	0.1	0.1	0.1
13	k	0.1	0.1	0.1	0.1
14	m	4.05E-06	4.05E-06	0.0705907	0.0705907
15	m1	0.0404977	0.0003976	0.0404977	0.0003976
16	m2	8.68E-06	8.68E-06	0.0825435	0.075392
17	m21	0.0867923	0.0813607	3.195E+04	3.195E+04
18	n	1.	1.	1.	1.
19	p	–0.5	–0.5	–0.5	–0.5
20	s	1.	1.	1.	1.
21	t	0.0001	0.0001	8.	8.
22	Y1	0.00019	0.00019	5.592051	3.8
23	z	0.00019	–0.0777359	2.021731	2.381E-05
24	z1	0	0	0.167872	0.1667625
25	z2	0	0	0.2548118	0.2548118
26	z3	0	0	5.169975	3.378686

Differential equations

1 d(z1)/d(t) = (exp(t/(1 + 0.067*t)))*((exp(0.001*m)))*m1*z^n

2 d(z2)/d(t) = (exp(t/(1 + 0.067*t)))*(A2*(z^s))*((exp(0.001*m)))*m1*z^n

3 d(z3)/d(t) = k*(exp(t/(1 + 0.067*t)))*A3*(f3*(0.1*t^p/(1 + 0.1*t^p)))*((exp(0.001*m2)))*m21*z^n

Explicit equations

```
1 m = (0.0405*t – 0.01126*t^2 + 0.001462*t^3 – 0.00006868*t^4)
2 m1 = (0.0405 – 0.02252*t^1 + 0.004386*t^2 – 0.0002747*t^3)
3 m2 = 0.0868*t – 0.0386*t^2 + 0.0080*t^3 – 0.000756*t^4 + 2.6*
  (10^ – 5)*t^5
4 fi = 0.1
5 k = 0.1
6 n = 1.0
7 s = 1.0
8 p = –0.5
9 E = (0.625 – 0.375*(tanh(5*(t – 3))))*(1 – fi) + (1/k)*
  (0.625 – 0.375*(tanh(5*(t – 5))))*(fi)
10 Y1 = t*E
11 A2 = 1
12 A3 = 1
13 f3 = if (t < 0) then (0) else (1)
14 f31 = (1/16)*t*(8 – t)
15 f32 = (exp(–0.1*t))
16 f33 = (sin(3.14*t/4))^2
17 f34 = if (t < 7.99) then (0) else (1)
18 f35 = (1/8)*t^1
19 f36 = if (t < 4) then (1) else (0)
20 f37 = if (t < 4) then (0) else (1)
21 z = t*E–(z1 + z2)*((exp(–0.001*m)))–z3*(exp(–0.001*m2))
22 m21 = 0.0868 – 0.0772*t + 0.024*t^2 – 0.003*t^3 + 1.3*(10 – 4)* t^4
```

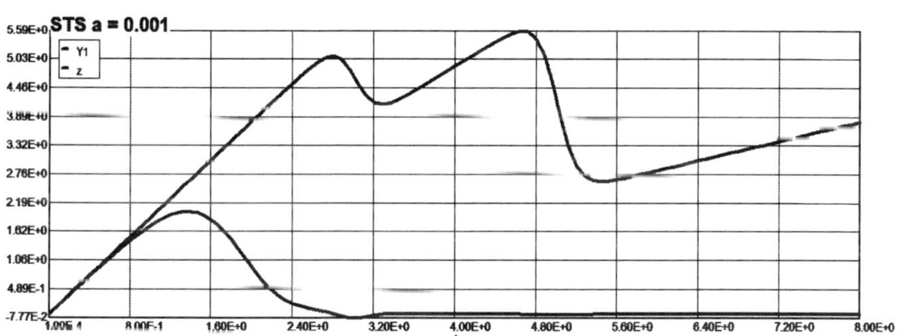

Figure 7.1: Stress-temperature-strain diagram (STS) α = 0.001.

Model: z = a1*t + a2*t^2 + a3*t^3 + a4*t^4 + a5*t^5

Variable	Value
a1	4.049795
a2	–3.163575
a3	0.8875755
a4	–0.106722
a5	0.0046657

$$\sigma = 4.05\theta - 3.16\theta^2 + 0.888\theta^3 - 0.107\theta^4 + 0.0047\theta^5 \quad (7.7)$$

Example 7.2 f31 = (1/16)*t*(8 – t) $\alpha = 0.001$

The high temperature effect on nanomaterials is increasing up to $\theta^* = 4$, and then decreasing.

Calculated values of DEQ variables

	Variable	Initial value	Minimal value	Maximal value	Final value
1	A2	1.	1.	1.	1.
2	A3	1.	1.	1.	1.
3	E	1.9	0.475	1.9	0.475
4	f3	1.	1.	1.	1.
5	f31	5.0E-05	0	0.9999954	0
6	f32	0.99999	0.449329	0.99999	0.449329
7	f33	6.162E-09	6.162E-09	0.9999968	1.015E-05
8	f34	0	0	1.	1.
9	f35	1.25E-05	1.25E-05	1.	1.
10	f36	1.	0	1.	0
11	f37	0	0	1.	1.
12	fi	0.1	0.1	0.1	0.1
13	k	0.1	0.1	0.1	0.1
14	m	4.05E-06	4.05E-06	0.0705907	0.0705907
15	m1	0.0404977	0.0003976	0.0404977	0.0003976
16	m2	8.68E-06	8.68E-06	0.0825435	0.075392
17	m21	0.0867923	0.0805237	3.195E+04	3.195E+04
18	n	1.	1.	1.	1.
19	p	–0.5	–0.5	–0.5	–0.5
20	s	1.	1.	1.	1.

21	t	0.0001	0.0001	8.	8.
22	Y1	0.00019	0.00019	5.591664	3.8
23	z	0.00019	–0.0824124	2.209326	0.0059897
24	z1	0	0	0.1903081	0.1891825
25	z2	0	0	0.3178923	0.3178923
26	z3	0	0	5.084099	3.287219

Differential equations

```
1 d(z1)/d(t) = (exp(t/(1 + 0.067*t)))*((exp(0.001*m)))*m1*z^n
2 d(z2)/d(t) = (exp(t/(1 + 0.067*t)))*(A2*(z^s))*((exp(0.001*m)))*m1*z^n
3 d(z3)/d(t) = k*(exp(t/(1 + 0.067*t)))*A3*(f31*(0.1*t^p/(1 + 0.1*t^p)))*
  ((exp(0.001*m2)))*m21*z^n
```

Explicit equations

```
1 m = (0.0405*t – 0.01126*t^2 + 0.001462*t^3 – 0.00006868*t^4)
2 m1 = (0.0405 – 0.02252*t^1 + 0.004386*t^2 – 0.0002747*t^3)
3 m2 = 0.0868*t – 0.0386*t^2 + 0.0080*t^3 – 0.000756*t^4 + 2.6*(10^ – 5)*t^5
4 fi = 0.1
5 k = 0.1
6 n = 1.0
7 s = 1.0
8 p = –0.5
9 E = (0.625 – 0.375*(tanh(5*(t – 3))))*(1 – fi) + (1/k)*
  (0.625 – 0.375*(tanh(5*(t – 5))))*(fi)
10 Y1 = t*E
11 A2 = 1
12 A3 = 1
13 f3 = if (t < 0) then (0) else (1)
14 f31 = (1/16)*t*(8 – t)
15 f32 = (exp(–0.1*t))
16 f33 = (sin(3.14*t/4))^2
17 f34 – if (t < 7.99) then (0) else (1)
18 f35 = (1/8)*t^1
19 f36 = if (t < 4) then (1) else (0)
20 f37 = if (t < 4) then (0) else (1)
21 z = t*E–(z1 + z2)*((exp(–0.001*m)))–z3*(exp(–0.001*m2))
22 m21 = 0.0868 – 0.0772*t + 0.024*t^2 – 0.003*t^3 + 1.3*(10 – 4)* t^4
```

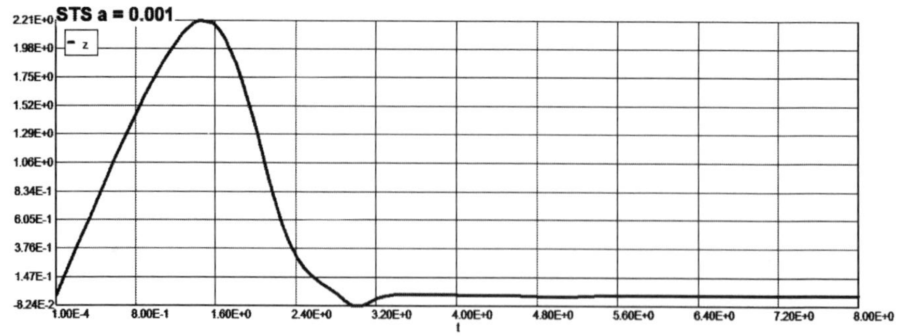

Figure 7.2: Stress-temperature-strain diagram (STS) f31 and α = 0.001.

Model: z = a1*t + a2*t^2 + a3*t^3 + a4*t^4 + a5*t^5

Variable	Value
a1	4.265851
a2	–3.237507
a3	0.8861352
a4	–0.104326
a5	0.0044799

$$\sigma = 4.26\theta - 3.24\theta^2 + 0.886\theta^3 - 0.104\theta^4 + 0.0045\theta^5 \quad (7.8)$$

Example 7.3 f34 = if (t < 7.99) then (0) else (1) α = 0.001

The high temperature does not affect the nanomaterials up to θ = 7.99 < 8.0

Calculated values of DEQ variables

	Variable	Initial value	Minimal value	Maximal value	Final value
1	A2	1.	1.	1.	1.
2	A3	1.	1.	1.	1.
3	E	1.9	0.475	1.9	0.475
4	f3	1.	1.	1.	1.
5	f31	5.0E-05	5.0E-05	0.9999548	0.0049938
6	f32	0.99999	0.4497785	0.99999	0.4497785
7	f33	6.162E-09	6.162E-09	0.9999622	0.0001218
8	f34	0	0	0	0
9	f35	1.25E-05	1.25E-05	0.99875	0.99875
10	f36	1.	0	1.	0

11	f37	0	0	1.	1.
12	fi	0.1	0.1	0.1	0.1
13	k	0.1	0.1	0.1	0.1
14	m	4.05E-06	4.05E-06	0.0705866	0.0705866
15	m1	0.0404977	0.0004482	0.0404977	0.0004482
16	m2	8.68E-06	8.68E-06	0.0825479	0.0754977
17	m21	0.0867923	0.0827145	3.179E+04	3.179E+04
18	n	1.	1.	1.	1.
19	p	–0.5	–0.5	–0.5	–0.5
20	s	1.	1.	1.	1.
21	t	0.0001	0.0001	8.	8.
22	Y1	0.00019	0.00019	5.596116	3.79525
23	z	0.00019	–0.4610609	3.705459	0.5433488
24	z1	0	0	0.8993469	0.8993469
25	z2	0	0	2.352784	2.352784
26	z3	0	0	0	0

Differential equations

1 d(z1)/d(t) = (exp(t/(1 + 0.067*t)))*((exp(0.001*m)))*m1*z^n
2 d(z2)/d(t) = (exp(t/(1 + 0.067*t)))*(A2*(z^s))*((exp(0.001*m)))*m1*z^n
3 d(z3)/d(t) = k*(exp(t/(1 + 0.067*t)))*A3*(f34*(0.1*t^p/(1 + 0.1*t^p)))*((exp(0.001*m2)))*m21*z^n

Explicit equations

1 m = (0.0405*t – 0.01126*t^2 + 0.001462*t^3 – 0.00006868*t^4)
2 m1 = (0.0405 – 0.02252*t^1 + 0.004386*t^2 – 0.0002747*t^3)
3 m2 = 0.0868*t – 0.0386*t^2 + 0.0080*t^3 – 0.000756*t^4 + 2.6*(10^ – 5)*t^5
4 fi = 0.1
5 k = 0.1
6 n = 1.0
7 s = 1.0
8 p = –0.5
9 E = (0.625 – 0.375*(tanh(5*(t – 3))))*(1 – fi) + (1/k)*(0.625 – 0.375*(tanh(5*(t – 5))))*(fi)
10 Y1 = t*E

```
11 A2 = 1
12 A3 = 1
13 f3 = if (t < 0) then (0) else (1)
14 f31 = (1/16)*t*(8 – t)
15 f32 = (exp(–0.1*t))
16 f33 = (sin(3.14*t/4))^2
17 f34 = if (t < 7.99) then (0) else (1)
18 f35 = (1/8)*t^1
19 f36 = if (t < 4) then (1) else (0)
20 f37 = if (t < 4) then (0) else (1)
21 z = t*E–(z1 + z2)*((exp(–0.001*m)))-z3*(exp(–0.001*m2))
22 m21 = 0.0868 – 0.0772*t + 0.024*t^2 – 0.003*t^3 + 1.3*(10 – 4)* t^4
```

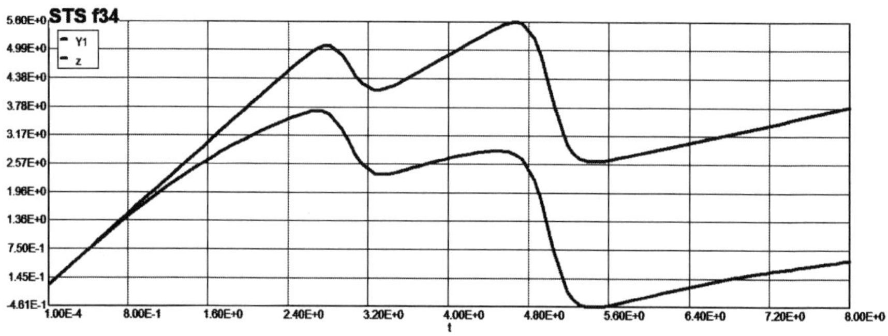

Figure 7.3: Stress-temperature-strain diagram (STS) f34 and $\alpha = 0.001$.

Model: z = a1*t + a2*t^2 + a3*t^3 + a4*t^4 + a5*t^5

Variable	Value
a1	1.301195
a2	0.8788917
a3	–0.5025695
a4	0.073129
a5	–0.0032978

$$\sigma = 1.3\theta + 0.879\theta^2 - 0.502\theta^3 + 0.073\theta^4 - 0.0033\theta^5 \qquad (7.9)$$

Example 7.4 f37 = if (t < 4) then (0) else (1) $\alpha = 0.001$

The high temperature does not affect the nanomaterials up to $\theta^* = 4$; after that the nanomaterials suddenly fail.

Calculated values of DEQ variables

	Variable	Initial value	Minimal value	Maximal value	Final value
1	A2	1.	1.	1.	1.
2	A3	1.	1.	1.	1.
3	E	1.9	0.475	1.9	0.475
4	f3	1.	1.	1.	1.
5	f31	5.0E-05	0	1.	0
6	f32	0.99999	0.449329	0.99999	0.449329
7	f33	6.162E-09	6.162E-09	0.9999969	1.015E-05
8	f34	0	0	1.	1.
9	f35	1.25E-05	1.25E-05	1.	1.
10	f36	1.	0	1.	0
11	f37	0	0	1.	1.
12	fi	0.1	0.1	0.1	0.1
13	k	0.1	0.1	0.1	0.1
14	m	4.05E-06	4.05E-06	0.0705907	0.0705907
15	m1	0.0404977	0.0003976	0.0404977	0.0003976
16	m2	8.68E-06	8.68E-06	0.0825436	0.075392
17	m21	0.0867923	0.0827145	3.195E+04	3.195E+04
18	n	1.	1.	1.	1.
19	p	–0.5	–0.5	–0.5	–0.5
20	s	1.	1.	1.	1.
21	t	0.0001	0.0001	8.	8.
22	Y1	0.00019	0.00019	5.591425	3.8
23	z	0.00019	–0.0095448	3.705459	2.382E-05
24	z1	0	0	0.578086	0.5777069
25	z2	0	0	1.618335	1.618335
26	z3	0	0	3.394816	1.60421

Differential equations

1 d(z1)/d(t) = (exp(t/(1 + 0.067*t)))*((exp(0.001*m)))*m1*z^n

2 d(z2)/d(t) = (exp(t/(1 + 0.067*t)))*(A2*(z^s))*((exp(0.001*m)))*m1*z^n

3 d(z3)/d(t) = k*(exp(t/(1 + 0.067*t)))*A3*(f37*(0.1*t^p/(1 + 0.1*t^p)))*((exp(0.001*m2)))*m21*z^n

Explicit equations

1 m = (0.0405*t – 0.01126*t^2 + 0.001462*t^3 – 0.00006868*t^4)

2 m1 = (0.0405 – 0.02252*t^1 + 0.004386*t^2 – 0.0002747*t^3)

3 m2 = 0.0868*t – 0.0386*t^2 + 0.0080*t^3 – 0.000756*t^4 + 2.6*(10^ – 5)*t^5

4 fi = 0.1

5 k = 0.1

6 n = 1.0

7 s = 1.0

8 p = –0.5

9 E = (0.625 – 0.375*(tanh(5*(t – 3))))*(1 – fi) + (1/k)*(0.625 – 0.375*(tanh(5*(t – 5))))*(fi)

10 Y1 = t*E

11 A2 = 1

12 A3 = 1

13 f3 = if (t < 0) then (0) else (1)

14 f31 = (1/16)*t*(8 – t)

15 f32 = (exp(–0.1*t))

16 f33 = (sin(3.14*t/4))^2

17 f34 = if (t < 7.99) then (0) else (1)

18 f35 = (1/8)*t^1

19 f36 = if (t < 4) then (1) else (0)

20 f37 = if (t < 4) then (0) else (1)

21 z = t*E–(z1 + z2)*((exp(–0.001*m)))–z3*(exp(–0.001*m2))

22 m21 = 0.0868 – 0.0772*t + 0.024*t^2 – 0.003*t^3 + 1.3*(10 – 4)* t^4

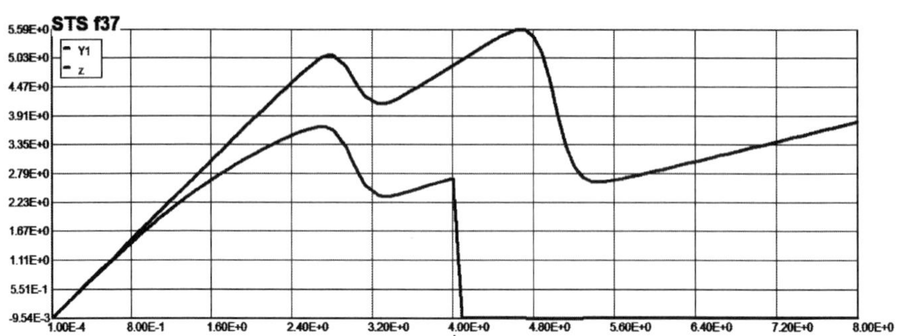

Figure 7.4: Stress-temperature-strain diagram (STS) f37 and α = 0.001.

Model: z = a1*t + a2*t^2 + a3*t^3 + a4*t^4 + a5*t^5

Variable	Value
a1	1.040597
a2	1.577153
a3	–0.911607
a4	0.1507622
a5	–0.0079519

$$\sigma = 1.04\theta + 1.577\theta^2 - 0.912\theta^3 + 0.151\theta^4 - 0.00795\theta^5 \qquad (7.10)$$

It can be seen from the above Examples 7.1–7.4 that the maximum allowable creep stress under high temperatures (σ = 3.7—see Examples 7.3 and 7.4) is obtained when the nanomaterials are so strong and rigid that they practically do not extend (shortened) under the action of high temperatures (within specified limits); the creep strain in this case occurs only as a result of matrix deformations and changes of some dimensionless parameters of the effective modulus of elasticity of the nanocomposite as a function of temperature (time). Examples for other types of functions are not given in this chapter, since they occupy an intermediate value between the extreme values of the functions (f3 in the sense of their effect on the creep process). To construct statistical data based on solutions of the corresponding deterministic formulation of nanocomposites creep process (similar to the previous model for conventional composites—see Chapter 6), it is assumed that the α_i—parameter in Equation (7.6) is a discrete random variable, and each deterministic solution represents the realization of a creep deformation random process of time (temperature). As before, the corresponding numerical dimensionless solutions are presented below in tabular form and do not require additional explanations. Consider now f37 = if (θ < 4) then (0) else (1):

7.3 Compilation of statistical data based on creep constitutive equation solutions

Example 7.5 $\alpha = 0.01$

Calculated values of DEQ variables

	Variable	Initial value	Minimal value	Maximal value	Final value
1	A2	1.	1.	1.	1.
2	A3	1.	1.	1.	1.
3	E	1.9	0.475	1.9	0.475
4	f3	1.	1.	1.	1.

5	f31	5.0E-05	0	1.	0
6	f32	0.99999	0.449329	0.99999	0.449329
7	f33	6.162E-09	6.162E-09	0.9999948	1.015E-05
8	f34	0	0	1.	1.
9	f35	1.25E-05	1.25E-05	1.	1.
10	f36	1.	0	1.	0
11	f37	0	0	1.	1.
12	fi	0.1	0.1	0.1	0.1
13	k	0.1	0.1	0.1	0.1
14	m	4.05E-06	4.05E-06	0.0705907	0.0705907
15	m1	0.0404977	0.0003976	0.0404977	0.0003976
16	m2	8.68E-06	8.68E-06	0.0825435	0.075392
17	m21	0.0867923	0.0827181	3.195E+04	3.195E+04
18	n	1.	1.	1.	1.
19	p	–0.5	–0.5	–0.5	–0.5
20	s	1.	1.	1.	1.
21	t	0.0001	0.0001	8.	8.
22	Y1	0.00019	0.00019	5.591468	3.8
23	z	0.00019	–0.0095458	3.705533	2.38E-05
24	z1	0	0	0.5783433	0.5779639
25	z2	0	0	1.619113	1.619113
26	z3	0	0	3.397341	1.60566

Differential equations

1 d(z1)/d(t) = (exp(t/(1 + 0.067*t)))*((exp(0.01*m)))*m1*z^n

2 d(z2)/d(t) = (exp(t/(1 + 0.067*t)))*(A2*(z^s))*((exp(0.01*m)))*m1*z^n

3 d(z3)/d(t) = k*(exp(t/(1 + 0.067*t)))*A3*(f37*(0.1*t^p/(1 + 0.1*t^p)))*((exp(0.01*m2)))*m21*z^n

Explicit equations

1 m = (0.0405*t – 0.01126*t^2 + 0.001462*t^3 – 0.00006868*t^4)

2 m1 = (0.0405 – 0.02252*t^1 + 0.004386*t^2 – 0.0002747*t^3)

3 m2 = 0.0868*t – 0.0386*t^2 + 0.0080*t^3 – 0.000756*t^4 + 2.6*(10^ – 5)*t^5

4 fi = 0.1

5 k = 0.1

6 n = 1.0

```
7 s = 1.0
8 p = –0.5
9 E = (0.625 – 0.375*(tanh(5*(t – 3))))*(1 – fi) + (1/k)*
  (0.625 – 0.375*(tanh(5*(t – 5))))*(fi)
10 Y1 = t*E
11 A2 = 1
12 A3 = 1
13 f3 = if (t < 0) then (0) else (1)
14 f31 = (1/16)*t*(8 – t)
15 f32 = (exp(–0.1*t))
16 f33 = (sin(3.14*t/4))^2
17 f34 = if (t < 7.99) then (0) else (1)
18 f35 = (1/8)*t^1
19 f36 = if (t < 4) then (1) else (0)
20 f37 = if (t < 4) then (0) else (1)
21 z = t*E–(z1 + z2)*((exp(–0.01*m)))–z3*(exp(–0.01*m2))
22 m21 = 0.0868 – 0.0772*t + 0.024*t^2 – 0.003*t^3 + 1.3*(10 – 4)* t^4
```

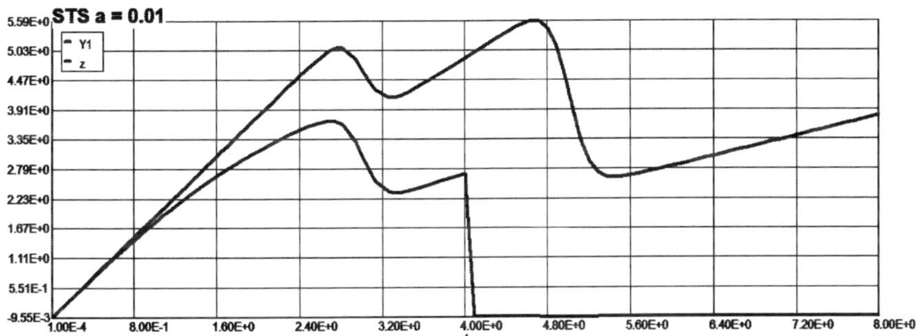

Figure 7.5: Stress-temperature-strain diagram (STS) α = 0.01.

Model: z = a1*t + a2*t^2 + a3*t^3 + a4*t^4 + a5*t^5

Variable	Valuc
a1	1.085611
a2	1.517337
a3	–0.8864206
a4	0.1465688
a5	–0.0077118

$$\sigma = 1.086\theta + 1.517\theta^2 - 0.886\theta^3 + 0.146\theta^4 - 0.00771\theta^5 \tag{7.11}$$

Example 7.6 α = 0.1

Calculated values of DEQ variables

	Variable	Initial value	Minimal value	Maximal value	Final value
1	A2	1.	1.	1.	1.
2	A3	1.	1.	1.	1.
3	E	1.9	0.475	1.9	0.475
4	f3	1.	1.	1.	1.
5	f31	5.0E-05	0	1.	0
6	f32	0.99999	0.449329	0.99999	0.449329
7	f33	6.162E-09	6.162E-09	0.9999968	1.015E-05
8	f34	0	0	1.	1.
9	f35	1.25E-05	1.25E-05	1.	1.
10	f36	1.	0	1.	0
11	f37	0	0	1.	1.
12	fi	0.1	0.1	0.1	0.1
13	k	0.1	0.1	0.1	0.1
14	m	4.05E-06	4.05E-06	0.0705907	0.0705907
15	m1	0.0404977	0.0003976	0.0404977	0.0003976
16	m2	8.68E-06	8.68E-06	0.0825436	0.075392
17	m21	0.0867923	0.0827568	3.195E+04	3.195E+04
18	n	1.	1.	1.	1.
19	p	–0.5	–0.5	–0.5	–0.5
20	s	1.	1.	1.	1.
21	t	0.0001	0.0001	8.	8.
22	Y1	0.00019	0.00019	5.592036	3.8
23	z	0.00019	–0.0095415	3.706267	2.374E-05
24	z1	0	0	0.5809223	0.5805411
25	z2	0	0	1.626912	1.626912
26	z3	0	0	3.422848	1.62022

Differential equations

1 d(z1)/d(t) = (exp(t/(1 + 0.067*t)))*((exp(0.1*m)))*m1*z^n

2 d(z2)/d(t) = (exp(t/(1 + 0.067*t)))*(A2*(z^s))*((exp(0.1*m)))*m1*z^n

3 d(z3)/d(t) = k*(exp(t/(1 + 0.067*t)))*A3*(f37*(0.1*t^p/(1 + 0.1*t^p)))*((exp(0.1*m2)))*m21*z^n

Explicit equations

1 m = (0.0405*t – 0.01126*t^2 + 0.001462*t^3 – 0.00006868*t^4)
2 m1 = (0.0405 – 0.02252*t^1 + 0.004386*t^2 – 0.0002747*t^3)
3 m2 = 0.0868*t – 0.0386*t^2 + 0.0080*t^3 – 0.000756*t^4 + 2.6*(10^ – 5)*t^5
4 fi = 0.1
5 k = 0.1
6 n = 1.0
7 s = 1.0
8 p = –0.5
9 E = (0.625 – 0.375*(tanh(5*(t – 3))))*(1 – fi) + (1/k)*(0.625 – 0.375*(tanh(5*(t – 5))))*(fi)
10 Y1 = t*E
11 A2 = 1
12 A3 = 1
13 f3 = if (t < 0) then (0) else (1)
14 f31 = (1/16)*t*(8 – t)
15 f32 = (exp(–0.1*t))
16 f33 = (sin(3.14*t/4))^2
17 f34 = if (t < 7.99) then (0) else (1)
18 f35 = (1/8)*t^1
19 f36 = if (t < 4) then (1) else (0)
20 f37 = if (t < 4) then (0) else (1)
21 z = t*E–(z1 + z2)*((exp(–0.1*m)))–z3*(exp(–0.1*m2))
22 m21 = 0.0868 – 0.0772*t + 0.024*t^2 – 0.003*t^3 + 1.3*(10 – 4)* t^4

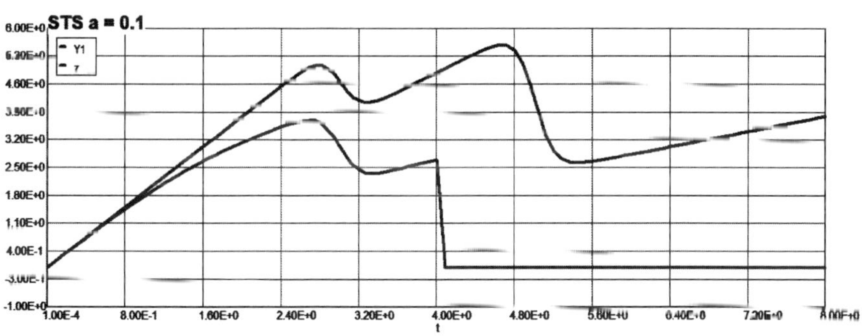

Figure 7.6: Stress-temperature-strain diagram (STS) α = 0.1.

Model: z = a1*t + a2*t^2 + a3*t^3 + a4*t^4 + a5*t^5

Variable	Value
a1	1.039517
a2	1.578584
a3	–0.9120835
a4	0.1508227
a5	–0.0079545

$$\sigma = 1.04\theta + 1.578\theta^2 - 0.912\theta^3 + 0.151\theta^4 - 0.008\theta^5 \tag{7.12}$$

Example 7.7 $\alpha = 1$

Calculated values of DEQ variables

	Variable	Initial value	Minimal value	Maximal value	Final value
1	A2	1.	1.	1.	1.
2	A3	1.	1.	1.	1.
3	E	1.9	0.475	1.9	0.475
4	f3	1.	1.	1.	1.
5	f31	5.0E-05	0	1.	0
6	f32	0.99999	0.449329	0.99999	0.449329
7	f33	6.162E-09	6.162E-09	0.9999951	1.015E-05
8	f34	0	0	1.	1.
9	f35	1.25E-05	1.25E-05	1.	1.
10	f36	1.	0	1.	0
11	f37	0	0	1.	1.
12	fi	0.1	0.1	0.1	0.1
13	k	0.1	0.1	0.1	0.1
14	m	4.05E-06	4.05E-06	0.0705907	0.0705907
15	m1	0.0404977	0.0003976	0.0404977	0.0003976
16	m2	8.68E-06	8.68E-06	0.0825437	0.075392
17	m21	0.0867923	0.0832385	3.195E+04	3.195E+04
18	n	1.	1.	1.	1.
19	p	–0.5	–0.5	–0.5	–0.5
20	s	1.	1.	1.	1.
21	t	0.0001	0.0001	8.	8.

22	Y1	0.00019	0.00019	5.59181	3.8
23	z	0.00019	–0.0095205	3.71359	2.296E-05
24	z1	0	0	0.6073765	0.6069772
25	z2	0	0	1.707014	1.707014
26	z3	0	0	3.681759	1.772413

Differential equations

1 d(z1)/d(t) = (exp(t/(1 + 0.067*t)))*((exp(1*m)))*m1*z^n
2 d(z2)/d(t) = (exp(t/(1 + 0.067*t)))*(A2*(z^s))*((exp(1*m)))*m1*z^n
3 d(z3)/d(t) = k*(exp(t/(1 + 0.067*t)))*A3*(f37*(0.1*t^p/(1 + 0.1*t^p)))*((exp(1*m2)))*m21*z^n

Explicit equations

1 m = (0.0405*t – 0.01126*t^2 + 0.001462*t^3 – 0.00006868*t^4)
2 m1 = (0.0405 – 0.02252*t^1 + 0.004386*t^2 – 0.0002747*t^3)
3 m2 = 0.0868*t – 0.0386*t^2 + 0.0080*t^3 – 0.000756*t^4 + 2.6*(10^ – 5)*t^5
4 fi = 0.1
5 k = 0.1
6 n = 1.0
7 s = 1.0
8 p = –0.5
9 E = (0.625 – 0.375*(tanh(5*(t – 3))))*(1 – fi) + (1/k)*(0.625 – 0.375*(tanh(5*(t – 5))))*(fi)
10 Y1 = t*E
11 A2 = 1
12 A3 = 1
13 f3 = if (t < 0) then (0) else (1)
14 f31 = (1/16)*t*(8 – t)
15 f32 = (exp(–0.1*t))
16 f33 = (sin(3.14*t/4))^2
17 f34 = if (t < 7.99) then (0) else (1)
18 f35 = (1/8)*t^1
19 f36 = if (t < 4) then (1) else (0)
20 f37 = if (t < 4) then (0) else (1)
21 z = t*E–(z1 + z2)*((exp(–1*m)))–z3*(exp(–1*m2))
22 m21 = 0.0868 – 0.0772*t + 0.024*t^2 – 0.003*t^3 + 1.3*(10 – 4)* t^4

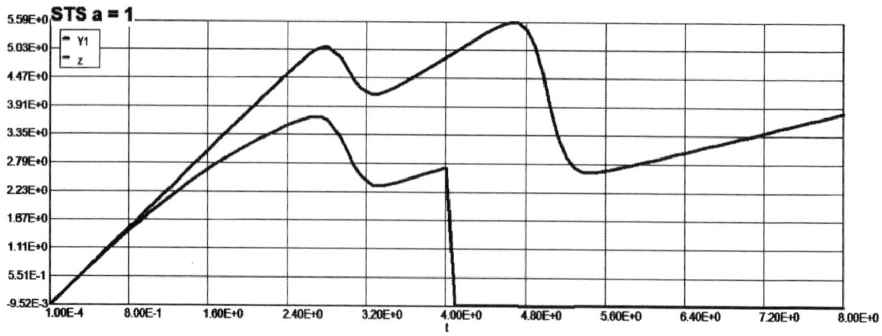

Figure 7.7: Stress-temperature-strain diagram (STS) $\alpha = 1$.

Model: z = a1*t + a2*t^2 + a3*t^3 + a4*t^4 + a5*t^5

Variable	Value
a1	1.074378
a2	1.532532
a3	–0.8916404
a4	0.1472547
a5	–0.0077429

$$\sigma = 1.074\theta + 1.532\theta^2 - 0.892\theta^3 + 0.147\theta^4 - 0.00774\theta^5 \quad (7.13)$$

Example 7.8 $\alpha = 10$

Calculated values of DEQ variables

	Variable	Initial value	Minimal value	Maximal value	Final value
1	A2	1.	1.	1.	1.
2	A3	1.	1.	1.	1.
3	E	1.9	0.475	1.9	0.475
4	f3	1.	1.	1.	1.
5	f31	5.0E-05	0	1.	0
6	f32	0.99999	0.449329	0.99999	0.449329
7	f33	6.162E-09	6.162E-09	0.9999977	1.015E-05
8	f34	0	0	1.	1.
9	f35	1.25E-05	1.25E-05	1.	1.
10	f36	1.	0	1.	0

11	f37	0	0	1.	1.
12	fi	0.1	0.1	0.1	0.1
13	k	0.1	0.1	0.1	0.1
14	m	4.05E-06	4.05E-06	0.0705907	0.0705907
15	m1	0.0404977	0.0003976	0.0404977	0.0003976
16	m2	8.68E-06	8.68E-06	0.0825436	0.075392
17	m21	0.0867923	0.080245	3.195E+04	3.195E+04
18	n	1.	1.	1.	1.
19	p	–0.5	–0.5	–0.5	–0.5
20	s	1.	1.	1.	1.
21	t	0.0001	0.0001	8.	8.
22	Y1	0.00019	0.00019	5.591845	3.8
23	z	0.00019	–0.0093329	3.778578	1.374E-05
24	z1	0	0	0.9516906	0.9510541
25	z2	0	0	2.763205	2.763205
26	z3	0	0	7.620206	4.179226

Differential equations

1 d(z1)/d(t) = (exp(t/(1 + 0.067*t)))*((exp(10*m)))*m1*z^n

2 d(z2)/d(t) = (exp(t/(1 + 0.067*t)))*(A2*(z^s))*((exp(10*m)))*m1*z^n

3 d(z3)/d(t) = k*(exp(t/(1 + 0.067*t)))*A3*(f37*(0.1*t^p/(1 + 0.1*t^p)))*((exp(10*m2)))*m21*z^n

Explicit equations

1 m = (0.0405*t – 0.01126*t^2 + 0.001462*t^3 – 0.00006868*t^4)

2 m1 = (0.0405 – 0.02252*t^1 + 0.004386*t^2 – 0.0002747*t^3)

3 m2 = 0.0868*t – 0.0386*t^2 + 0.0080*t^3 – 0.000756*t^4 + 2.6*(10^ – 5)*t^5

4 fi = 0.1

5 k = 0.1

6 n = 1.0

7 s = 1.0

8 p = –0.5

9 E = (0.625 – 0.375*(tanh(5*(t – 3))))*(1 – fi) + (1/k)*

(0.625 – 0.375*(tanh(5*(t –5))))*(fi)
10 Y1 = t*E
11 A2 = 1
12 A3 = 1
13 f3 = if (t < 0) then (0) else (1)
14 f31 = (1/16)*t*(8 – t)
15 f32 = (exp(–0.1*t))
16 f33 = (sin(3.14*t/4))^2
17 f34 = if (t < 7.99) then (0) else (1)
18 f35 = (1/8)*t^1
19 f36 = if (t < 4) then (1) else (0)
20 f37 = if (t < 4) then (0) else (1)
21 z = t*E–(z1 + z2)*((exp(–10*m)))–z3*(exp(–10*m2))
22 m21 = 0.0868 – 0.0772*t + 0.024*t^2 – 0.003*t^3 + 1.3*(10 – 4)* t^4

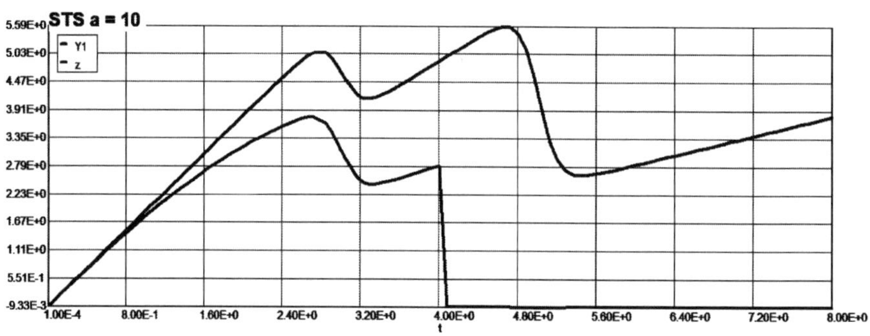

Figure 7.8: Stress-temperature-strain diagram (STS) α = 10.

Model: z = a1*t + a2*t^2 + a3*t^3 + a4*t^4 + a5*t^5

Variable	Value
a1	1.003661
a2	1.641596
a3	–0.9323027
a4	0.1529798
a5	–0.0080192

$$\sigma = 1.00\theta + 1.642\theta^2 - 0.932\theta^3 + 0.153\theta^4 - 0.008\theta^5 \quad (7.14)$$

Example 7.9 $\alpha = 100$

Calculated values of DEQ variables

	Variable	Initial value	Minimal value	Maximal value	Final value
1	A2	1.	1.	1.	1.
2	A3	1.	1.	1.	1.
3	E	1.9	0.475	1.9	0.475
4	f3	1.	1.	1.	1.
5	f31	5.0E-05	0	1.	0
6	f32	0.99999	0.449329	0.99999	0.449329
7	f33	6.162E-09	6.162E-09	0.9999969	1.015E-05
8	f34	0	0	1.	1.
9	f35	1.25E-05	1.25E-05	1.	1.
10	f36	1.	0	1.	0
11	f37	0	0	1.	1.
12	fi	0.1	0.1	0.1	0.1
13	k	0.1	0.1	0.1	0.1
14	m	4.05E-06	4.05E-06	0.0705907	0.0705907
15	m1	0.0404977	0.0003976	0.0404977	0.0003976
16	m2	8.68E-06	8.68E-06	0.0825437	0.075392
17	m21	0.0867923	0.0804735	3.195E+04	3.195E+04
18	n	1.	1.	1.	1.
19	p	–0.5	–0.5	–0.5	–0.5
20	s	1.	1.	1.	1.
21	t	0.0001	0.0001	8.	8.
22	Y1	0.00019	0.00019	5.591571	3.8
23	z	0.00019	–0.007445	4.172044	–0.0001558
24	z1	0	0	107.0953	107.0294
25	z2	0	0	368.3238	368.3238
26	z3	0	0	1.027E+04	6377.224

Differential equations

1 d(z1)/d(t) = (exp(t/(1 + 0.067*t)))*((exp(100*m)))*m1*z^n

2 d(z2)/d(t) = (exp(t/(1 + 0.067*t)))*(A2*(z^s))*((exp(100*m)))*m1*z^n

3 d(z3)/d(t) = k*(exp(t/(1 + 0.067*t)))*A3*(f37*(0.1*t^p/(1 + 0.1*t^p)))*((exp(100*m2)))*m21*z^n

Explicit equations

1 m = (0.0405*t – 0.01126*t^2 + 0.001462*t^3 – 0.00006868*t^4)

2 m1 = (0.0405 – 0.02252*t^1 + 0.004386*t^2 – 0.0002747*t^3)

3 m2 = 0.0868*t – 0.0386*t^2 + 0.0080*t^3 – 0.000756*t^4 + 2.6*(10^ – 5)*t^5

4 fi = 0.1

5 k = 0.1

6 n = 1.0

7 s = 1.0

8 p = –0.5

9 E = (0.625 – 0.375*(tanh(5*(t – 3))))*(1 – fi) + (1/k)*(0.625 – 0.375*(tanh(5*(t – 5))))*(fi)

10 Y1 = t*E

11 A2 = 1

12 A3 = 1

13 f3 = if (t < 0) then (0) else (1)

14 f31 = (1/16)*t*(8 – t)

15 f32 = (exp(–0.1*t))

16 f33 = (sin(3.14*t/4))^2

17 f34 = if (t < 7.99) then (0) else (1)

18 f35 = (1/8)*t^1

19 f36 = if (t < 4) then (1) else (0)

20 f37 = if (t < 4) then (0) else (1)

21 z = t*E–(z1 + z2)*((exp(–100*m)))–z3*(exp(–100*m2))

22 m21 = 0.0868 – 0.0772*t + 0.024*t^2 – 0.003*t^3 + 1.3*(10 – 4)* t^4

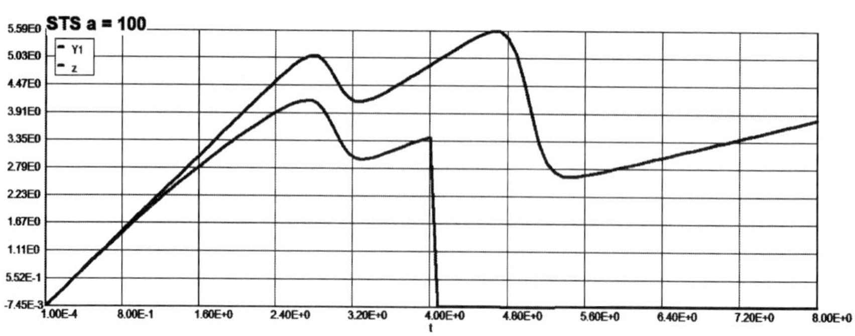

Figure 7.9: Stress-temperature-strain diagram (STS) $\alpha = 100$.

Model: z = a1*t + a2*t^2 + a3*t^3 + a4*t^4 + a5*t^5

Variable	Value
a1	0.4808137
a2	2.398604
a3	–1.210207
a4	0.1920256
a5	–0.0099137

$$\sigma = 0.481\theta + 2.400\theta^2 - 1.210\theta^3 + 0.192\theta^4 - 0.010\theta^5 \qquad (7.15)$$

Example 7.10 $\alpha = 1000$

Calculated values of DEQ variables

	Variable	Initial value	Minimal value	Maximal value	Final value
1	A2	1.	1.	1.	1.
2	A3	1.	1.	1.	1.
3	E	1.9	1.224997	1.9	1.224997
4	f3	1.	1.	1.	1.
5	f31	5.0E-05	5.0E-05	1.	1.
6	f32	0.99999	0.67032	0.99999	0.67032
7	f33	6.162E-09	6.162E-09	0.999771	2.537E-06
8	f34	0	0	0	0
9	f35	1.25E-05	1.25E-05	0.5	0.5
10	f36	1.	1.	1.	1.
11	f37	0	0	0	0
12	fi	0.1	0.1	0.1	0.1
13	k	0.1	0.1	0.1	0.1
14	m	4.05E-06	4.05E-06	0.0578259	0.0578259
15	m1	0.0404977	0.0030152	0.0404977	0.0030152
16	m2	8.68E-06	8.68E-06	0.074688	0.074688
17	m21	0.0867923	0.0794636	1996.77	1996.77
18	n	1.	1.	1.	1.
19	p	–0.5	–0.5	–0.5	–0.5
20	s	1.	1.	1.	1.
21	t	0.0001	0.0001	4.	4.

22	Y1	0.00019	0.00019	5.091451	4.899986
23	z	0.00019	0.00019	4.841059	4.459477
24	z1	0	0	1.086E+24	1.086E+24
25	z2	0	0	4.634E+24	4.634E+24
26	z3	0	0	0	0

Differential equations

1 d(z1)/d(t) = (exp(t/(1 + 0.067*t)))*((exp(1000*m)))*m1*z^n

2 d(z2)/d(t) = (exp(t/(1 + 0.067*t)))*(A2*(z^s))*((exp(1000*m)))*m1*z^n

3 d(z3)/d(t) = k*(exp(t/(1 + 0.067*t)))*A3*(f37*(0.1*t^p/(1 + 0.1*t^p)))*((exp(1000*m2)))*m21*z^n

Explicit equations

1 m = (0.0405*t – 0.01126*t^2 + 0.001462*t^3 – 0.00006868*t^4)

2 m1 = (0.0405 – 0.02252*t^1 + 0.004386*t^2 – 0.0002747*t^3)

3 m2 = 0.0868*t – 0.0386*t^2 + 0.0080*t^3 – 0.000756*t^4 + 2.6*(10^ – 5)*t^5

4 fi = 0.1

5 k = 0.1

6 n = 1.0

7 s = 1.0

8 p = –0.5

9 E = (0.625 – 0.375*(tanh(5*(t – 3))))*(1 – fi) + (1/k)*(0.625 – 0.375*(tanh(5*(t – 5))))*(fi)

10 Y1 = t*E

11 A2 = 1

12 A3 = 1

13 f3 = if (t < 0) then (0) else (1)

14 f31 = (1/16)*t*(8 – t)

15 f32 = (exp(–0.1*t))

16 f33 = (sin(3.14*t/4))^2

17 f34 = if (t < 7.99) then (0) else (1)

18 f35 = (1/8)*t^1

19 f36 = if (t < 4) then (1) else (0)

20 f37 = if (t < 4) then (0) else (1)

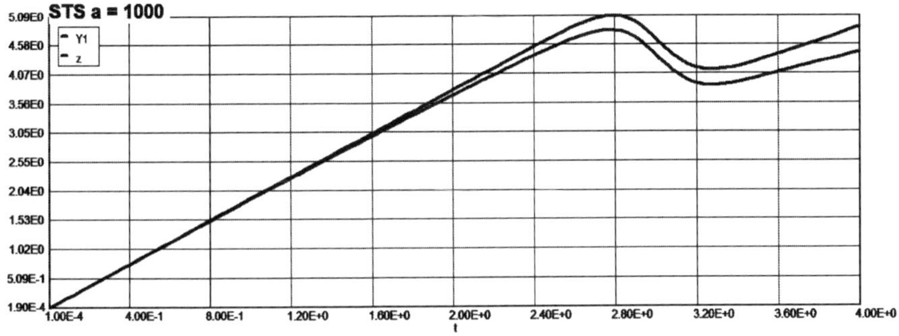

Figure 7.10: Stress-temperature-strain diagram (STS) $\alpha = 1000$.

21 z = t*E–(z1 + z2)*((exp(–1000*m)))–z3*(exp(–1000*m2))

22 m21 = 0.0868 – 0.0772*t + 0.024*t^2 – 0.003*t^3 + 1.3*(10 – 4)* t^4

Model: z = a1*t + a2*t^2 + a3*t^3 + a4*t^4 + a5*t^5

Variable	Value
a1	2.910287
a2	–2.987717
a3	2.711108
a4	–0.9450679
a5	0.1066385

$$\sigma = 2.91\theta - 2.99\theta^2 + 2.71\theta^3 - 0.945\theta^4 + 0.107\theta^5 \quad (7.16)$$

The results are summarized in Table 7.2.

Table 7.2: Allowable creep stress vs. parameter α.

α	$\alpha = 0.001$	$\alpha = 0.01$	$\alpha = 0.1$	$\alpha = 1$	$\alpha = 10$	$\alpha = 100$	$\alpha = 1000$	Aver.
σ_{max}	3.70	3.71	3.71	3.71	3.78	4.17	4.84	3.95

7.4 Creep deformation process of nanocomposites as an ergodic random process

One can see now that the realizations of random creep process present the ergodic process, therefore use $\alpha = 50$ for further analysis (as it has been done before).

Example 7.11 f37 = if (t < 4) then (0) else (1) $\alpha = 50$

Calculated values of DEQ variables

	Variable	Initial value	Minimal value	Maximal value	Final value
1	A2	1.	1.	1.	1.
2	A3	1.	1.	1.	1.
3	E	1.9	0.475	1.9	0.475
4	f3	1.	1.	1.	1.
5	f31	5.0E-05	0	1.	0
6	f32	0.99999	0.449329	0.99999	0.449329
7	f33	6.162E-09	6.162E-09	0.9999963	1.015E-05
8	f34	0	0	1.	1.
9	f35	1.25E-05	1.25E-05	1.	1.
10	f36	1.	0	1.	0
11	f37	0	0	1.	1.
12	fi	0.1	0.1	0.1	0.1
13	k	0.1	0.1	0.1	0.1
14	m	4.05E-06	4.05E-06	0.0705907	0.0705907
15	m1	0.0404977	0.0003976	0.0404977	0.0003976
16	m2	8.68E-06	8.68E-06	0.0825436	0.075392
17	m21	0.0867923	0.082811	3.195E+04	3.195E+04
18	n	1.	1.	1.	1.
19	p	–0.5	–0.5	–0.5	–0.5
20	s	1.	1.	1.	1.
21	t	0.0001	0.0001	8.	8.
22	Y1	0.00019	0.00019	5.591774	3.8
23	z	0.00019	–0.0085222	3.998674	–5.18E-05
24	z1	0	0	7.463017	7.458004
25	z2	0	0	23.95511	23.95511
26	z3	0	0	185.3801	124.8441

Differential equations

1 d(z1)/d(t) = (exp(t/(1 + 0.067*t)))*((exp(50*m)))*m1*z^n

2 d(z2)/d(t) = (exp(t/(1 + 0.067*t)))*(A2*(z^s))*((exp(50*m)))*m1*z^n

3 d(z3)/d(t) = k*(exp(t/(1 + 0.067*t)))*A3*(f37*(0.1*t^p/(1 + 0.1*t^p)))*((exp(50*m2)))*m21*z^n

Explicit equations

1 m = (0.0405*t – 0.01126*t^2 + 0.001462*t^3 – 0.00006868*t^4)
2 ml = (0.0405 – 0.02252*t^1 + 0.004386*t^2 – 0.0002747*t^3)
3 m2 = 0.0868*t – 0.0386*t^2 + 0.0080*t^3 – 0.000756*t^4 + 2.6*(10^ – 5)*t^5
4 fi = 0.1
5 k = 0.1
6 n = 1.0
7 s = 1.0
8 p = –0.5
9 E = (0.625 – 0.375*(tanh(5*(t – 3))))*(1 – fi) + (1/k)*(0.625 – 0.375*(tanh(5*(t – 5))))*(fi)
10 Y1 = t*E
11 A2 = 1
12 A3 – 1
13 f3 = if (t < 0) then (0) else (1)
14 f31 = (1/16)*t*(8 – t)
15 f32 = (exp(–0.1*t))
16 f33 = (sin(3.14*t/4))^2
17 f34 = if (t < 7.99) then (0) else (1)
18 f35 = (1/8)*t^1
19 f36 = if (t < 4) then (1) else (0)
20 f37 = if (t < 4) then (0) else (1)
21 z = t*E–(z1 + z2)*((exp(–50*m)))–z3*(exp(–50*m2))
22 m21 = 0.0868 – 0.0772*t + 0.024*t^2 – 0.003*t^3 + 1.3*(10 – 4)* t^4

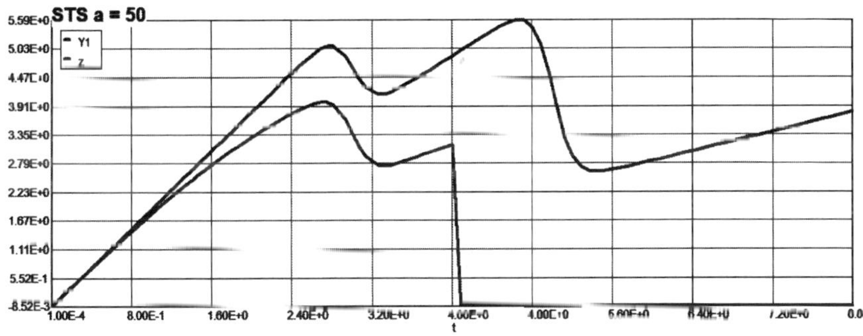

Figure 7.11: Stress-temperature-strain diagram (STS) α = 50.

Model: z = a1*t + a2*t^2 + a3*t^3 + a4*t^4 + a5*t^5

Variable	Value
a1	0.696548
a2	2.083282
a3	−1.095301
a4	0.1760928
a5	−0.0091536

$$\sigma = 0.697\theta + 2.083\theta^2 - 1.095\theta^3 + 0.176\theta^4 - 0.0092\theta^5 \tag{7.17}$$

7.5 Creep of the nanocomposite in the framework of the correlation theory of probability

7.5.1 *Mean value of allowable creep stress and strain*

The mean value can be calculated now as follows [24]:

$$\mu_\sigma = \frac{1}{\theta_{max}} \int_0^\theta \sigma(\tau)d\tau = \frac{1}{8}\int_0^8 \mathbf{(0.697\theta + 2.083\theta^2 - 1.095\theta^3 + 0.176\theta^4 - 0.0092\theta^5)d\theta}$$

$\mu_x = 8.0/8 = 1.0;$

7.5.2 *Standard deviation and autocorrelation function of allowable creep stress and strain*

In order to calculate the standard deviation in this case the chosen function (7.17) has to be centered. Again, after using the POLYMATH software we have:

$$\sigma_\sigma^2 = \frac{1}{\theta_{max}} \int_0^{\theta_{max}} \sigma(\tau)d\tau = \frac{1}{8}\int_0^8 \mathbf{(0.697\theta + 2.083\theta^2 - 1.095\theta^3 + 0.176\theta^4 - 0.0092\theta^5 - 1)^2 d\theta}$$

$\sigma = [23.64 /8]^{1/2} = 1.72$

In order to calculate the correlation function the chosen function (7.17) has to be centered. The correlation function of a steady time process can be computed now as follows [24]:

$$K_\sigma(\theta) = \frac{1}{\sigma - \tau} \int_0^{\theta-\tau} [\hat{\sigma}(\tau)\hat{\sigma}(\theta + \tau)]dt \tag{7.18}$$

$$\text{where} : \tau = 0;1;2;\ldots\ldots\ldots\ldots\ldots\ldots, 8$$

$$\hat{\sigma}(\tau) = 0.697\theta + 2.083\theta^2 - 1.095\theta^3 + 0.176\theta^4 - 0.0092\theta^5 - 1.0 \tag{7.19}$$

The correlation function of a steady time process can be computed now as follows:

After substituting (7.19) into (7.18) we have:

$$K_\sigma(\theta) = \frac{1}{\theta-\tau}\int_0^{\theta-\tau}[\hat{\sigma}(\theta)\hat{\theta}(\theta+\tau)]d\tau =$$

$$= \frac{1}{\theta-\tau}\int_0^{\theta-\tau}[(\mathbf{0.697\tau + 2.083\tau^2 - 1.095\tau^3 + 0.176\tau^4 - 0.0092\tau^5 - 1)(0.697(\tau+x) +}$$
$$\mathbf{+2.083(\tau+x)^2 - 1.095(\tau+x)^3 + 0.176(\tau+x)^4 - 0.0092(\tau+x)^5 - 1)]dx} \qquad (7.20)$$

Now calculate $K_\sigma(\tau)$ for each $x = 0;1;2;\ldots\ldots\ldots\ldots\ldots\ldots,8$. The results are presented in Table 7.3 below:

Table 7.3: Autocorrelation function.

τ	0	1	2	3	4	5	6	7	8
$K_\sigma(\theta)$	2.955	2.176	0.493	–1.428	–0.2675	–2.81	–1.86	0.173	0.0

The normalized correlation function is:

Table 7.4: Normalized autocorrelation function.

Temp. θ	0	1	2	3	4	5	6	7	8
$\rho_\sigma(\tau)$	1	0.736	0.167	–0.483	–0.09	–0.951	–0.629	0.058	0

Let's approximate the data from Table 7.4 by the formulae:

Model: y = (exp(–A*t))*(cos(B*t))

Variable	Initial guess	Value
A	0.1	0.0878
B	2.	0.624

$$\rho_\sigma(\theta) = (\exp(-0.0878*\theta))*(\cos(0.624*\theta) \qquad (7.21)$$

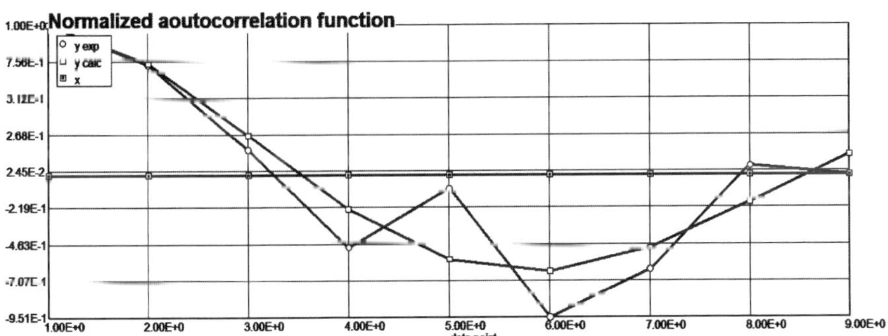

Figure 7.12: Normalized correlation function $\rho_\theta(\tau)$.

The ergodic function method is much simpler then the "classical" method of obtaining the correlation functions and it will be used in analysis of other fire creep deformation scenarios.
The corresponding spectral function is:

$$S(\omega) = \frac{1}{2}\frac{\sigma_\sigma^2}{\pi}\left[\frac{a}{(\omega - b)^2 + a^2} + \frac{a}{(\omega + b)^2 + a^2}\right] = \\ = 0.214\left[\frac{1}{(\omega - 0.624)^2 + 0.0878^2} + \frac{1}{(\omega + 0.624)^2 + 0.0878^2}\right] \quad (7.22)$$

The graphic presentation of spectral density is (see Fig. 7.13)

7.6 The first-occurrence time problem and the probability density P (a, t)

The average first-occurrence duration of time "t" above a given level "a" for stationary processes is defined as follows [25]:

$$\bar{t}_a = T\int_a^\infty f(x)dx \quad (7.23)$$

where: f(x)—probability density of the maximum temperature ordinates, which are not dependent on time for a stationary processes.

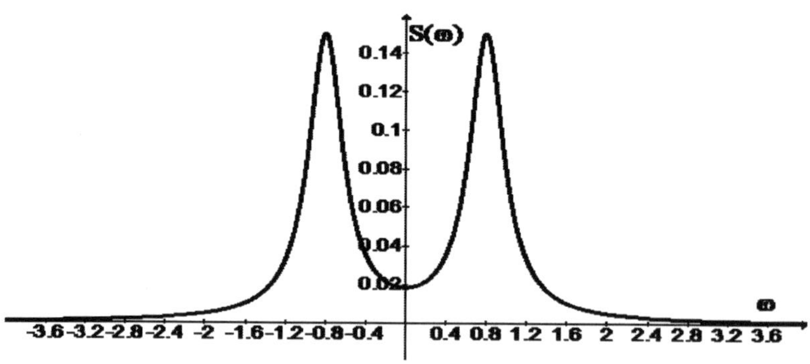

Figure 7.13: Spectral density function.

The average number of the occurrences above a given level "a" for stationary processes during the same period of time "T" is defined as follows [25]:

$$\bar{n}_a = T\int_0^\infty vf(a, v)dv \quad (7.24)$$

The average duration of time for all occurrences above a given level "a" for stationary processes during the same period of time "T" is defined as follows [25]:

$$\overline{\tau} = \frac{\int_a^{\infty} f(x)dx}{\int_0^{\infty} vf(a,v)dv} \quad (7.25)$$

The average area between the stationary random curve and the horizontal line y = a for the average first-occurrence duration of time "t" above a given level "a" is given as [25]:

$$\overline{s\mathbf{v}}_a = \overline{x} - \int_{-\infty}^{a} xf(x)dx - a\overline{\tau \mathbf{v}}_a$$

$$\text{Where:} \quad (7.26)$$

$$\overline{\mathbf{v}}_a = \frac{\overline{n}_a}{T}$$

Formulas (7.24); (7.25) and (7.26) are reduced as follows in our case of normal stationary process:

$$\overline{t}_a = T\int_a^{\infty} f(x)dx = \frac{T}{\sigma_\theta\sqrt{2\pi}}\int_a^{\infty}\exp(-\frac{(x-\overline{x})^2}{2\sigma_\theta^2})dx \quad (7.27)$$

$$\overline{n}_a = T\int_0^{\infty} vf(a,v)dv = \frac{T\sigma_v}{2\pi\sigma_\theta}\exp(-\frac{(a-\overline{x})^2}{2\sigma_\theta^2})$$

$$\text{Where:} \quad (7.28)$$

$$\sigma_v^2 = -\frac{d^2}{d\tau^2}K_\theta(\tau)|_{\tau=0}$$

$$\overline{\tau} = \frac{\int_a^{\infty} f(x)dx}{\int_0^{\infty} vf(a,v)dv} = \pi\frac{\sigma_\theta}{\sigma_v}\exp(+\frac{(a-\overline{x})^2}{2\sigma_\theta^2})[1-\Phi^*(\frac{a-\overline{x}}{\sigma_\theta})] \quad (7.29)$$

$$\overline{s} = \frac{\sigma_\theta^2\sqrt{2\pi}}{\sigma_v} + \frac{\pi(\overline{x}-a)\sigma_\theta}{\sigma_v}[1-\Phi^*(\frac{a-\overline{x}}{\sigma_0})][\exp(\frac{(a-\overline{x})^2}{2\sigma_\theta^2}] \quad (7.30)$$

For correlation function given by (8.47):

$$\sigma_{v\sigma}^2 = K_{\dot{\sigma}}(0) = -\frac{d^2}{d\theta^2}K_{\dot{\sigma}}(\theta)|_{\theta=0} = \sigma_{\sigma}^2(b^2-a^2)$$

$$\overline{n}_a = \theta_{max}\frac{\sqrt{b^2-a^2}}{2\pi}\exp(-\frac{a_0^2}{2\sigma_\sigma^2}) = 8\frac{\sqrt{0.624^2-0.0878^2}}{2\pi}\exp(-\frac{1.0^2}{2(1.72)^2}) = 0.0526$$

$$\overline{\mathbf{v}}_a = \frac{\overline{n}_a}{\theta_{max}} = \frac{0.0526}{8} = 0.00657$$

Now based on Poisson formulae of the probability to not having the minimum allowable stress ordinates crossing downwards the level "a = 4.0" is: $P_{rel} = P_o = \exp(-\bar{n}_a) = \exp(-0.0526) = 0.95$. This probability might characterize the reliability of the structure (one element or the whole structure).

7.7 Allowable creep stress vs. volumetric content of nanoparticles

In many nanocomposite structures the volumetric parameter fi is small (fi << 1), then the effective modulus of elasticity E in Equation 7.1 can be approximately presented as the matrix modulus of elasticity, i.e., $E_m = 0.625 - 0.375*(\tanh(5*(t-3)))$. Consider for example fi = 0.01, then the solution of Equation (7.1) is as follows:

Example 7.12 fi = 0.01

Calculated values of DEQ variables

	Variable	Initial value	Minimal value	Maximal value	Final value
1	A2	1.	1.	1.	1.
2	A3	1.	1.	1.	1.
3	E	1.09	0.2725	1.09	0.2725
4	f	0.9090909	0.034148	0.9090909	0.034148
5	f3	1.	1.	1.	1.
6	f31	5.0E-05	0	0.9999998	0
7	f32	0.99999	0.449329	0.99999	0.449329
8	f33	6.162E-09	9.062E-10	0.9999951	1.015E-05
9	f34	0	0	1.	1.
10	f35	1.563E-10	1.563E-10	1.	1.
11	f36	1.	0	1.	0
12	f37	0	0	1.	1.
13	fi	0.01	0.01	0.01	0.01
14	k	0.1	0.1	0.1	0.1
15	m	4.05E-06	4.05E-06	0.0705907	0.0705907
16	m1	0.0404977	0.0003976	0.0404977	0.0003976
17	m2	8.68E-06	8.68E-06	0.0825437	0.075392
18	m21	0.0867923	0.0802282	3.195E+04	3.195E+04
19	n	1.	1.	1.	1.

20	p	–0.5	–0.5	–0.5	–0.5
21	s	1.	1.	1.	1.
22	t	0.0001	0.0001	8.	8.
23	Y1	0.000109	0.000109	2.848447	2.18
24	z	0.000109	–0.1076569	1.25464	–4.327E-05
25	z1	0	0	0.7065149	0.6795097
26	z2	0	0	0.6644044	0.6644044
27	z3	0	0	112.6157	92.82404

Differential equations

1 d(z1)/d(t) = (exp(t/(1 + 0.067*t)))*((exp(50*m)))*m1*z^n

2 d(z2)/d(t) = (exp(t/(1 + 0.067*t)))*(A2*(z^s))*((exp(50*m)))*m1*z^n

3 d(z3)/d(t) = k*(exp(t/(1 + 0.067*t)))*A3*f3*f*((exp(50*m2)))*m21*z^n

Explicit equations

1 m = (0.0405*t – 0.01126*t^2 + 0.001462*t^3 – 0.00006868*t^4)

2 m1 = (0.0405 – 0.02252*t^1 + 0.004386*t^2 – 0.0002747*t^3)

3 m2 = 0.0868*t – 0.0386*t^2 + 0.0080*t^3 – 0.000756*t^4 + 2.6*(10^ – 5)*t^5

4 fi = 0.01

5 k = 0.1

6 n = 1.0

7 s = 1.0

8 p = –0.5

9 E = (0.625 – 0.375*(tanh(5*(t – 3))))*(1 – fi)+(1/k)*(0.625 – 0.375*(tanh(5*(t – 5))))*(fi)

10 Y1 = t*E

11 A2 = 1

12 A3 = 1

13 f = (0.1*t^p/(1 + 0.1*t^p))

14 f3 = if (t < 0) then (0) else (1)

15 f31 = (1/16)*t*(8 – t)

16 f32 = (exp(–0.1*t))

17 f33 = (sin(3.14*t/4))^2

18 f34 = if (t < 7.99) then (0) else (1)

19 f35 = (1/64)*t^2

20 f36 = if (t < 4) then (1) else (0)
21 f37 = if (t < 4) then (0) else (1)
22 z = t*E–(z1 + z2)*((exp(–50*m)))–z3*(exp(–50*m2))
23 m21 = 0.0868 – 0.0772*t + 0.024*t^2 – 0.003*t^3 + 1.3*(10 – 4)* t^4

Example 7.13 fi → 0

Calculated values of DEQ variables

	Variable	Initial value	Minimal value	Maximal value	Final value
1	A2	1.	1.	1.	1.
2	A3	1.	1.	1.	1.
3	E	1.	0.25	1.	0.25
4	f	0.9090909	0.034148	0.9090909	0.034148
5	f3	1.	1.	1.	1.
6	f31	5.0E-05	0	0.9999999	0
7	f32	0.99999	0.449329	0.99999	0.449329
8	f33	6.162E-09	6.162E-09	0.9999962	1.015E-05
9	f34	0	0	1.	1.
10	f35	1.563E-10	1.563E-10	1.	1.
11	f36	1.	0	1.	0
12	f37	0	0	1.	1.
13	fi	0	0	0	0
14	k	0.1	0.1	0.1	0.1
15	m	4.05E-06	4.05E-06	0.0705907	0.0705907
16	m1	0.0404977	0.0003976	0.0404977	0.0003976
17	m2	8.68E-06	8.68E-06	0.0825436	0.075392
18	m21	0.0867923	0.080441	3.195E+04	3.195E+04
19	n	1.	1.	1.	1.
20	p	–0.5	–0.5	–0.5	–0.5
21	s	1.	1.	1.	1.
22	t	0.0001	0.0001	8.	8.
23	Y1	1.0E-04	1.0E-04	2.60162	2.
24	z	1.0E-04	–0.1131457	1.154044	–3.972E-05
25	z1	0	0	0.6489842	0.6198773
26	z2	0	0	0.5619788	0.5619788
27	z3	0	0	103.4172	85.22463

Differential equations

1 d(z1)/d(t) = (exp(t/(1 + 0.067*t)))*((exp(50*m)))*m1*z^n
2 d(z2)/d(t) = (exp(t/(1 + 0.067*t)))*(A2*(z^s))*((exp(50*m)))*m1*z^n
3 d(z3)/d(t) = k*(exp(t/(1 + 0.067*t)))*A3*f3*f*((exp(50*m2)))*m21*z^n

Explicit equations

1 m = (0.0405*t – 0.01126*t^2 + 0.001462*t^3 – 0.00006868*t^4)
2 m1 = (0.0405 – 0.02252*t^1 + 0.004386*t^2 – 0.0002747*t^3)
3 m2 = 0.0868*t – 0.0386*t^2 + 0.0080*t^3 – 0.000756*t^4 + 2.6*(10^ – 5)*t^5
4 fi = 0.0
5 k = 0.1
6 n = 1.0
7 s = 1.0
8 p = –0.5
9 E = (0.625 – 0.375*(tanh(5*(t – 3))))*(1 – fi) + (1/k)*(0.625 – 0.375*(tanh(5*(t – 5))))*(fi)
10 Y1 = t*E
11 A2 = 1
12 A3 = 1
13 f = (0.1*t^p/(1 + 0.1*t^p))
14 f3 = if (t < 0) then (0) else (1)
15 f31 = (1/16)*t*(8 – t)
16 f32 = (exp(–0.1*t))
17 f33 = (sin(3.14*t/4))^2
18 f34 = if (t < 7.99) then (0) else (1)
19 f35 = (1/64)*t^2
20 f36 = if (t < 4) then (1) else (0)
21 f37 = if (t < 4) then (0) else (1)
22 z = t*E–(z1 + z2)*((exp(–50*m)))–z3*(exp(–50*m2))
23 m21 = 0.0868 – 0.0772*t + 0.024*t^2 – 0.003*t^3 + 1.3*(10 – 4)* t^4

The maximum allowable creep stress $\sigma_{max} = 1.25 < 4.0$. Therefore by decreasing volumetric concentration of nanoparticles from fi = 0.1 to fi = 0.01 the allowable creep stress decreased also from 4.0 to 1.25 and in the limit, if: fi → 0, then $\sigma_{max} = 1.15$ (see Fig. 7.14).

One can see from Fig. 7.14 that the allowable creep stress increases rapidly when 0.01 < fi < 0.05 and decreases rapidly if 0 < fi < 0.01.

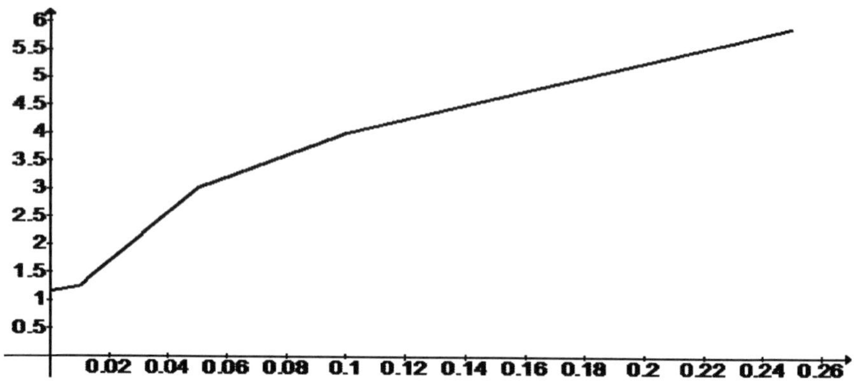

Figure 7.14: Allowable creep stress as a function of volumetric ratio.

It should also be noted that an increase in the volume content of nanoparticles does not lead to an increase in the deformability of the nanocomposite, since the deformability of the system as a whole is dictated by the elasticity of the matrix, rather than nanoparticles (assumed to be uniformly distributed in the homogeneous body of the nanocomposite structural element).

To achieve optimal design of a nanocomposite, it is necessary in most practical applications that the they simultaneously satisfy, two conditions: high strength and sufficiently large elasticity. However, it is known that nanocomposite structural elements in most cases, although they have high strength, but exhibit brittleness at failure (for example, at high temperatures), thereby reducing their applicability in many areas of industry. In order to solve the problems of optimal designing of nanocomposites, the creep model (in the third integral expression of Equation 7.6, which is a creep deformation due to a change in the properties of nanoparticles under the action of high temperatures), includes few additional types of functional dependencies f3. In addition to all above there is a series of "logical" functions f3; f31,..., f37 that characterizes the qualities of nanoparticles participation in the creep deformation process. For example, the function f3 indicates that high temperature affects nanoparticles from the very beginning of creep deformations at the base temperature θ^* (the nanoparticles are growing and that reduces the allowable creep stress in accordance with the HP law). The function f34 has an opposite effect: the nanoparticles are "good" that is they are not affected at all by the applied high temperature (it just increases the effective modulus of elasticity and allowing to have greater elasticity). For example:

Example 7.14

Data: fi = 0.1; k = 0.1 and f3

Calculated values of DEQ variables

	Variable	Initial value	Minimal value	Maximal value	Final value
1	A2	1.	1.	1.	1.
2	A3	0.01	0.01	0.01	0.01
3	E	1.9	0.475	1.9	0.475
4	f	0.9090909	0.034148	0.9090909	0.034148
5	f3	1.	1.	1.	1.
6	f31	5.0E-05	0	0.9999568	0
7	f32	0.99999	0.449329	0.99999	0.449329
8	f33	6.162E-09	6.162E-09	0.9996823	1.015E-05
9	f34	0	0	1.	1.
10	f35	1.563E-10	1.563E-10	1.	1.
11	f36	1.	0	1.	0
12	f37	0	0	1.	1.
13	fi	0.1	0.1	0.1	0.1
14	k	0.1	0.1	0.1	0.1
15	m	4.05E-06	4.05E-06	0.0705907	0.0705907
16	m1	0.0404977	0.0003976	0.0404977	0.0003976
17	m2	8.68E-06	8.68E-06	0.0825458	0.075392
18	m21	0.0867923	0.0817689	3.195E+04	3.195E+04
19	n	1.	1.	1.	1.
20	p	–0.5	–0.5	–0.5	–0.5
21	s	1.	1.	1.	1.
22	t	0.0001	0.0001	8.	8.
23	Y1	0.00019	0.00019	6.931025	3.8
24	z	0.00019	–0.6514727	3.803557	–0.0052029
25	z1	0	0	7.610594	7.180105
26	z2	0	0	23.95266	23.95266
27	z3	0	0	180.5908	125.4239

Differential equations

1 d(z1)/d(t) = (exp(t/(1 + 0.067*t)))*((exp(50*m)))*m1*z^n

2 d(z2)/d(t) = (exp(t/(1 + 0.067*t)))*(A2*(z^s))*((exp(50*m)))*m1*z^n

3 d(z3)/d(t) = k*(exp(t/(1 + 0.067*t)))*A3*f3*f*((exp(50*m2)))*m21*z^n

Explicit equations

1 m = (0.0405*t – 0.01126*t^2 + 0.001462*t^3 – 0.00006868*t^4)

2 m1 = (0.0405 – 0.02252*t^1 + 0.004386*t^2 – 0.0002747*t^3)

3 m2 = 0.0868*t – 0.0386*t^2 + 0.0080*t^3 – 0.000756*t^4 + 2.6*(10^ – 5)*t^5

4 fi = 0.1

5 k = 0.1

6 n = 1.0

7 s = 1.0

8 p = –0.5

9 E = (0.625 – 0.375*(tanh(5*(t – 4))))*(1 – fi) + (1/k)*(0.625 – 0.375*(tanh(5*(t – 5))))*(fi)

10 Y1 = t*E

11 A2 = 1

12 A3 = 0.01

13 f = (0.1*t^p/(1 + 0.1*t^p))

14 f3 = if (t < 0) then (0) else (1)

15 f31 = (1/16)*t*(8 – t)

16 f32 = (exp(–0.1*t))

17 f33 = (sin(3.14*t/4))^2

18 f34 = if (t < 7.99) then (0) else (1)

19 f35 = (1/64)*t^2

20 f36 = if (t < 4) then (1) else (0)

21 f37 = if (t < 4) then (0) else (1)

22 z = t*E–(z1 + z2)*((exp(–50*m)))–z3*(exp(–50*m2))

23 m21 = 0.0868 – 0.0772*t + 0.024*t^2 – 0.003*t^3 + 1.3*(10 – 4)* t^4

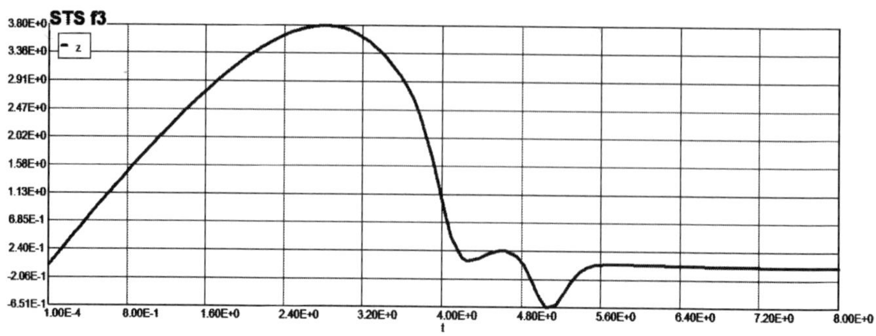

Figure 7.15: Stress-temperature-strain diagram (STS) f3: $\theta_{max} = 2.8$ $\sigma = 3.8$.

Compare now:

Example: fi = 0.1; k = 0.1 and f34

Calculated values of DEQ variables

	Variable	Initial value	Minimal value	Maximal value	Final value
1	A2	1.	1.	1.	1.
2	A3	0.01	0.01	0.01	0.01
3	E	1.9	0.475	1.9	0.475
4	f	0.9090909	0.0341687	0.9090909	0.0341687
5	f3	1.	1.	1.	1.
6	f31	5.0E-05	5.0E-05	0.9999506	0.0049938
7	f32	0.99999	0.4497785	0.99999	0.4497785
8	f33	6.162E-09	6.162E-09	0.9999494	0.0001218
9	f34	0	0	0	0
10	f35	1.563E-10	1.563E-10	0.9975016	0.9975016
11	f36	1.	0	1.	0
12	f37	0	0	1.	1.
13	fi	0.1	0.1	0.1	0.1
14	k	0.1	0.1	0.1	0.1
15	m	4.05E-06	4.05E-06	0.0705866	0.0705866
16	m1	0.0404977	0.0004482	0.0404977	0.0004482
17	m2	8.68E-06	8.68E-06	0.082548	0.0754977
18	m21	0.0867923	0.0828108	3.179E+04	3.179E+04
19	n	1.	1.	1.	1.
20	p	–0.5	–0.5	–0.5	–0.5
21	s	1.	1.	1.	1.
22	t	0.0001	0.0001	8.	8.
23	Y1	0.00019	0.00019	6.930487	3.79525
24	z	0.00019	–0.2264161	4.721589	1.043084
25	z1	0	0	30.45152	30.45152
26	z2	0	0	63.40035	63.40035
27	z3	0	0	0	0

Differential equations

1 d(z1)/d(t) = (exp(t/(1 + 0.067*t)))*((exp(50*m)))*m1*z^n
2 d(z2)/d(t) = (exp(t/(1 + 0.067*t)))*(A2*(z^s))*((exp(50*m)))*m1*z^n
3 d(z3)/d(t) = k*(exp(t/(1 + 0.067*t)))*A3*f34*f*((exp(50*m2)))*m21*z^n

Explicit equations

1 m = (0.0405*t – 0.01126*t^2 + 0.001462*t^3 – 0.00006868*t^4)
2 m1 = (0.0405 – 0.02252*t^1 + 0.004386*t^2 – 0.0002747*t^3)
3 m2 = 0.0868*t – 0.0386*t^2 + 0.0080*t^3 – 0.000756*t^4 + 2.6*(10^ – 5)*t^5
4 fi = 0.1
5 k = 0.1
6 n = 1.0
7 s = 1.0
8 p = –0.5
9 E = (0.625 – 0.375*(tanh(5*(t – 4))))*(1 – fi) + (1/k)*
(0.625 – 0.375*(tanh(5*(t – 5))))*(fi)
10 Y1 = t*E
11 A2 = 1
12 A3 = 0.01
13 f = (0.1*t^p/(1 + 0.1*t^p))
14 f3 = if (t < 0) then (0) else (1)
15 f31 = (1/16)*t*(8 – t)
16 f32 = (exp(–0.1*t))
17 f33 = (sin(3.14*t/4))^2
18 f34 = if (t < 7.99) then (0) else (1)
19 f35 = (1/64)*t^2
20 f36 = if (t < 4) then (1) else (0)

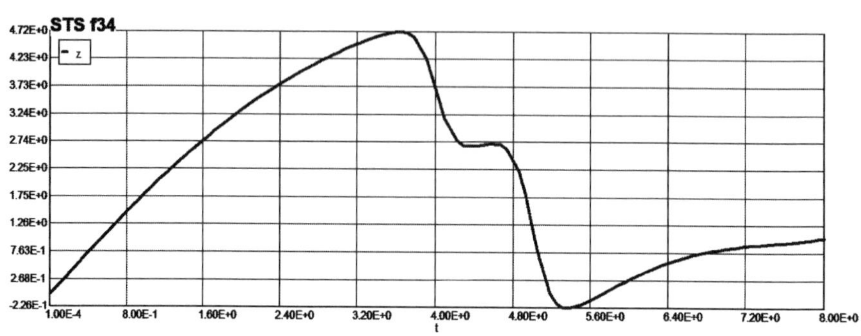

Figure 7.16: Stress-temperature-strain diagram (STS) f34: $\theta_{max} = 3.8$ $\sigma = 4.72$.

21 f37 = if (t < 4) then (0) else (1)

22 z = t*E–(z1 + z2)*((exp(–50*m)))–z3*(exp(–50*m2))

23 m21 = 0.0868 – 0.0772*t + 0.024*t^2 – 0.003*t^3 + 1.3*(10 – 4)* t^4

$\theta_{max} = 3.8$ $\sigma = 4.72$ (approximately ≈ 35% increase in elasticity and ≈ 22% in strength).

Let's say the final choice (as a realization of a random ergodic function selected) is as follows:

Data: f31; fi = 0.1 and k = 0.3; $T^* = 400°$ K and $\beta = 0.067$. $E_0 = 100$ [GPa]; $\alpha = 4(10^{-6})$ [1/°C]

Example 7.15

Calculated values of DEQ variables

	Variable	Initial value	Minimal value	Maximal value	Final value
1	A2	1.	1.	1.	1.
2	A3	1.	1.	1.	1.
3	E	1.233333	0.3083333	1.233333	0.3083333
4	f	0.9090909	0.034148	0.9090909	0.034148
5	f3	1.	1.	1.	1.
6	f31	5.0E-05	0	1.	0
7	f32	0.99999	0.449329	0.99999	0.449329
8	f33	6.162E-09	6.162E-09	0.9999947	1.015E-05
9	f34	0	0	1.	1.
10	f35	1.563E-10	1.563E-10	1.	1.
11	f36	1.	0	1.	0
12	f37	0	0	1.	1.
13	fi	0.1	0.1	0.1	0.1
14	k	0.3	0.3	0.3	0.3
15	m	4.05E-06	4.05E-06	0.0705907	0.0705907
16	m1	0.0404977	0.0003976	0.0404977	0.0003976
17	m2	8.68E-06	8.68E-06	0.0825436	0.075392
18	m21	0.0867923	0.0815046	3.195E+04	3.195E+04
19	n	1.	1.	1.	1.
20	p	–0.5	–0.5	–0.5	–0.5
21	s	1.	1.	1.	1.
22	t	0.0001	0.0001	8.	8.

23	Y1	0.0001233	0.0001233	4.442098	2.466667
24	z	0.0001233	–0.0082849	1.325117	–0.0071119
25	z1	0	0	0.5914846	0.5877002
26	z2	0	0	0.5669104	0.5669104
27	z3	0	0	179.4735	105.8019

Differential equations

1 d(z1)/d(t) = (exp(t/(1 + 0.067*t)))*((exp(50*m)))*m1*z^n

2 d(z2)/d(t) = (exp(t/(1 + 0.067*t)))*(A2*(z^s))*((exp(50*m)))*m1*z^n

3 d(z3)/d(t) = k*(exp(t/(1 + 0.067*t)))*A3*f31*f*((exp(50*m2)))*m21*z^n

Explicit equations

1 m = (0.0405*t – 0.01126*t^2 + 0.001462*t^3 – 0.00006868*t^4)

2 m1 = (0.0405 – 0.02252*t^1 + 0.004386*t^2 – 0.0002747*t^3)

3 m2 = 0.0868*t – 0.0386*t^2 + 0.0080*t^3 – 0.000756*t^4 + 2.6*(10^ – 5)*t^5

4 fi = 0.1

5 k = 0.3

6 n = 1.0

7 s = 1.0

8 p = –0.5

9 E = (0.625 – 0.375*(tanh(5*(t – 4))))*(1 – fi) + (1/k)*(0.625 – 0.375*(tanh(5*(t – 5))))*(fi)

10 Y1 = t*E

11 A2 = 1

12 A3 = 1

13 f = (0.1*t^p/(1 + 0.1*t^p))

14 f3 = if (t < 0) then (0) else (1)

15 f31 = (1/16)*t*(8 – t)

16 f32 = (exp(–0.1*t))

17 f33 = (sin(3.14*t/4))^2

18 f34 = if (t < 7.99) then (0) else (1)

19 f35 = (1/64)*t^2

20 f36 = if (t < 4) then (1) else (0)

21 f37 = if (t < 4) then (0) else (1)

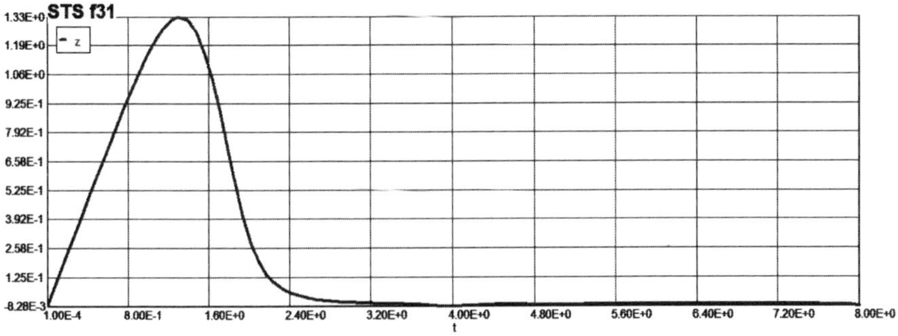

Figure 7.17: Stress-temperature-strain diagram (STS) f31.

22 z = t*E–(z1 + z2)*((exp(–50*m)))–z3*(exp(–50*m2))
23 m21 = 0.0868 – 0.0772*t + 0.024*t^2 – 0.003*t^3 + 1.3*(10 – 4)* t^4

Model: z = a1*t + a2*t^2 + a3*t^3 + a4*t^4 + a5*t^5

Variable	Value
a1	2.605806
a2	–2.10892
a3	0.6099215
a4	–0.0752943
a5	0.0033688

$$\sigma = 2.6\theta - 2.1\theta^2 + 0.61\theta^3 - 0.075\theta^4 + 0.00337\theta^5 \quad (7.31)$$

The mean value can be calculated now is:

$$\mu_\sigma = \frac{1}{\theta_{max}}\int_0^\theta \sigma(\tau)d\tau = \frac{1}{8}\int_0^8 \mathbf{(2.6\theta - 2.1\theta^2 + 0.61\theta^3 - 0.075\theta^4 + 0.00337\theta^5)d\theta} \quad (7.32)$$

$\mu_x = 5.16/8 = 0.645$; $\sigma = [1.58/8]^{1/2} = 0.444$

In order to calculate the correlation function the chosen function (7.31) has to be centered. The correlation function of a steady time process can be computed now as follows:

$$\hat{\sigma}(t) = 2.6\theta - 2.1\theta^2 + 0.61\theta^3 - 0.075\theta^4 + 0.00337\theta^5 - 0.645 \quad (7.33)$$

The correlation function of a steady time process can be computed now as follows:
After substituting (7.33) into (7.18) we have:

$$K_{\sigma}(\theta)=\frac{1}{\theta-\tau}\int_{0}^{\theta-\tau}[\hat{\sigma}(\theta)\hat{\theta}(\theta+\tau)]d\tau=$$

$$=\frac{1}{\theta-\tau}\int_{0}^{\theta-\tau}[(\mathbf{2.6\theta-2.1\theta^2+0.61\theta^3-0.075\theta^4+0.00337\theta^5-0.645})(\mathbf{2.6(\theta+x)}-$$

$$\mathbf{-2.1(\theta+x)^2+0.61(\theta+x)^3-0.075(\theta=x)^4+0.00337(\theta+x)^5-0.645})]d\theta \tag{7.34}$$

Now calculate $K_{\sigma}(\tau)$ for each $x = 0;1;2;......................,8$. The results are presented in Table 7.5 below:

Table 7.5: Autocorrelation function.

τ	0	1	2	3	4	5	6	7	8
$K_{\sigma}(\theta)$	0.2	0.627/7 = 0.0896	–0.166/6 = –0.0277	–0.618/5 = –0.124	–0.598/4 = –0.15	–0.216/3 = –0.072	0.186/2 = 0.093	0.138/1 = 0.138	0.0

The normalized correlation function is:

Table 7.6: Normalized autocorrelation function.

Temp. θ	0	1	2	3	4	5	6	7	8
$\rho_{\sigma}(\tau)$	1	0.448	–0.138	–0.62	–0.75	–0.36	0.465	0.69	0

Let's approximate the data from Table 7.6 by the same type of formulae:

Model: C04 = (exp(–A*t))*(cos(B*t))

Variable	Initial guess	Value
A	0.1	0.079
B	2.	0.909

A = 0.079; B = 0.909

$$\rho_{\sigma}(\theta) = (\exp(-0.079*\theta))*(\cos(0.909*\theta)) \tag{7.35}$$

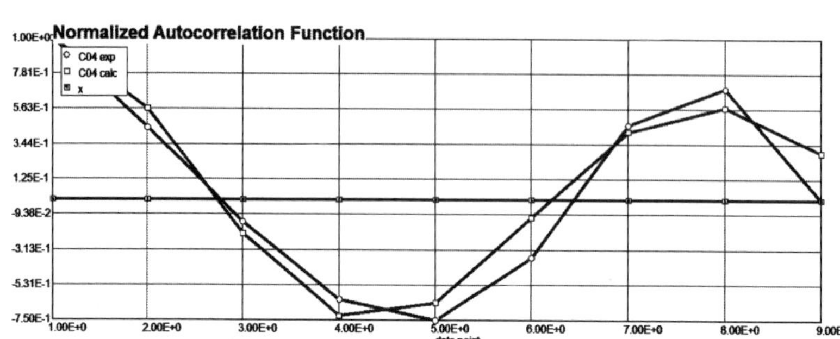

Figure 7.18: Normalized autocorrelation function f31.

For the normalized correlation function given by (7.35):

$$\sigma_{v\sigma}^2 = K_{\dot{\sigma}}(0) = -\frac{d^2}{d\theta^2} K_{\dot{\sigma}}(\theta)|_{\theta=0} = \sigma_{\dot{\sigma}}^2 (b^2 - a^2)$$

$$\bar{n}_a = \theta_{max} \frac{\sqrt{b^2 - a^2}}{2\pi} \exp(-\frac{a_0^2}{2\sigma_\sigma^2}) = 8 \frac{\sqrt{0.909^2 - 0.079^2}}{2\pi} \exp(-\frac{1.32^2}{2(0.444)^2}) = 0.012$$

$$\bar{v}_a = \frac{\bar{n}_a}{\theta_{max}} = \frac{0.012}{8} = 0.0015$$

Now based on the Poisson formulae of the probability to not having the minimum allowable stress ordinates crossing downwards the level "a = 1.32" is: $P_{rel} = P_o = \exp(-\bar{n}_a) = \exp(-0.012) = 0.988$. This probability might characterize the reliability of the structure (one element or the whole structure).

Conclusions

1. This chapter is devoted to the study of creep of nanocomposite materials based on fillers in the form of carbon nanotubes and ultra-dispersed particles. The performed mathematical simulation of the behavior of nanocomposites using the special creep function, based on the theory of hereditary mechanics and *the kinetics of a phase transformation in a nanocomposite material* with the defining equation in the form of Volterra integral equation of the second kind, showed a satisfactory result, which indicates the possibility of using the above technique to predict the creep of nanocomposite materials.
2. The increase in rigidity (instantaneous modulus of elasticity) of the nanocomposite material rises with the volumetric concentration increase of nanoparticles, but on the other hand it is accompanied by a sufficiently intensive reduction of the ultimate deformation to failure, therefore the optimal concentration of nanomaterials introduced into the matrices should be chosen taking both of these processes into account.
3. Another positive consequence of the introduction of nanoparticles in the matrices is a decrease in the intensity of the creep processes in the nanocomposite material. This effect is important from a practical point of view, since it allows stabilizing the creep process during the long-term action of mechanical loads.

References

[1] Razdolsky, L. 2014. Probability-Based Structural Fire Load. Cambridge University Press, UK.

[2] Purvis. 1992. Handbook of Industrial Materials. Oxford: Elsevier Advanced Technology. 803 p.

[3] Ajayan, P.M., Schadler, L.S. and Braun, P.V. 2003. Nanocomposite Science and Technology. Wiley: New York, NY, USA.

[4] Jordan, J., Jacob, K.I., Tannenbaum, R., Sharaf, M.A. and Jasiuk, I. 2005. Experimental trends in polymer nanocomposites—A review. Mater. Sci. Eng. 393: 1–11.

[5] Thostenson, E., Li, C. and Chou, T. 2005. Nanocomposites in context. Composites Science and Technology 65(3-4): 491–516.

[6] Yu, M.-F., Lourie, O., Dyer, M.J., Moloni, K., Kelly, T.F. and Ruoff, R.S. 2000. Strength and breaking mechanism of multiwalled carbon nanotubes under tensile load. Science 287(5453): 637–640.

[7] Peng, B., Locascio, M., Zapol, P., Li, S., Mielke, S.L., Schatz, George, C. and Espinosa, H.D. 2008. Measurements of near-ultimate strength for multiwalled carbon nanotubes and irradiation-induced crosslinking improvements. Nature Nanotechnology 3(10): 626–631.

[8] Chen, L., Xie, X., Wang, B., Wang, K. and Xie, J. 2006. Spherical nanostructured Si/C composite prepared by spray drying technique for lithium ion batteries. Materials Science and Eng. B 131: 186–190.

[9] Morgan, J.D., Greegor, R.B., Ackerman, P.K. and Le, Q.N. 2013. Thermal simulation and testing of expanded metal foils used for lightning protection of composite aircraft structures. SAE Int. J. Aerosp. 6(2): 371–377.

[10] Greegor, R.B., Morgan, J.D., Le, Q.N. and Ackerman, P.K. 2013. Finite element modeling and testing of expanded metal foils used for lightning protection of composite aircraft structures. Proceedings of ICOLSE Conference; Seattle, WA, September 18–20, 2013.

[11] Hall, E.O. 1951. The deformation and ageing of mild steel: III discussion of results. Proc. Phys. Soc. London 64: 747–753.

[12] Petch, N.J. 1953. The cleavage strength of polycristals. J. Iron Steel Inst. London 173: 25–28.

[13] Avrami, M. 1939. Kinetics of phase change (I). General theory. J. Chem. Phys. 7: 1103–1112.

[14] Avrami, M. 1940. Kinetics of phase change (II). Transformation-time relations for random distribution of nuclei. J. Chem. Phys. 8: 212–224.

[15] Avrami, M. 1941. Granulation, phase change and microstructure, Kinetics of phase change. III. J. Chem. Phys. 9: 177–184.

[16] Johnson, W. and Mehl, K. 1939. Reaction kinetics in processes of nucleation and growth. Trans. Am. Inst. Min. Met. Eng. 195: 416–458.

[17] Kolmogorov, A. 1937. Static theory of metals crystallization. Izvestia Academia Nauk SSSR. Ser. Mater. 1: 355–359.

[18] Razdolsky, L. 2012. Structural Fire Loads: Theory and Principles. McGraw—Hill Co. N.Y., N.Y.

[19] Prout, E.G. and Tompkins, F.C. 1944. The thermal decomposition of potassium permanganate. Trans. Faraday Soc. 40: 488–498.

[20] Razdolsky, L. 2017. Probability Based High Temperature Engineering. Springer Nature Publishing Co., AG Switzerland.

[21] Cornell, C.A. 1967. Bounds on the reliability of structural systems. Journal of the Structural Division. ASCE, Vol. 93, No. ST., February.

[22] Lind, N.C. 1973. The design of structural design norms. Journal of Structural Mechanics 1(3).
[23] Hasofer, A.M. and Lind, N.C. 1974. Exact and invariant second-moment code format. Journal of the Engineering Mechanics Division, ASCE, Vol. 100, No. EM1, February 1974.
[24] Sveshnokov, A.A. 1978. Problems in Probability Theory, Mathematical Statistics and Theory of Random Functions. Dover Publications, New York.
[25] Rice, J.R. and Beer, F.P. 1966. First-occurrence time of high-level crossings in a continuous random process. Journal of the Acoustical Society of America 39(2).

Index

F

G

H

I

K

L

M

N

O

P

Printed and bound by PG in the USA

USA2019PGIL